Broadband Networking

Broadband Networking

GLEN **CARTY**

McGraw-Hill/Osborne
New York Chicago San Francisco
Lisbon London Madrid Mexico City Milan
New Delhi San JuanSeoul Singapore Sydney Toronto

McGraw-Hill/Osborne
2600 Tenth Street
Berkeley, California 94710
U.S.A.

To arrange bulk purchase discounts for sales promotions, premiums, or fund-raisers, please contact **McGraw-Hill**/Osborne at the above address. For information on translations or book distributors outside the U.S.A., please see the International Contact Information page immediately following the index of this book.

Broadband Networking

1234567890 CUS CUS 0198765432

ISBN 0-07-219510-X

Publisher
  Brandon A. Nordin
Vice President & Associate Publisher
  Scott Rogers
Acquisitions Editor
  Francis Kelly
Project Editor
  Jennifer Malnick
Acquisitions Coordinator
  Emma Acker
Technical Editor
  Tony Ryan
Copy Editor
  Robert Campbell

Proofreader
  Susie Elkind
Indexer
  Claire Splan
Computer Designers
  George Toma Charbak
  Michelle Galicia
Illustrators
  Michael Mueller
  Lyssa Wald
  Richard Coda
  Jackie Sieben
Cover Series Design
  Amparo Del Rio
Series Design
  Peter F. Hancik

This book was composed with Corel VENTURA™ Publisher.

To Karen, Jade-Rae, and Cody

AT A GLANCE

CONTENTS

Part II

Access Networks

ACKNOWLEDGMENTS

The first edition of this book would not have been possible without the support of my wife, Karen, and my daughters, Jade-Rae and Cody. A project of this size is time-consuming and presents challenges in maintaining a work-life balance. They encouraged me when it appeared most daunting, and understood when deadlines were looming.

Thanks to the Grant County Public Utility District, Washington State, for taking the time to answer questions and explain what they are doing in the area of FTTx networks. Their story is one of vision and success. In many ways, they are pioneers in the role of Fiber to the home; they have utilized technology to breathe new life into the community.

It was also a pleasure working with the team at McGraw-Hill/Osborne, especially the book's editor, Franny Kelly, and his assistant, Emma Acker. To project editor Jennifer Malnick, I owe a debt of gratitude. To copyeditor Robert Campbell, all I can say is that he is a magician with words. Finally, to technical editor Tony Ryan, thank you.

INTRODUCTION

▼

The writing of this book was a lot of fun because, unlike other books, *Broadband Networking* is not a discussion on a single technology or vendor; instead, it is a broad topic that addresses multiple technologies.

The broadband concept is not new; it has been around since the 1980s, when the promise of Broadband ISDN was seen as the way to deliver converged services. However, it now appears that we have finally gotten traction, and the promise of broadband networks is finally being realized. We are not quite there yet, and instead of a single technology, many different technologies and access networks are being deployed to deliver on this promise.

We have seen many changes sine 1990. The rate of technological achievements, innovation, and expansion is unprecedented. During this time, the way we traditionally viewed the world and its telecommunication, data, and broadcasting infrastructure significantly changed. For the first time, there is a distinct blurring of lines between the telephone, data, and audio/video broadcast network industries. The providers of services, who once operated within the confines of their respective industries, now compete across industries to deliver all three services.

As a result of this change, there is evidence of convergence. It has brought us a lot closer than we have ever been to the promise of an integrated network, and to all the benefits to be gained from this arrangement. For the consumer, it means a single provider will be able to provide an integrated service—as well as the potential for lower prices for a combined service—as a result of competition. To the provider, it means additional revenues and a single infrastructure from which multiple services can be delivered.

I believe we are at the cusp of a wave of exciting change. The advent of Digital TV means a whole new breadth of services—some obvious, such as Interactive TV, and some not yet imagined. The promise of the smart home and intelligent appliances means a new wave of home automation that can be controlled from any corner of the world.

We have seen more than a hundred years of technology based on copper wires and electricity. In contrast, fiber optic communication has a much shorter history—the first useable commercial optical fiber was developed in 1970. This means that we are at the early stages of fiber optics and have only begun to realize its potential. The most significant years of technological change is not behind us but ahead. In previous years, technological change benefited the business enterprise more than the consumer; future years will see a change in this trend, and the consumer will be the main beneficiary of technology.

The Intended Audience

The objective of this book is to provide a thorough understanding of current and emerging broadband technologies—features of the technology, operations, architecture, and issues of deployment. It is structured to be the first—and hopefully the only—book needed for anybody new to internetworking, the professional who needs a deeper understanding of broadband, and marketing professionals who need to understand how each technology is positioned and the applications behind their growth.

Special care has been taken to present the subject in a clear and easy-to-understand format. The assumption is that the reader is new to broadband networking, and so any reference to a new term or acronym is described in context when first mentioned. If the term reappears in a later chapter, the reader is reminded of its meaning for ease of understanding. By taking this approach, the book appeals to a wider audience and the reader is able to understand and follow from the basic to the more advanced details of a technology regardless of his or her background or training.

It is also assumed that the availability of the different broadband access solutions is in and of itself confusing. For this reason, the differentiating features of each technology are presented to help remove some of the confusion created by too much choice. Some technologies inherently provide better performance for specific applications. Armed with this knowledge, the reader is better able to decide the technology that is best suited to his or her particular application or circumstance.

What This Book Covers

This book consists of twelve chapters, each logically divided into four different parts.

Part I: The Move to Broadband

In Part I, the question "What is Broadband?" is defined and positioned in context to Baseband, Narrowband, and Wideband. We examine the applications and trends that are fueling the growth of broadband are examined, a few of which include Digital TV, the Internet, and new corporate initiatives such as E-learning.

Finally, the networking solutions that are available for home, multiunit buildings, and corporate offices are presented. Not all the solutions reviewed support broadband services, but by taking a look at each, the reader gains a better and broader appreciation of how it all fits together.

Part II: The Access Networks

In Part II, the technology and details of current broadband access solutions are presented. To most people, this is the only part of the network they ever see or care about because it represents the broadband solution and is the gateway that connects the home and office to multiple services.

The technologies presented in Part II include:

- ▼ DSL Access Networks
- ■ Cable Access Networks
- ■ Wireless Access Networks
- ■ FTTx Access Networks
- ▲ Digital Transmission and Frame Relay

Each technology is covered with enough detail to provide a thorough understanding of how it works, its architecture, standards, and its differentiating features. Emerging broadband access solutions are also presented, and Gigabit Ethernet (1Gbe and 10GbE) and Ethernet Passive Optical Networks are examined in the same detailed manner.

Part III: Core Technologies

The technologies that make up the metro and core network are presented in Part III. At the heart of this network are technologies that function as high-speed, high-performance bit pumps, pumping data through high-capacity pipes between different points in the network. The plumbing of the core network moves from the electrical to the optical domain, so instead of metallic wire, fiber optic is the predominant medium.

A technical foundation on fiber optics and the technologies that are used to squeeze more bandwidth from a single fiber strand are presented. This sets the stage for a detailed review of:

▼   DWDM

■   ATM

■   SONET/SDH

▲   MPLS

Part IV: Convergence and Content Services

In Part IV, we look beyond broadband to the emergence of content services and issues of convergence. We also attempt to raise some of the more frequently asked questions that are typically presented in discussions of next-generation networks. Our best interpretation of the issues and our best ideas on how things may play out are presented. The topics covered in Part IV include:

▼   ATM and SONET Versus IP and Ethernet

■   Why MPLS and Not IP Over ATM

■   The Future of Data

■   Convergence and its Wider Security Implications

▲   Flash Crowds and Bottlenecks

Finally, in this part we examine the following technologies:

▼   Content Delivery Networks (CDN)

▲   Layer 4 Switching

PART I

The Move to Broadband

Part 1 of this book explains some the technologies and trends that are driving the growth of broadband. In the first chapter, various terminologies and technologies are explained, and broadband is compared with other technologies such as baseband, narrowband, and wideband to help you understand the differences. The technologies and issues—regulatory, social, or business—are also discussed to give you an understanding of the drivers that are influencing the growth and acceptance of broadband.

Chapter 2 takes a more practical view of home, multitenant, and corporate networks. It aims to provide a broad understanding of the various types of networks found in the home, a description of how they function, and the standards behind each. It also focuses on multitenant dwellings and gives examples, where possible, of some of the ways in which building managers are building network infrastructure that allows tenants to gain access to broadband facilities. Finally, this chapter offers a review of the networks typically found within the corporate environment. New trends for remote access are also reviewed and explained.

CHAPTER 1

Introduction to Broadband

The term *broadband* is widely used today in many different contexts, the most common being a way to describe speeds greater than 200 Kbps. This, however, is a very narrow description, as broadband actually embodies a gateway to new possibilities. It is the means by which we can now gain access to new methods of instantaneous communication, an always-on pipeline to freely available information and new forms of commerce and recreation.

During World War II, General Eisenhower saw the advantages Germany enjoyed because of the autobahn network. He also noted the enhanced mobility of the Allies when they fought their way into Germany. These experiences shaped Eisenhower's views on highways. "The old convoy," he said, "had started me thinking about good, two-lane highways, but Germany had made me see the wisdom of broader ribbons across the land." During the State of the Union address on January 7, 1954, President Eisenhower made clear that he was ready to turn his attention to the nation's highway problems. He considered it important to "protect the vital interest of every citizen in a safe and adequate highway system."

In much the same way the Interstate system opened up the possibilities of exploration and commerce, broadband provides the pipeline that allows multiple digital services to enter the home and business.

WHAT IS BROADBAND?

Broadband refers to the wide bandwidth characteristics of a transmission medium and its ability to transport multiple signals and traffic types simultaneously. The medium can be coax, fiber, twisted pair, or wireless. With broadband, a wide range of frequencies is available to transmit information. This allows for multiple signals of possibly different traffic types to be sent concurrently using different frequencies or channels, thereby allowing more information to be carried at one time. In terms of speed, the FCC defines broadband as the capability of supporting speeds in excess of 200 Kbps upstream (consumer to the provider) and downstream (provider to consumer) in the last mile. This definition is based on the reasoning that speeds of 200 Kbps and greater would be enough to enable users to surf through Web pages as fast as one could flip through pages of a normal book and to transmit full-motion video.

Broadband Versus Baseband

The computer and communication industries began as two distinct and separate industries. In the late seventies, the two began to converge as computers began communicating with one another through the emerging medium of local area networks. This convergence created a debate over what was the more efficient wiring technology. The two available choices were baseband and broadband.

Baseband was based on the Ethernet developments, which used a 20 MHz, 50Ω cable. Designed to operate in half-duplex mode, Ethernet allowed a user to transmit at speeds up to 10 Mbps (or 10 million bits per second). This was a lot more desirable than the alternative, twisted pair telephone wires, which had a limit of 1 Mbps. Another reason for the appeal of this new medium was the fact that the transmission was all digital instead of analog modulation. The data was digitally applied to the wire through a DC signal.

Broadband, which was the other choice, operates using frequency division multiplexing (FDM) on a 75Ω cable with a bandwidth of about 350 MHz. As a technology, it was well known as the means by which cable television was delivered. Second, it had been around for many years and, as a result, carried commodity pricing. Its capacity was also a strong argument for the technology.

Ultimately, baseband LANs became the predominant type of LANs deployed in most offices. The political ramifications of the two technologies played a role in this choice: baseband was deemed to be the domain of data processing specialists, while broadband, because of its origins and use, was more in the realm of the voice types. So in fact, the debate was also about who maintained control.

Baseband LANs

When used in terms of telecommunication, baseband describes a communication system in which information is transported in a digital format across a single channel, which occupies the entire bandwidth of the transmission medium. In other words, it is a one-channel band. Examples of baseband networks are Ethernet, Token Ring, and Arcnet. In each case, digital information is directly placed on the wire, and the full width of the transmission medium is used. As a result, at any given instant, only one station has full access to the medium, as you can see from Figure 1-1. Baseband networks typically operate in a half-duplex mode, which means that data flows in one direction at a time. Variants of the

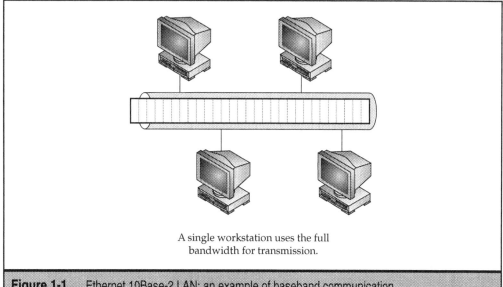

A single workstation uses the full
bandwidth for transmission.

Figure 1-1.   Ethernet 10Base-2 LAN: an example of baseband communication

original Ethernet now allow for full-duplex operation, made possible by removing the contention-based scheme by giving each station its own dedicated medium to a switch.

Broadband LANs

Even though baseband LANs far outnumbered their broadband cousins, broadband found itself a market in places where multiple services—for instance, data and closed circuit TV—had to be carried across the same infrastructure. Factories and campuses are examples of some of these installations.

The way broadband LANs operate is best described by using the analogy of the U.S. interstate highway system. A highway is of a finite width and capacity, but it can be subdivided into multiple lanes to accommodate different types of traffic; faster traffic uses the outside lane and slower vehicles, the inside. Broadband networks operate in much the same way. The width of the connection is measured in Hertz, which is a measure of cycles per second of a frequency. The greater the difference between the highest and lowest frequencies, the greater the capacity or bandwidth of the connection.

Frequency division multiplexing, or FDM, uses different frequencies to create different channels on the cable just as there are different lanes on a highway. Signals carried along a cable are modulated into radio frequency (RF) channels that are typically 6 MHz wide, which is the bandwidth required for a standard TV channel (in the U.S.). Dividing services into separate channels permits different types of signals to coexist and travel in opposite directions "inbound and outbound" along the same cable without concern for one signal interfering with another.

FDM, being an analog technique, requires a modem to convert digital information to analog and analog to digital at the other end. Unlike a baseband connection, which places the digital information directly onto the cable, broadband converts the digital information to an analog signal and places the analog signal on its own channel or set of frequencies. This allows different traffic types to share the same cable and not interfere with each other. The ability to use a set of frequencies, one for sending and the other for receiving, also differentiates broadband from baseband. This capability allows for full-duplex operation or the ability to send and receive data simultaneously. A baseband connection uses the full bandwidth of the cable so only one device can transmit at a time.

The core of a broadband network is the *headend* through which all transmission must pass. As Figure 1-2 shows, even if two devices are adjacent to each other, information from one to the other must flow to the headend, which translates the transmitting device frequency to the frequency of the receiving device, in a process that is called *remodulation*.

It is highly possible that the signal from the device to the headend passes through several other components, such as splitters and couplers, that degrade the strength of the signal or through amplifiers that boost the signal strength. The signal, however, regardless of the components that it may have traversed, must arrive at the headend with the correct signal strength, usually a lower level than the transmission strength.

The objective of a well-designed network is to ensure that the amount of loss is equal to the amount of gain provided by the amplifiers. This relationship is called *unity gain*. The challenge, however, lies in the fact that signal loss is not uniform over different frequencies, some being affected more than others by different components. This aspect of

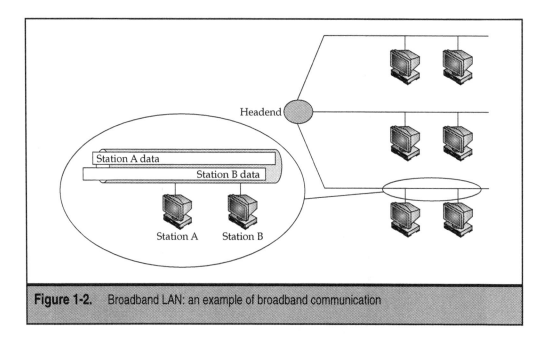

Figure 1-2.   Broadband LAN: an example of broadband communication

broadband design makes it more complex than designing a baseband local area network and requires specialized knowledge to ensure optimum performance.

Narrowband Versus Wideband

The term *narrowband* generally describes telecommunication that carries voice information in a narrow band of frequencies. More loosely, it has been used to describe "not broadband" or sometimes a band that is just wide enough to carry voice. More specifically, the term has been used to describe a specific frequency range set aside by the U.S. FCC for mobile or radio services, including paging systems, from 50 characters per second to 64 Kbps.

Wideband, a synonym of broadband, refers to a transmission medium or channel that has a wider bandwidth than one voice channel (with a carrier wave of a certain modulated frequency).

MARKET DRIVERS: THE APPEAL OF BROADBAND

The Yankee Group, a telecommunications research firm, estimates there will be over thirty-six million broadband subscribers connections in the U.S. by year-end 2005 (see Figure 1-3). ARS Inc., a competitive market intelligence company, research has determined that the growth rate of broadband subscribers in the third quarter of 2001 was 14.2 percent over the previous quarter. The rate of increase from first quarter to the second was 14.9 percent, and 25.8 percent from the end of 2000 to the end of the first quarter 2001.

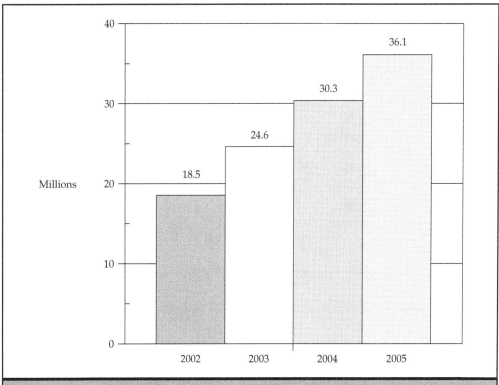

Figure 1-3. Estimated broadband subscriber growth (source: Yankee Group)

The adoption of high-speed broadband connections is driven in part by the types of services that are made available as a result of the higher speeds. In fact, it is somewhat of a "what comes first, the chicken or the egg" dilemma. In order to justify the cost of new services, developers need evidence of a growing market, but the speed of growth is directly impacted by the availability of the services.

Today, the cost of providing broadband services to the home and business is very high. Service providers who invest in the build-out of the infrastructure necessary to provide broadband services are in fact making big bets that broadband will be widely adopted and will eventually result in a return on their investments.

How, then, can a broadband provider offset the cost of building the network? One answer is to retain an interest in the content and services offered, and to limit the user only to these choices. As an example, today RoadRunner is the Internet service provider for users of Time Warner's broadband network. Earthlink, a competitor of RoadRunner, is not able to provide ISP services to a user of the Time Warner broadband network—their only choice is RoadRunner. This question of choice has gotten a lot of attention from some local government and industry regulators who fear this practice could stifle competition and lead to higher prices. This issue is widely knows as the issue of "open access."

The openNET coalition, a group of over 900 providers of consumer Internet service, contends that customers deserve a choice and would like to have access to cable providers' high-speed broadband networks. They also contend that choice eventually leads to the other benefits of competition, including affordable prices and improved quality of service. The cable providers' response is that it would be a major technical challenge and a significant investment would be needed to facilitate this access. They have, however, agreed to review the possibilities of allowing access by multiple ISPs onto their cable systems. The issue is still unresolved and the answer is yet to be determined, but it underscores some of the issues that consumers, businesses, and industry regulators face as we embark on the journey to broadband.

The Convenience of Broadband

If you always had to rely on connecting a hose to an outdoor tap to get water into your home, then you would probably wonder what the big deal was about indoor plumbing. Apart from the obvious convenience of not having to go outside to turn on and off the tap whenever you needed water, the extra benefits of an internal source would probably not be too obvious. Unless, of course, you had a vivid imagination and could see the new possibilities an inside source brings. One obvious consequence of not having indoor plumbing is that you would probably use less water, and only when necessary, because of the inconvenience and hassle of having to go outside to turn it on and off. One not-so-obvious benefit that would probably be lost on you without some exercise of the imagination would be the convenience of an ice maker. So too is the concept of an always-on connection to the Internet. An obvious convenience is the fact that it's always there when you need it. A not-so-obvious benefit is the new services that are spawned as a result.

The availability of an "always on" high-speed connection to most businesses, and residences, opens a whole new market for new services that can be developed and sold. In this environment, the only constant is change. So what are some of these services that would entice a user to begin using high-speed broadband access? Some possibilities include:

▼   Digital television and all the services it brings

■   The Internet—news ways of interacting and emerging content

▲   New corporate initiatives—telecommuting, video conferencing, and e-learning

Digital TV

Digital television, or DTV, is a new type of television service that is now becoming available. TV stations in the U.S. use an analog format for broadcasting television programs that was created by the National Television Standards Committee (NTSC). Stations will continue to broadcast on their current channels and format for the next several years but will eventually begin to broadcast on a new channel that has been provided for broadcasts in a new digital format. By the year 2006, all U.S. television stations will cease to broadcast in the analog format and will begin broadcasting in the new digital format. Currently this is the planned date, but factors such as market readiness could cause the date to slip.

The new digital format provides several improvements over the old analog format, which include:

- ▼ A much clearer and brilliant picture
- ■ No ghosting or snowing
- ■ Increased capacity of each channel through multicasting (multiplexing)
- ■ Dolby Digital CD-quality audio
- ■ Wide-screen pictures much like the cinema
- ▲ On-screen text and graphics

With DTV, a single channel can be used to broadcast multiple programs with accompanying Dolby Digital Surround Sound as well as download data for interactive programs. About 84 percent of all television stations have been granted a DTV construction license, and about 212 stations are currently on the air with some type of digital content.

Will we need to replace our old television sets? The good news is no, set-top converter boxes will be available to allow us to use our old TVs to receive digital signals. The bad news, however, is that in order to enjoy the full benefits of HDTV (high-definition television), a DTV receiver, which currently is still relatively expensive, is needed.

Television Basics

To fully appreciate the features and benefits of digital television, a review of some of the terminology, history, and technology is appropriate.

The Standards   The three main worldwide standards for encoding and transmitting analog television are:

- ▼ **National Television Standards Committee (NTSC)**   The standard adopted in the United States
- ■ **Phase Alternation Line (PAL)**   Adopted in countries such as the United Kingdom
- ▲ **SEquential Coleur Avec Memoire (SECAM)**   Adopted by France and the Eastern Bloc countries

The major features of each standard are outlined in Table 1-1.

Two main power frequencies are used throughout the world, 50 Hz and 60 Hz. The countries that use a 60 Hz frequency generally adopted NTSC, and those with a 50 Hz frequency adopted PAL or SECAM. Nevertheless, the underlying reasons for choosing SECAM over PAL were rooted more in politics than technology.

	NTSC	PAL	SECAM
Aspect ratio	4:3	4:3	4:3
Field rate/second	59.94	50	50
Scan lines	525	625	625
Active lines	480	575	575
Pixels per line	640	580	580
Frame rate/second	29.97	25	25
Size uncompressed signal (Mbps)	221.2	400.2	400.2
Channel bandwidth	6 MHz	8 MHz	8 MHz

Table 1-1.   Analog Television Broadcast Standards

Aspect Ratio   Before the advent of Cinemascope, movie studios produced movies in a 4:3 aspect ratio. This means that the height of the picture was 3/4 its width, or the width was four units of measurement for every three units of height. When the NTSC decided on the original TV specifications, they decided to adopt the same ratio that was being used by the movie studios.

Over time, the movie studios began using a wider format to better fit the human field of vision in an effort to enhance the viewing experience. When this happened, movies had to be modified to fit the screen of the television or be shown in a letterbox format. *Letterboxing* is the process where black lines—above and below the picture—are used to retain the original aspect ratio of the film. The drawback with this is that it makes the picture smaller.

Display Formats   A number of resolutions are available with DTV. *Resolution* refers to the number of lines of information displayed on a screen from top to bottom. The NTSC and PAL standards use interlaced scanning as their display format. The purpose of interlacing was to minimize flicker that was inherent in pictures transmitted over low bandwidth. Another way to display the same information is through a process called *progressive scanning,* the method used in computer monitors.

Interlacing works by broadcasting only half of an image, what is called a *field,* at any instant. As Figure 1-4 shows, the picture is painted onto the screen one half at a time by illuminating pixels of odd-numbered lines on the first pass and then pixels of even-numbered lines on the next pass. On the first pass of the scan, pixels are successively illuminated starting on the first line of the screen and progressing from left to right. The second line is skipped, the third line is illuminated, and then the fifth, the seventh, and so on. On the second pass, scanning begins at line two, then line four, line six, and so on.

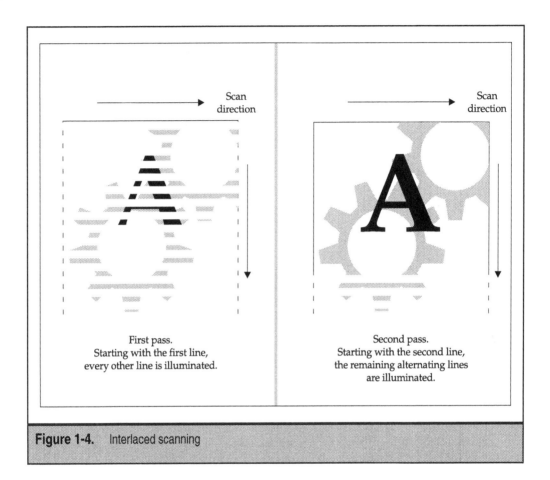

Figure 1-4.   Interlaced scanning

Each field takes one sixtieth of a second to paint onto the screen, with the entire frame (consists of two fields) taking one thirtieth a second. The effect of this is to refresh the entire picture about 30 times every second. What appear to the eyes as 30 fps (frames per seconds) is actually 60 interlaced fields per second. An advantage of this technique is that only one half of the frame is being transmitted at any one moment, and as a result less information is being painted onto the screen, so less bandwidth and less expensive circuitry are needed to give the appearance of high resolution.

Progressive scanning, also known as "noninterlaced scanning" or "sequential scanning," is the means by which the entire frame is painted onto the screen by illuminating every line sequentially. Computer monitors use progressive scanning, and as a result, a video card must do a scan conversion if a monitor is to be used to display television signals.

The number of scan lines varies among the standards. For NTSC, the standard is 525 lines at an interlace of 30 fps. Of the 525 lines, only 480 lines are used for the picture; the remaining lines are used for broadcasting information. As Table 1-2 shows, digital TV consists of 18 different combinations of resolution, frame rate, and aspect ratio.

Resolution	Aspect Ratio	Frames per Second
Horizontal × Vertical		(Interlaced = i; Progressive = p)
1920×1080 (HDTV)	16:9	30i, 30p, 24p
1280×720 (HDTV)	16:9	60p, 30p, 24p
704×480 (SDTV)	16:9	30i, 60p, 30p, 24p
704×480 (SDTV)	4:3	30i, 60p, 30p, 24p
640×480 (SDTV)	4:3	30i, 60p, 30p, 24p

Table 1-2.   Digital TV Formats

Bandwidth

In the U.S., the FCC has allotted 6 MHz of bandwidth for each broadcast channel. As Table 1-3 shows, all TV channels do not occupy a contiguous space within the frequency spectrum; rather, blocks of channels are interspersed with other services.

In 1987, the FCC issued a ruling indicating that the HDTV standards that were to be issued would be compatible with existing NTSC service and would be confined to the

Frequency Range (MHz)	Type of Service
54–72	TV Channels 2 through 4
72–76	Manufacturing, remote control, eavesdropping bugs, etc.
76–88	TV Channels 5 and 6
88–108	Standard FM radio
108–136	Aeronautical communications
136–138	Satellites
138–144	Military communications
144–148	Amateur radio
148–150.8	Military use
150.8–174	Business, highway, law enforcement, weather, maritime

Table 1-3.   Radio Frequency Spectrum

Frequency Range (MHz)	Type of Service
174–216	TV Channels 7 through 13
216–220	Maritime and aeronautical
220–222	Land mobile communications
222–225	Amateur radio
225–400	Military aviation and space
400–406	Military and government
406–420	U.S. Government
420–450	Amateur radio
450–470	Ultra High Frequency band. Business, industry, military, fire government
470–806	TV Channels 14 through 69

Table 1-3.   Radio Frequency Spectrum *(Continued)*

existing VHF and UHF frequency bands. The challenge to digital TV was that the signal must fit into a 6 MHz bandwidth and take no more channel space than today's NTSC broadcast signal.

Somehow, more than five times as much information had to be accommodated into the same bandwidth of 6 MHz. That doesn't include the additional bandwidth required for the planned audio or data feed. How then could this be accomplished if DTV also meant the capability to deliver a bigger and better-quality picture, improved sound, and less distortion?

The answer was found in a compression technique developed by the Moving Picture Experts Group (MPEG) called MPEG-2. It works by recording just enough detail within each frame to make it look as if nothing is missing. It also compares adjacent frames and records only the section of the picture that has moved or changed. Using this technique, a compression ratio of up to 55:1 is achievable, and HDTV, which promises the best that DTV has to offer, fits in a 19.39 Mbps channel.

HDTV and SDTV

High-definition television, HDTV, is the highest level of digital television. The signal must be 1080i (using interlaced scanning) or 780p (progressive scanning), more than double the scan lines of analog TV and six times the resolution. HDTV attempts to capture the movie viewer's experience by increasing the aspect ratio to 16:9—the width of the image is almost twice that of the height. HDTV uses the Dolby Digital/AC-3 audio encoding system—the digital sound used in most movie theaters, DVDs, and many home theaters since the early 1990's. As shown in Figure 1-5, it can include up to 5.1 channels of sound: three in front (left, center, and right), two in back (left and right), and a subwoofer bass (that's the .1 channel).

SDTV, or standard-definition TV, has a lower bandwidth requirement than that of HDTV. It uses NTSC pixel resolution and has 12 different authorized video scanning formats (see earlier Table 1-2). It is possible to multicast four or more different SDTV programs on one broadcast channel. What this means in reality is that when you tune in to a specific channel, instead of watching a single program you may be presented with an option of viewing a do-it-yourself program, a documentary, the local news, or some other programming. All these choices would be available from a single broadcaster on a single channel.

A program can be television, nontelevision data, or a combination of both. Using excess bandwidth, a broadcaster also has the potential to deliver data over the connection. Web

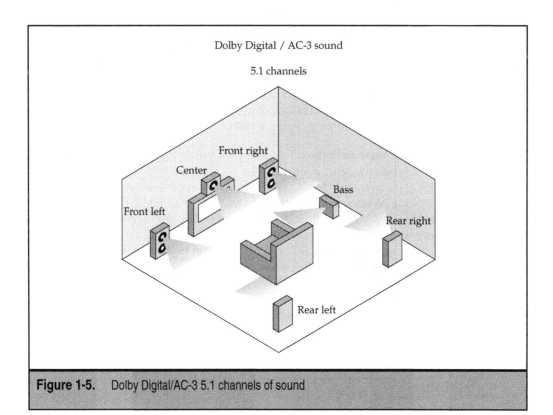

Figure 1-5.   Dolby Digital/AC-3 5.1 channels of sound

pages, interactive programs, audio, images, and computer programs can all be down-loaded along with broadcasts. This capability is beginning to be known as *enhanced TV*.

Interactive TV

Interactive TV, once just a dream, is now becoming more widespread, especially among cable operators. Interactivity has always posed a challenge to television programming because television broadcasting has always been a one-way communication. In order to exercise interactivity, the user must have a way of communicating back to the broadcaster.

For cable operators, this simply means using some of the bandwidth for the return path and upgrading the infrastructure by installing additional amplifiers—and laser transmitters for any fiber links—for the upstream communication. In the case of over-the-air broadcasters, however, it's a bit trickier. In many instances, the telephone is being used for the return path to the broadcaster. This is not only a possibility for over-the-air broadcasting but is also used by cable operators who choose to delay the up-grade of their infrastructure.

In addition to true interactivity, different ways of simulating interaction are being explored. In these types of applications, the interaction happens between the user and the set-top box instead of the broadcaster. An example of how this may be achieved is by enabling the user to select different camera angles during a Superbowl playoff. The broadcaster would transmit the different camera angles on different program streams. To the users, it would appear as if the communication is happening between them and the broadcaster, but in fact the set-top box would be doing the switching. With some imagination, it's not hard to see how this feature could appear even more functional if the set-top box had some storage—instant replay would be at your fingertips.

Video on Demand and Near Video on Demand   Video on Demand (VoD) is probably what most people think of when the term interactive TV is mentioned. In fact, most people interpret VoD as Movie on Demand (MoD), which is just one application of VoD. Another application may include News and Weather on Demand. The point being, as the application matures and is more widely deployed, more uses will be developed, and they may include additional services that go beyond the common definition of the term today. As an example, the ability to record your favorite program on the broadcaster's servers for later playback could be offered as a service that replaces the need for a personal VCR.

VoD allows subscribers to purchase the viewing of a movie whenever they wish and offers features similar to a VCR—like pause, rewind, and forward. It eliminates the hassle of renting a movie. Just think, you no longer have to get dressed, jump in the car, drive 3–5 miles, enter a store, rent a video, return home, watch the movie, and then try to get the movie back before the rental period ends. Instead, you sit before your television, select the movie, order it, enjoy it, and that's it.

The broadcaster, however, incurs heavy costs to provide this convenience, which has been one of the major reasons for the slow roll out of the service. In spite of this, broadcasters do not wish to ignore this opportunity, as it means passing on a piece of the home video rental market—estimated to be in excess of $10 billion per year. As shown in Figure 1-6, the typical components of a VoD network consist of:

1. **Video servers**   These are the heart of, and an expensive part of, the network, as it is the server that serves up the content. The falling prices per stream and per gigabit of storage have contributed greatly to lowering server cost. Scalability, fault tolerance, and streaming efficiency are critical factors in the efficient operation of video servers.

2. **VoD server and streaming manager**   As with all networking systems, a good management system is necessary to ensure the smooth operation and availability of the network. For each movie ordered, the video stream must be established and torn down in an orderly fashion, and a system is needed to ensure that this happens the way it should and that any service disruptions, malfunctions, or degradation get immediate attention.

3. **Application server**   The headend of the VoD network, this is usually dedicated to a specific type of service—Weather on Demand, Movie on Demand, and so on. Multiple application servers may exist in a network, each interfacing to the application client that lives in the set-top box.

4. **Subscriber management systems (SMS) and billing**   Previously used to manage and track the viewing of prescheduled programming, these systems now need to be more flexible as programming becomes more dynamic and determined by user requests. Another challenge is the addition of services that may not be movie based. A request for a paid data feed, for example, represents a new type of service that would need to be tracked and billed.

5. **Session and resource manager**   This system manages the topology of the network and the video streaming radio frequency spectrum.

6. **Asset Manager**   This system manages the different types of content.

7. **Application clients**   These reside in set-top boxes and interface with application servers. The set-top box must be able to accommodate multiple application client interfaces for each type of service. An operating system is usually installed within the set-top box to handle any requirement for data exchange between different application client interfaces.

A major contributor to the high cost of delivering VoD services is the impact it has on the infrastructure itself. Every user of a VoD network must have a dedicated connection back to the server that cannot be shared. This is a very expensive proposition to the broadcaster and one that has led to development of other types of service that make better use of bandwidth.

Near Video on Demand (nVoD) tries to solve the problem (and at least partly succeeds) by providing a service that allows for the sharing of bandwidth but gives the user some

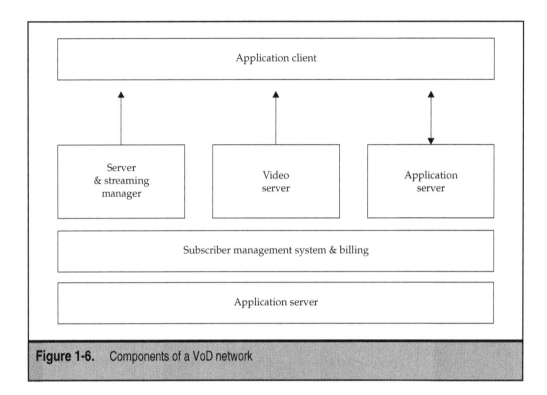

Figure 1-6. Components of a VoD network

limited flexibility to jump backward or forward within the program, at set time intervals. With nVoD, a program is broadcast on several channels with multiple start times, as you can see in Figure 1-7. For instance, a one and one-half hour movie is broadcast on six channels starting 15 minutes apart. To get the benefit of "rewind," the user would jump to a channel that started later than the one they are watching. To "fast forward," they would jump to the channel that started before the one they are watching.

This does not provide for the full flexibility of rewind and fast forward, but with the right pricing it may be an inconvenience that is acceptable to the consumer and would solve the dedicated bandwidth requirement problem. The staggered start times could also be decreased, though this means using more channels, but that may not be a problem, as multiple program streams on one channel are possible with Digital TV. The benefits to the broadcaster lie in the fact that the bandwidth is shared by as many users who may wish to view the program—a fact that keeps infrastructure and server costs down and allows the programs to start at prescheduled times—which is great for existing SMS systems.

The Internet

The potential scope of the Internet, its uses and capabilities, is limited only by what the most creative minds can conceive. The World Wide Web has experienced an exponential growth in both the number of servers that are connected daily and the number of users that are connecting to use these servers.

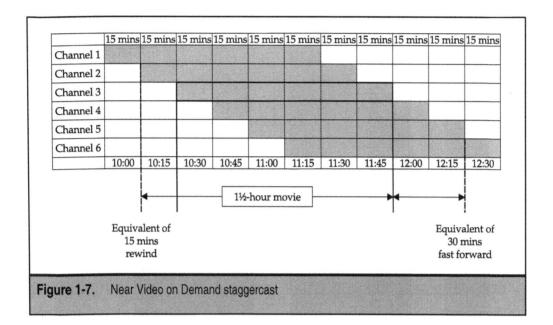

Figure 1-7. Near Video on Demand staggercast

The growth rate has slowed in recent years, but the Web's content and the way it's being used are evolving in ways that require lots more bandwidth. In the past, a dial connection was more than adequate to send e-mail and browse web sites which were predominantly text based. Today's web sites have a lot more graphics and utilize applications like Macromedia Shockwave to provide multimedia and interactive content to the user. Before delving much further into why the Internet is viewed as another key market driver for broadband networks, we will review some terminology.

Latency and *bandwidth* are two aspects of network performance that are independent of each other (see Figure 1-8). Latency is the measure of time it takes for a packet to travel from points A to B in a network and is usually measured in milliseconds, which are thousandths of a second. A packet traveling between the end user and the target application will pass through several different devices before it reaches its destination, each different device introducing delays. Latency affects all components of the network across which the packet travels.

Latency is also affected by distance, most evidently with satellite communications. An orbiting satellite is so distant that the limit imposed by the speed of light becomes an appreciable source of delay, adding perhaps 239 ms of latency for the signal to reach a geosynchronous satellite and the same amount of time for it to return. The lower the latency, the better the user experience, especially when using interactive applications. High latency appears to the user as long delays that degrade the user experience.

In networking, the perfect world would be one where latency is almost nonexistent and bandwidth was in abundance. Bandwidth is not a measure of speed but of capacity. In analog transmissions, bandwidth or capacity is measured in Hertz (a measure of frequency) and is the difference between the highest and lowest frequencies available on the medium.

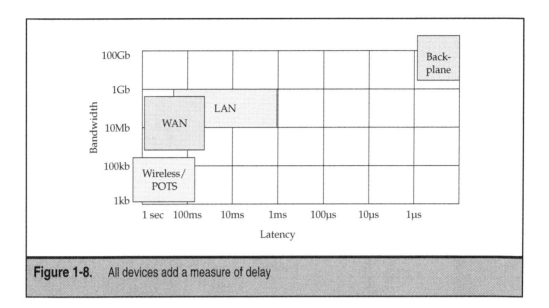

Figure 1-8. All devices add a measure of delay

For digital services like T-1 and ATM, bandwidth or capacity is measured in bits per second (bps), or the number of bits that can be transmitted in a second.

So which is more important, latency or bandwidth? The answer is: both are important. To appreciate the relationship between bandwidth and latency, let's revisit the analogy of a highway. A connection with high bandwidth (good) and high latency (bad) would be akin to a 100-lane highway with each lane traveling at 15 mph. A connection with low bandwidth (potentially bad) and low latency (good) would be analogous to a single-lane highway with a speed limit of 100 mph. So the more data you wish to move, the more you benefit from bandwidth. The greater the urgency to get it from point A to point B, the more you will benefit from lower latency.

The fastest analog modems—those that operate at close to 56 Kbps—introduce latency up to 100 ms or more. This is a result of having to do modulation and demodulation of the signal—digital to analog and analog to digital. An ISDN modem, in contrast, has a delay of about 2 ms, which is about 50 times faster than the analog version, a direct result of having not to do the analog-to-digital conversion.

The Internet: An Evolving Entity

During the heyday of Napster, the file-trading web site was so popular—it had nearly 70 million registered users—that universities across the country erected firewalls to restrict student access to the program that was clogging their networks with music files.

There are several interesting points about the Napster story that highlight the way the use of the Internet evolves and why it is helping to drive the requirement for bandwidth and speed. The web site will probably be remembered for the copyright infringement lawsuit brought by the Recording Industry Association of America (RIAA). Its defense was that its users were merely engaging in a larger-scale version of behavior that is acceptable today—that of making a copy of a song or VCR recording for a friend.

Napster is an application that allows people to share MP3 audio files. Users connect to a centralized server to search for an MP3 file they want. The server to which they connect contains a database with the names of all files that other users have made freely available for download. The application then makes a connection directly to the server that has the file, and the user downloads it from there. The server containing the file can be another user's machine, and in the same way, the machine that connects to the Napster database can be a server in its own right.

Many, if not most, of the songs being downloaded are copyrighted material recorded by major bands and record labels who deem the practice to be music piracy, an infringement of copyright laws. Central to the issue is the concept of fair use, which describes what you are allowed to do with material that you have purchased. Under fair use, you are allowed to lend a book to a friend or tape a song from a CD. You are not allowed, however, to sell the material or claim it as your own.

Now all of this makes good reading, especially if you are interested in the interpretation of copyright laws, but what is relevant to the subject at hand is the way in which Napster grew and distributed its material. Whether or not Napster will succeed is still to be determined, but the trend toward digital openness will undoubtedly continue.

Napster does not store the MP3 files that users want; instead, it holds a master index of all the files that other users have made freely available for download. The actual files reside on the users' computers, which automatically become a server for downloads. This approach raises two interesting points: First, most statistics given about the growth of the number of servers on the Web include web servers that are associated with registered domain names but do not include the machines that serve up some of these MP3 files. Second, the distribution of the files across multiple machines follows more of a distributed model than a straight client/server relationship.

MP3.com, another music site, is an example of a more client/server approach. All files are stored on their servers, which handle all requests for downloads. Their approach—in the early years—was to offer a service where unsigned bands had an opportunity to host their original works on MP3.com servers. In return, MP3.com would provide a web page dedicated to the artist and would handle any sales and distribution of the work.

A user of the service was given a choice to either download a free copy of the file or purchase a CD that would be burned and mailed by MP3.com. The company and the artist would share the proceeds. Using this means of distribution, the need for adequate bandwidth lay mainly between the user and MP3.com. In the Napster model, the bandwidth requirement went beyond the inquiring user and Napster to include the machine hosting the file—which might have been just another user. Figure 1-9 shows the difference in approach.

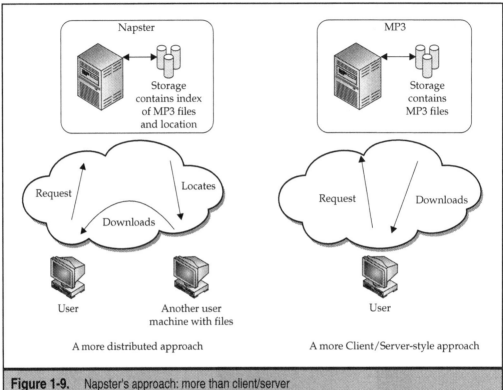

Figure 1-9.   Napster's approach: more than client/server

One weakness in the Napster approach is the dependence on the centralized server. Without it, there would not be a service. To prove yet again that the uses and capabilities of the Internet are only limited by the most creative minds, it was not long before something came along that improved on the Napster concept.

Gnutella—originally developed by Nullsoft—is a peer-to-peer networking application, as sketched in Figure 1-10. Each computer on the network can be a client or a server. The network is not dependent on a centralized server to operate and is not limited to any particular file type. It operates as both a mini–search engine and a file serving system. Here is how it works: First, you specify the IP address of the machine to which you would like to connect—a friend maybe. Once connected, you have access to all the machines that are running Gnutella. When you search for a file, it does a recursive search through the entire domain of Gnutella machines, starting with the one to which you are connected. Finally, if you close the application in the middle of a download, it will resume where you left off (assuming the system from which you were downloading is still connected).

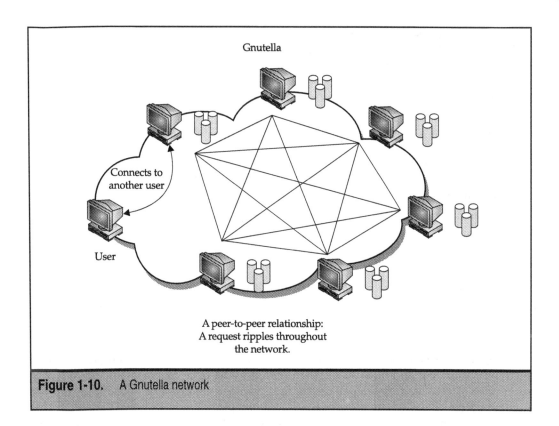

Gnutella

Connects to
another user

User

A peer-to-peer relationship:
A request ripples throughout
the network.

Figure 1-10.   A Gnutella network

In a Gnutella-like network, it is difficult to isolate any one area that benefits the most from increased bandwidth, because the entire network benefits. The reality is, however, that the connections will vary from dial-up connections to T1s to cable or DSL connections. As these types of networks gain traction and become more pervasive, they become another market driver for broadband networks.

The Internet: Expanding Content

According to a Yankee Group survey, 93 percent of households that own PCs now have Internet access. In addition, the top three uses for home PCs are: online services/Internet access (76 percent), games/entertainment (58 percent), and personal/household finances (30 percent).

Web developers are constantly seeking ways to draw more eyes to a web site, keep them there, and have them tell someone about it. Multimedia is being used more and more to improve the user experience and to make the visit more memorable. According to Macromedia—a developer of dynamic Web content applications—as of 2001, over

270 million Web users have installed their Shockwave multimedia player, with over 200,000 new installs per day. The number of new installs underscores the adoption and growth of rich media content that is tailored for fast networks. The following table gives examples of the types of content and their typical bandwidth requirement for efficient operation:

Application	Typical Data Rate	Comments
E-mail	up to 56 Kbps	Requires more when attachment is sent
File transfer	56 Kbps to 1.5 Mbps	
Remote database access	1.5 Mbps to 10 Mbps	
Imaging	1.5 Mbps to 45 Mbps	Digitization and storage of documents, drawings, photographs, X rays, etc.
Multimedia	3 Mbps to 45 Mbps	Streaming audio, video, MP3, interactive games, etc.
LAN interconnect	10 Mbps+	

The growth of computers' processing power has been the main fuel behind growth of the types of content and how they are presented to the user. In 1965, Gordon Moore—cofounder of Intel and developer of the first microprocessor chip—observed that the number of transistors on a microprocessor doubles approximately every 12 months. This held true until the late 1970s, when the doubling interval slowed to about 18 months, which is what it has remained. In other words, processing power doubles every 18 months. Gordon Moore has gone on to further state that he sees no barriers to Intel achieving a microprocessor with one billion transistors and an operating speed of 10 GHz by 2012, effectively achieving a performance rate of 100,000 MIPs (million instructions per second). The price and size of microprocessors have also been decreasing at a rate that is proportional to the increase in processing power.

The storage industry is also experiencing similar growth rates and technological advancements. The amount of information that can be packed into a square inch of space is increasing at a rate that parallels the microprocessor growth rate. The price per megabyte of storage is also decreasing at a rate that is proportional to the increase in capacity. It is this relationship between power, price, and size of microprocessors and storage that has continued to provide the catalyst for the development of new types of content and new ways for the consumer to use that content.

Types of Content and Issues Affecting Their Use

Audio Files   In its uncompressed format, every second of CD-quality audio (44.1 kHz, 16-bit, stereo) requires about 172 kilobytes, or about 10MB per minute, so 30 megabytes of storage would be needed for a three-minute song. Most early audio file formats were uncompressed, examples of which include Windows wave files (WAV) and the Mac's Audio

Interchange File Format (AIFF). Compression techniques were required to reduce the size of audio files and not lose sound quality. Without compression, the files were just too large for any practical application.

As with video, the solution was found in compression techniques developed by MPEG (the Moving Picture Experts Group) called MP3—short for MPEG-1 Audio Layer 3. The higher the audio layer, the greater the level of complexity of the encoding software, and thus the greater the sound quality for a given level of compression. A highly compressed file sounds better using Layer 3 encoding than using Layer 1 encoding. Some common audio file formats are listed in Table 1-4.

Format	Description/Features	Compression Ratio
WAV	Developed by Microsoft, the wave format offers the highest-quality 16-bit uncompressed audio. Both the sample rate and the number of channels, however, can be adjusted at the expense of quality.	1:1
AU	Developed by Sun Microsystems, this format stores any number of audio channels, yielding theater-like sound.	1:1
AIFF	The Audio Interchange File Format is the audio format commonly used on the Mac. It produces larger files because the format is largely uncompressed.	1:1
MP3	Developed by the German company, Fraunhofer-Gesellschaft, MP3 retains most of the quality of the original. Expanded use of the Internet as a data delivery medium and increased processor power capable of decoding MP3 in real time, as well as its early adoption by the college student community, have contributed to its widespread popularity.	11:1
SHOUTCast	This format was created by NullSoft, the creators of Winamp MP3 player. Essentially, the encoding compression is MP3, but the format delivers the file through streaming. Internet radio broadcasters are its main users.	10:1

Table 1-4.   Audio File Formats

Format	Description/Features	Compression Ratio
RealAudio	Audio streaming technology developed by Progressive Networks, this is used by many Internet radio stations and web sites selling music for small streaming audio samples. Quality is less than MP3 at lower levels of compression.	16:1
VQF or TwinVQ	Developed by NTT (Nippon Telegraph and Telephone) for Yamaha, this requires a special VQF player. It produces a smaller file than MP3, but the quality is not as good.	18:1
Liquid Audio	Developed by Liquid Audio in direct competition to MP3, this format has the ability to use different compression methods: MPEG-1 Audio Level 2, Dolby Digital/AC3, or AA3. It can include images, artist information, and playback restrictions (digital watermarks) in addition to the audio information. The security features have gained the attention of the recording industry, and the company has aggressively sought out partnerships that could extend the format's reach and increase its popularity. The format also provides for pay-for-play. The Dolby Digital/AC3 format allows for 5.1 audio channels.	>12:1

Table 1-4. Audio File Formats *(Continued)*

MP3 uses a process called "perceptual coding." The audio data is analyzed and compared to a model of human hearing. Any information that fits into the model—that is, any sound that the human ear can detect—is encoded. The MP3 file format has the largest user base of all compression techniques and file formats. Several different MP3 players now on the market afford a user the ability to download or digitize an audio CD and store it on the player for playback.

Internet broadcasting   Audio and video streaming allow a user to play an audio or video file direct from the network. Streaming an audio feed gives the user the capability to listen to large files such as the entire recording of a concert or a live broadcast without downloading. This is also true of video content; you can watch a movie directly from the network. Many of the features and challenges of video have already been discussed in the section on digital TV.

Sites that provide streaming audio and video are now more commonplace, and entire programming lineups are now being developed for airing on the Internet. Yahoo's FinanceVision (http://vision.yahoo.com) is an example; it features a full lineup of programming each day and includes interviews with audience participation, reports, live market updates, and advertising.

Regulation FD   On October 23, 2000, the SEC Fair Disclosure Rule went into effect. It was the SEC's attempt to level the playing field for all investors by requiring all U.S. public companies to fully disclose information that may influence investment decisions to all investors at the same time, not just selectively to analysts and insiders before events occur. As a result of this ruling, companies began using the Internet as a medium to distribute information.

This mandate had a significant impact. At the time of the announcement, the Internet was already being used by corporations to post text-based information on their web sites. A few companies had begun trials with streaming audio and video, but the uptake appeared to be somewhat cautious. The new regulation appeared to remove any hesitancy, and corporations began adopting and rolling out streaming content at a more rapid pace. Live and recorded broadcasts of company earnings conference calls are now commonplace. Many years from now, historians will probably look back and name Regulation FD as one of the turning point for the adoption of streaming content.

New Corporate Initiatives

Technology is advancing rapidly, and corporations' budgets are continually being adjusted to maintain earnings growth. Corporations with the best bottom lines are rewarded through rising stock prices; those that fail are dealt with in a decisive manner. The competitive nature of business forces companies to constantly find new ways to improve productivity while keeping costs low. Companies are changing the way they view their workforce. Telecommuting, for instance, is more prevalent than it was in the 1990s. More employees are working from home, and this trend is fueling the need for more bandwidth to enable them to connect back to the office at the same level of service they would have if they were physically present.

Another change that is becoming more commonplace is in the area of learning. Not long ago, corporate training consisted of reading, on-the-job training, or classroom tuition. Most teaching was done face-to-face. Today, a lot of teaching is being accomplished remotely. Distance learning, or e-learning as it is called, is fast becoming the norm. The benefit of this approach to the corporation is that less time is spent away from the office, as it eliminates the need for travel and usually means a more flexible schedule. All these points provide a cost benefit to the company.

Video Conferencing

Video conferencing is the transmission and reception of video images and speech between two or more people. This is accomplished through the use of cameras to capture the image and microphones that capture the voice for transmission. Incoming images and speech are presented on the computer screen and computer speaker. The costs of video conferencing equipment—which have traditionally been a key inhibitor to its adoption—have declined in recent years. This fact, coupled with the emergence of a standard called H.323, promulgated by the International Telecommunication Union (ITU) for supporting audio/video conferencing over IP, is encouraging more corporations to take an interest in the technology. The standard was first approved in 1996 and has been through several revisions; it has also been incorporated into equipment from multiple vendors.

Facilitating Meetings   The simplest and most popular application of video conferencing is hosting meetings. This reduces travel costs as well as travel time and makes meeting attendance more convenient. In most businesses today, a considerable number of meetings are conducted via telephones. Video conferencing provides remote participants with much of the face-to-face familiarity that comes with physical presence, including elements of facial expression, body language, and eye contact. The integration of video conferencing with other collaborative electronic tools—data transfer, shared whiteboards, and shared applications—can significantly enhance productivity.

Microsoft NetMeeting is an example of an application that combines video conferencing with useful collaborative tools. Aimed at the collaborative market, NetMeeting offers the following features:

Video and Audio Conferencing	Facilitates communication between any users on the network
Whiteboard	Allows the sharing of ideas through diagrams that are drawn on a shared screen space
Chat	Allows for real-time conversations via text with all meeting participants
Internet Directory	The Microsoft Internet Directory is a web site provided and maintained by Microsoft to locate people to call on the Internet
File Transfer	Allows for the exchange of files between participants. The transfer happens in the background and does not interfere with the flow of the meeting
Program Sharing	Lets you share multiple programs during a conference
Remote Desktop Sharing	Lets you operate a computer from a remote location

Security	Uses three types of security measures to protect your privacy
Advanced Calling	Gives you the flexibility to send a mail message to a NetMeeting user or initiate a NetMeeting call directly from your mail address book

eLearning   The growing use of networked technology to deliver training to workers has spawned an entirely new industry—one known as eLearning. In late 1999, Cisco Systems CEO John Chambers brought attention to eLearning by calling it the next "killer application." He further stated that education over the Internet was going to be so big it would make e-mail look like a rounding error.

The typical company today has a growing percentage of its workforce spread across a wide geographical area. It is not unusual to find a single department of less than ten people all located in as many as five different cities. In addition to the challenges such a dispersed work unit presents, it also makes it difficult to provide training to those who are away from the main office campus.

Research on telecommuting has shown that it enhances productivity, but it can be a drain on the department training budget if the employee needs to travel long distances to obtain training. Companies faced with this problem are turning to eLearning as an alternative approach. With eLearning, the course is brought to the employee anywhere he or she is without the need for travel. Training is delivered to employees when they need it and where they need it.

eLearning classes are becoming more sophisticated and are much more than presenting pages of a presentation. They incorporate the best elements of multimedia and provide the features that are typically only available in a classroom setting plus some that are unique to eLearning. Some of these features include:

▼ A live virtual classroom with a professional instructor

■ Multimedia features including 3-D graphics, audio, and advanced animation, making the concepts and lessons easier to understand and more engaging

■ The ability to raise your hand—get the instructor's attention—to ask questions

■ Students' ability to change seats and notify the instructor when they have left the room

■ Students' ability to chat with the instructors and fellow classmates

■ Students' ability to participate in quizzes

■ Students' ability to participate in a hands-on lab

■ Visual aids that are better than most of those presented in a formal classroom setting

■ Collaborative learning through the use of whiteboards, etc.

▲ Up-to-date information

Having sat through several well-structured e-learning courses, I have found them to be a very efficient way to impart training. The structure of the course and the technology behind it make for easier understanding of the subject matter. Interaction with the instructor and other students helps to set the pace of the lesson and provides a setting where everyone benefits from each other's experience much as you would in a regular classroom. Hands-on labs help you apply the knowledge and reinforce learning through trying the concepts.

SUMMARY

This chapter has tried to present several market drivers behind the need for, and growth of, broadband networks. We have touched on just a few of the technologies, services, and issues. In fact, I am sure that within a year of this book's being published, new services will be in development that were not even on the horizon at the time of writing. Hopefully, the technologies that were covered were a good enough incentive to take a closer look at broadband and its promise in the rest of this book.

CHAPTER 2

Commercial and Residential Networks

In Chapter 1, we examined some of the technologies, applications, services, and issues that are influencing the growth of broadband. In this chapter, we will take a look at some of the networking solutions that are available for home, multiunit buildings, and corporate offices. Not all the solutions reviewed support broadband services, but by taking a look at each, we gain a better and broader appreciation of how it all fits together.

HOME NETWORKING

Home networking, as a concept, is an entirely different animal from networks found in the office. In a home network, there is a lot to network, a fact that is not too obvious because it is an area that has been predominantly overlooked. There are four basic types of home networks:

▼ Data networks that connect PCs

■ Communication networks that connect phones

■ Entertainment networks that connect TVs, stereos, and game consoles

▲ Control networks that tie in lighting, security, home automation, heating, and cooling

Additionally, home network solutions fall into two broad categories, those that require new wires to be installed and those that are designed to be installed without the need to pull any new wires—a concept that is known as *"no new wires."*

Networking computers can be a daunting task. Unlike offices, consumer households don't have system administrators to configure, install, maintain, and protect their networks; and running wires can be difficult to impossible. The most important aspect of any home networking solution is ease of use, and as a result, this aspect very often become the final deciding factor on selecting a solution, even at the expense of performance. In an office environment, the opposite is true: performance is typically the first of several criteria considered, long before any thought for ease of use.

Growth of Home PCs

Home networking has been largely ignored due in part to the fact that until a few years ago most households had only a single PC—and there is no reason to network a single PC. Times are changing, and more households are purchasing a second computer. According to the U.S. Census Bureau, 51 percent of U.S. households had at least one computer in August 2000, and four in five of these households had at least one member using the Internet (see Figure 2-1).

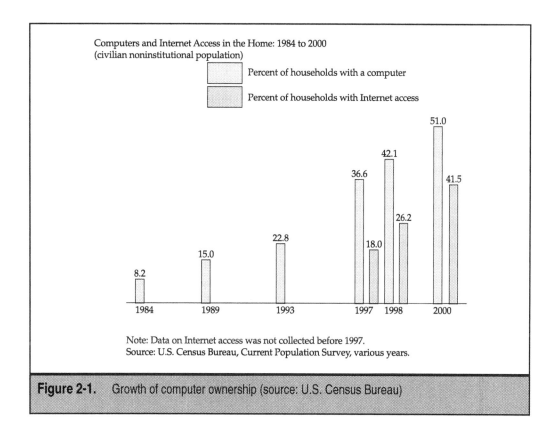

Computers and Internet Access in the Home: 1984 to 2000
(civilian noninstitutional population)

☐ Percent of households with a computer

☐ Percent of households with Internet access

Note: Data on Internet access was not collected before 1997.
Source: U.S. Census Bureau, Current Population Survey, various years.

Figure 2-1.   Growth of computer ownership (source: U.S. Census Bureau)

This data represents a greater than five-fold growth in home computer ownership since 1984. Following the path of previous inventions such as the television, more households are now installing a second PC—sometimes more—to meet the needs of the family. In May 2001, more than 20 million U.S. households had two or more PCs installed in the home. Figure 2-2 shows a Dataquest projection for the growth of home networks, which is projected to reach 7.9 million in 2003.

Driving these trends are several factors. Internet use is on the increase, PC prices have fallen below $1000, home offices are becoming more commonplace, and the PC is being used for more tasks. With PC use quickly becoming a national pastime, how does a parent choose between doing the family finances and letting their kid do research for a school project? One way is to develop a schedule, but this becomes inconvenient very quickly. The moment a second PC is installed, the competition for resources shifts to another device, a printer or access to the Internet, for instance. The next question becomes

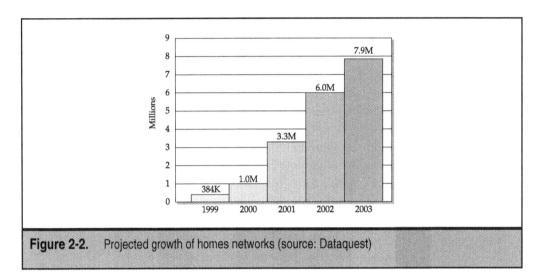

Figure 2-2. Projected growth of homes networks (source: Dataquest)

how to link both computers to share files and resources such as printers and high-speed Internet connections.

A home network provides flexibility and function—its immediate benefits include:

▼ **Printer sharing**   Printers become accessible from any place in the house.

■ **Simultaneous Internet access**   All users of the home network can share a single connection to the Internet, and in some cases, a single account.

■ **File sharing**   Files can be exchanged between computers without the need to resort to swapping floppy disks.

▲ **Multiplayer games**   Family members can participate in games designed for two or more players from two or more PCs in the house at the same time.

Several options exist for building a home network. We will examine some of these options, most of which fall within the category of "no new wires," the concept of providing a home networking solution without the need to add additional wiring. In this section, we will look at the following technologies:

▼ Phone line

■ Power line

■ Wireless

▲ Ethernet

Phone-Line Networking

Phone-line networking—commonly called HomePNA, short for Home Phoneline Networking Alliance (also the name of the group that created the standard)—is a networking solution that utilizes preexisting telephone wires in a building. As shown in Figure 2-3,

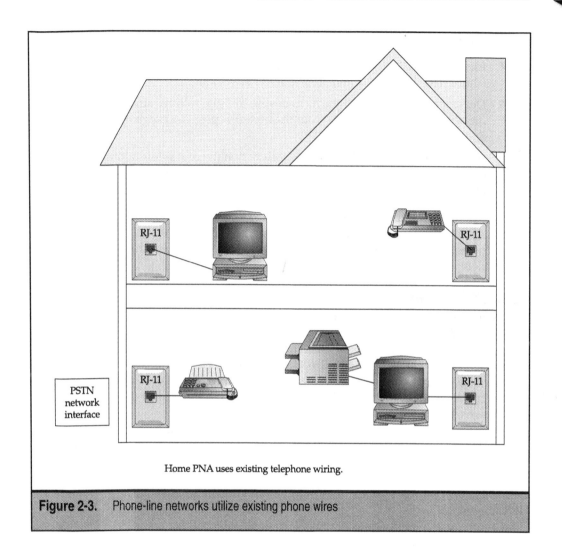

Home PNA uses existing telephone wiring.

Figure 2-3.   Phone-line networks utilize existing phone wires

the data signals travel on frequencies different from those of voice, and as a result, the wiring infrastructure is leveraged to carry both voice and data. The ability to utilize existing wires reduces the cost per node of this solution.

The Home Phoneline Networking Alliance

The Home Phoneline Networking Alliance, an association of industry-leading companies founded in 1998, developed the specification for HomePNA. The promoter members of the group include 3Com, Agere Systems, AMD, AT&T Wireless Services, Broadcom, Compaq, Conexant, Hewlett-Packard, Intel, Motorola, and Tut Systems. Since its formation, the Alliance's membership has grown to include over 150 companies, spanning the networking, telecommunications, hardware, software, and consumer electronics industries.

The primary objective of the Alliance is to:

▼ Ensure mass deployment of a consumer-friendly, low-cost, "no new wires" solution for in-home, phone line–based networking.

◼ Develop certification standards to ensure interoperability among HomePNA member company products from the broadest possible range of technology and equipment vendors.

▲ Achieve industry standardization both nationally and internationally through deployment and acceptance by appropriate standards bodies such as the ITU and the IEEE.

To gain the HomePNA's "seal of approval," manufacturers are required to go through a certification program to ensure interoperability. An independent body—CEBus Industry Council's (CIC) Pluglab at Purdue University—conducts the testing and certification process.

Network Features

The original version of HomePNA (version 1.0) was designed to operate with computers and devices up to 500 feet apart at a rate of 1 Mbps and is ideal for homes of up to 10,000 square feet, or 99.5 percent of homes in the U.S. The current version, 2.0, is a significant improvement over version 1.0. Based upon Broadcom chips, it operates at 10 Mbps, extends the distance between nodes to 1000 feet, is fully backward compatible. The version 2.0 specification has received approval by the ITU as a global standard, a fact that has opened up overseas markets.

At the Comdex 2001 show, the Home Phoneline Networking Alliance announced the availability of the Voice over HomePNA (VoHPNA) protocol—a version 2 enhancement. VoHPNA extends the 2.0 specification to include support for digital voice services, enabling eight simultaneous high-quality voice streams within the home (see Figure 2-4). With this new feature, you are able to integrate your plain old analog phone into the HPNA network.

A standard analog phone is plugged into a special VoHPNA adapter. The adapter is then plugged into the wall using the standard phone RJ-11 wall jack. Once connected, voice travels as packets over the wire along with any data traffic generated by your computers. Incorporate this feature with a broadband connection and you gain the ability to communicate using VoIP (Voice over IP) to any VoIP user in the world or another VoHPNA end point.

The third generation of HPNA is now being ratified and is slated to be finalized and ready for release in the fourth quarter of 2002. HomePNA 3.0 will be optimized for broadband entertainment including video on demand, streaming video and audio, head-to-head gaming, voice, and multi-PC file and Internet sharing applications without disruption of regular phone service. It will operate at a throughput rate of up to 100 Mbps. At 100 Mbps, the last major competitive edge of Ethernet is removed and HomePNA becomes a serious challenger for the home network. Version 3.0 will be compatible with

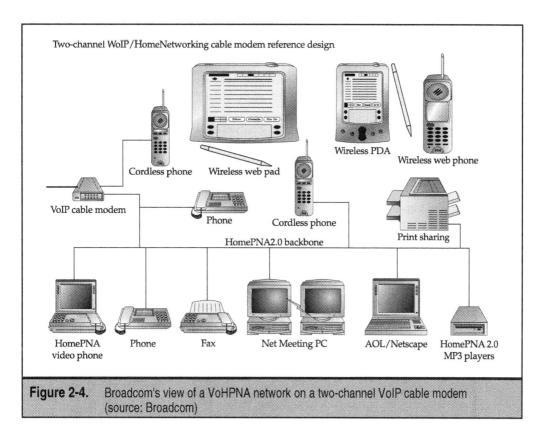

Two-channel WoIP/HomeNetworking cable modem reference design

Cordless phone   Wireless web pad

Wireless PDA

Wireless web phone

VoIP cable modem

Phone

Cordless phone

Print sharing

HomePNA2.0 backbone

HomePNA
video phone

Phone

Fax

Net Meeting PC

AOL/Netscape

HomePNA 2.0
MP3 players

Figure 2-4.   Broadcom's view of a VoHPNA network on a two-channel VoIP cable modem
(source: Broadcom)

other services, such as POTS, ISDN, and xDSL, and will be backward compatible with
HomePNA 1.0 and 2.0. It will also support VoHPNA.

Under the Covers of HomePNA

HomePNA version 2.0 operates as a multipoint CSMA/CD packet network that supports
unicast, multicast, and broadcast transmissions. A *unicast* frame is any frame that is addressed to a specific receiver; this is in contrast to a *multicast* frame, which is addressed to
a group of receivers, and a *broadcast* frame, which is addressed to receivers.

HomePNA operates in a similar way to Ethernet with the following exception: it
places no restriction on the wiring type, wiring topology, or termination. It has more in
common with Ethernet 10Base-2 in that it uses a common shared physical medium without the need for a hub or a switch. In contrast, 10Base-T uses a discrete wiring between
devices back to a hub or switch.

At the physical layer, frequency division multiplexing is used to create different channels for the different traffic types. Analog telephone service uses the lower part of the
available frequency spectrum—below 4 KHz. ADSL uses the spectrum up to 1.1 MHz,
and HPNA the frequencies between 4 and 10 MHz. The lower limit of 4 MHz makes it

feasible to implement the filters needed to reduce out-of-band interference between ADSL and HPNA traffic. The upper limit of 10 MHz was chosen to ensure broad applicability across different grades of wiring.

The protocol uses a CSMA/CD just like Ethernet, but it uses a new collision resolution algorithm called *distributed fair priority queuing (DFPQ)*. Eight levels of priority queuing were also implemented.

The brains behind the HomePNA v2.0 network are two Broadcom chips—the BCM4100 iLine10 analog front end and the BCM4210 iLine10 PCI/MSI controller.

The *BCM4100 iLine10 analog front end* acts as a transceiver that interfaces with the phone line and is responsible for sending and receiving analog signals from devices over 1000 feet away. It converts the analog signals received from the phone line into a digital format for the BCM4210 to process, and the digital output of the BCM4210 into an analog format for transmission onto the wiring medium. Key features include the following:

▼ An integrated iLine10 analog front-end transceiver for 10 Mbps home networking

■ A digital multiplexed transceiver interface (DMTI) to iLine10 MAC/PHY controllers

▲ Support for both iLine10 and HPNA 1.0 operation

The *BCM4210 iLine10 PCI/MSI controller* is a chip of many talents. It features a highly adaptive and integrated Internet communications processor that transmits and receives data at speeds up to 10 Mbps and above while overcoming any noise introduced by wiring impairments. The maximum data rate for the BCM4210 controller is 16 Mbps. Basically, the BCM4210 takes the unfiltered digital signal it received from the BCM4100 chip and removes any noise that may be present before passing it to the processor for processing. The processor then sends an acknowledgment back to the sending computer to let it know that it successfully received the data.

The BCM4210 and BCM4100 combination functions on existing home telephone wiring (up to 1000 feet) without disturbing normal telephone operation or other existing services such as V.90 modems, ISDN, or G.lite DSL (see Figure 2-5).

HomePNA at a Glance

HomePNA offers these basic features and advantages to the home networker:

▼ It is inexpensive, reliable, and easy to install—it is possible to get a two-computer kit for under $100.

■ It uses FDM—frequency-division multiplexing technology.

■ It is based on a standard that ensures interoperability (always ensure that the equipment has the HomePNA certification seal).

■ It operates at 1 Mbps for v1.0 and 10 Mbps v2.0, even when the house phone is being used—fast enough for bandwidth-intensive applications.

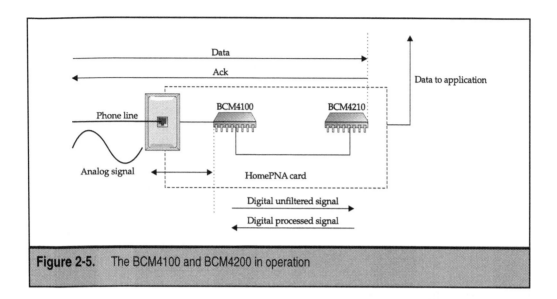

Figure 2-5. The BCM4100 and BCM4200 in operation

- It utilizes existing wiring and requires no additional network equipment—no hubs or routers.

- It supports up to 25 devices with a distance limitation of 500 feet for v 1.0 and 1000 feet for v 2.0.

- It can be connected to other types of networks.

- It works in Macs and PCs.

▲ It supports digital voice streams.

Power-Line Networking

Power-line communications, also known as PLC, are, like phone-line networks, based on the concept of no new wires. These networks have an edge over phone-line networks in that every room in a home typically has at least one or two power outlets. In fact, the average U.S. home has an average of three power outlets per room, resulting in greater choice of location and mobility—as long as there is a power socket, there is a connection to the network, as shown in Figure 2-6. In contrast, telephone jacks are usually limited to a few rooms in the home, with some rooms having just one—thereby limiting the places from which a connection to the network may be made.

The technology has tremendous market potential. In the coming years, we're likely not only to use PLC to network PCs over existing residential power-line infrastructure, but also to network anything that is electrically powered. It is conceivable that consumer electronics equipment (stereos, TVs), home automation, control and security systems, PCs, and white goods (refrigerators and dishwashers) could all be networked using the

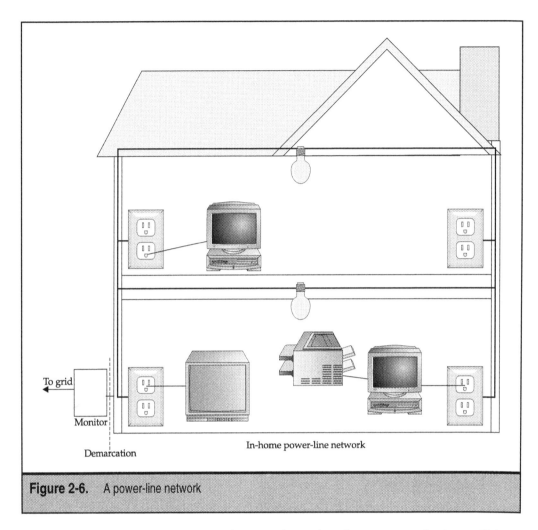

To grid

Monitor

Demarcation

In-home power-line network

Figure 2-6. A power-line network

existing AC power infrastructure. What's tricky is that these various devices will have varying throughput requirements ranging from 100 bps to 10 Mbps or more.

Technical Challenges

Having ubiquitous network access points throughout the home is good, but as a transmission medium, power lines are horrendous. Of the different media types used for networking, the wiring used for powering the home is certainly the most difficult—and a very harsh environment—for communication. At some frequencies, the transmitted signal may arrive at the receiver with relatively little loss, while at other frequencies it may be overwhelmed by noise.

The worst aspect of using power lines is their unpredictable and inconsistent characteristics. The transfer function of the wiring changes with the different types of devices that are being powered at any given instance. Switching on a motor, for instance, wreaks havoc on the transfer capabilities of the wire because of the time-varying impedance characteristics of the motor. In fact, just the act of plugging in a new device may change the transfer function of the line.

Without a consistent baseline for communication, the challenge of using the wires for data communications greatly increases. And as if that were not bad enough, consider other issues, such as interference. Interference in this environment is like riding into an ambush; just when you think all is well, someone jumps you—so much for the payload. The net result is that there is no guarantee that the packet sent will be the same as the packet received.

Types of Power-Line Network

Unlike its phone-line counterpart, power-line networking has been slow in advancing a single standard for interoperability for networks operating over 1 Mbps. Until recently, the only existing power-line networking specifications were for low-bandwidth applications. An area that is receiving some debate is whether or not a single standard for both low- and high-bandwidth applications is practical. The components and engineering necessary to provide speeds of 1 Mbps or more are more expensive than those required for applications that run in the kilobit per second range.

If the application is exclusively for features that require only low speeds, it may not make economic sense to purchase a system that supports greater speeds. Until prices fall enough to make the point moot, it is highly probable that we will continue to have multiple standards for low- and high-bandwidth solutions. The speed of market adoption and technology enhancements will determine when this will change.

High-Bandwidth Networks   There are two competing standards for high-speed (1 Mbps and above) power-line networks. In June 2001, the HomePlug Powerline Alliance announced the completion of its version 1.0 specification for power line–based communications at a data rate of up to 14 Mbps. Several vendors, including Cayman Systems, Netgear, and Phoenix Broadband Corporation, have already announced their intent to begin producing compliant products by late 2001 to early 2002.

A competing group, the Consumer Electronics Association's (CEA) R7.3 Committee, is working to develop a scalable power-line networking standard that will meet the throughput and QoS requirements of all power-line networking products. They are still in the process of developing their version of a specification and have given a projected completion date of mid 2002.

The HomePlug Powerline Alliance is composed of 90 diverse companies from different industries including services, retail, hardware and software, semiconductors, and technology sectors. Based on the PowerPacket INT5130 chip set by Intellon, the 1.0 specification supports a data rate of 14 Mbps.

Critics of the new specification—some members of the opposing CEA camp—believe the technology is too PC-centric and is not optimized for home entertainment use. They contend that a true home network should support multiple nodes, Quality of Service with little latency, multicasting and broadcasting—all features that are promised to be addressed in the CEA R7.3 version of the specification. The CEA also argue that HomePlug's speed makes it too expensive to use for slower applications such as home control and security, and that R7.3 will be more suited to this market.

HomePlug, in response, has been urging vendors who intend to target the lower-speed market to use the HomePlug spec as a basis for building HomePlug-compliant products. In the absence of a competing specification, HomePlug has an edge and should most likely benefit from being first to market. The vendors addressing the higher-speed markets will probably be first to adopt the new specification, as a few have already confirmed, but for the lower-speed markets only time will tell.

Under the Covers of HomePlug 1.0 Intellon's PowerPacket INT5130 chipset forms the basis of the new HomePlug 1.0 specification. It includes both a robust physical layer (PHY) and a media access control (MAC) protocol. The PHY specifies the modulation, coding, and basic packet formats, while the MAC protocol controls the sharing of the medium among multiple clients.

The HomePlug PHY uses an enhanced form of orthogonal frequency-division multiplexing (OFDM) with forward error correction (similar to the technology found in DSL modems). OFDM splits the available range of frequencies (4.3 MHz to 20.9 MHz) on the electrical subsystem into 84 separate carriers. The high-speed data stream is then divided into multiple parallel bit streams with each bit being placed on some of the carriers. If a transmission is disrupted because of noise, the PowerPacket chip senses the fact and switches the data to another set of carriers (Chapter 3 provides more details on modulation and error correction techniques).

The MAC protocol is a variant of the well-known Carrier Sense Multiple Access with Collision Avoidance (CSMA/CA) protocol. Features have been added to support priority classes, provide fairness, and allow the control of latency. In order to support CSMA/CA, the PHY must support burst transmission and reception just like an Ethernet network; in other words, each client enables its transmitter only when it has data to send and, upon finishing, turns off its transmitter and returns to the receive mode.

The HomePlug frame format shown in Figure 2-7 uses a virtual carrier sense (VCS) methodology and contention resolution to minimize collision. The frame format uses *delimiters* to indicate the start of a frame, the end of a frame, and the response; it consists of a preamble and a frame control. Start of frame, end of frame, and response delimiters all have the same symbol structure, with the frame control being used to indicate whether the delimiter is the start of the frame, the end of the frame, or a response delimiter.

If the delimiter is indicating the start of the frame, it specifies the duration of the data to follow. If it is used to indicate the end of the frame or a response, it defines where the end of the transmission lies. So the receiver is able to determine, from the frame control,

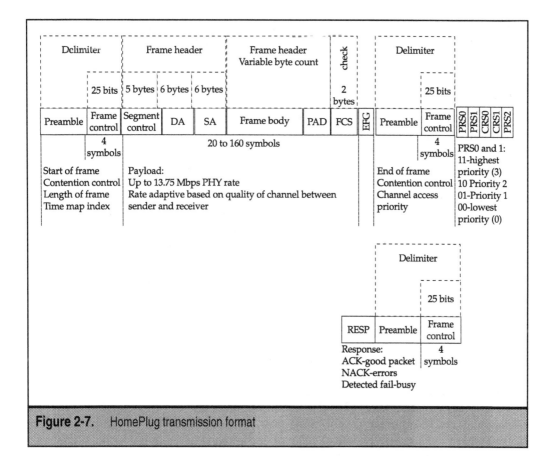

Figure 2-7.   HomePlug transmission format

the duration for which the channel will be occupied by the transmission; it then synchro-nizes its VCS to this time interval. If the frame control cannot be decoded, it must be as-sumed that a frame of the maximum allowed length is being transmitted, and the VCS is set accordingly. The VCS can be subsequently corrected the next time a delimiter is seen, whether it is an end-of-frame or response delimiter.

As with all collision avoidance schemes, all unicast frames are always acknowledged by sending a response delimiter. If the sender does not receive a response, it assumes a collision has occurred and the frame is retransmitted. The receiving station also has the option of sending a FAIL message if it does not have the resources to handle the frame, or it may send a NACK to indicate that the packet contained errors that could not be cor-rected by FEC.

As with Ethernet, the contention method has a random back-off algorithm whenever a collision is assumed (the word "assumed" is used because in collision avoidance, the assumption of a collision is made if an ACK response is not received). Upon completion of a transmission, other nodes wishing to transmit signal their priority in a priority resolution interval indicated by PRS0 and PRS1 in Figure 2-7.

Low-Bandwidth Networks   There are several specifications for low-bandwidth networks—X10 and LonWorks to name a couple. You may have seen the ads for the little X10 surveillance cameras available from X10.com. Don't be fooled by the name; X10 is a lot more than a camera—it embodies a whole suite of command and control devices for the home. It was developed by a company in Scotland—Pico Electronics—with the first shipped product hitting the market in 1978. The patent on the standard has since expired, and prices have fallen sharply. X10's ActiveHome lets you control the lights, temperature, stereo, garage door, television, and more using your PC.

X10 technology is a good example of simplicity and ease of use. It transmits signals over electrical wiring by using 120 KHz signal bursts, each 1 millisecond long. Each bit transmitted consists of two bursts; a binary 1 is represented by a burst followed by a no-burst of (1 ms duration), while a 0 is a no-burst followed by a burst. A preamble that consists of the sequence "burst, burst, burst, no-burst" is used to indicate the start of a packet. The first two bursts are neither a 0 nor a 1, so this makes for an easy identifier of the start of the preamble sequence. The addition of a 1 (burst—then no-burst) completes the sequence and further helps to identify the sequence as a preamble. The logic of the preamble is based on the assumption that if spurious noise inadvertently generates a double burst, it would be highly improbable that the noise would have followed the double burst with a binary 1 (burst—no burst).

The data rate is a slow 18.62 bits/sec. The X10 protocol transmits each bit at the change of the voltage cycle, so it takes 13 cycles to transmit an entire frame (shown in Figure 2-8). Additionally, each command is sent twice to ensure reliability, for a total of 26 cycles. An additional 3 cycles are required after each pair of commands, bringing the total to 29 cycles for a single 9-bit command string (data portion of the frame). The resulting speed in a 60 Hz circuit is about 18.62 bits/second (60 / 29 * 9 – number of data bits).

The X10 ActiveHome works with a computer to run customized home lighting and appliance routines. The software allows for the creation of macros or groups of events that can be triggered by motion, time, or remote control. Up to 256 devices may be controlled.

LonWorks (a group of 36 member companies) formed the LonMark Interoperability Association in 1994. The membership has since grown to over 200 companies. The mission of the group is to enable the easy integration of multivendor systems based on LonWorks networks. LonWorks networks are aimed at higher-end industrial applications.

In a LonWorks network—a creation of Echelon Corporation—intelligent control devices, or nodes, communicate with each other using a common protocol. Each node in the network implements the protocol and performs control functions. In addition, each node includes a physical interface that couples the node's microcontroller with the communications medium. Each node in the network—switches, sensors, motors, motion detectors, etc.—performs a simple task. The overall network performs a complex control application such as automating a building.

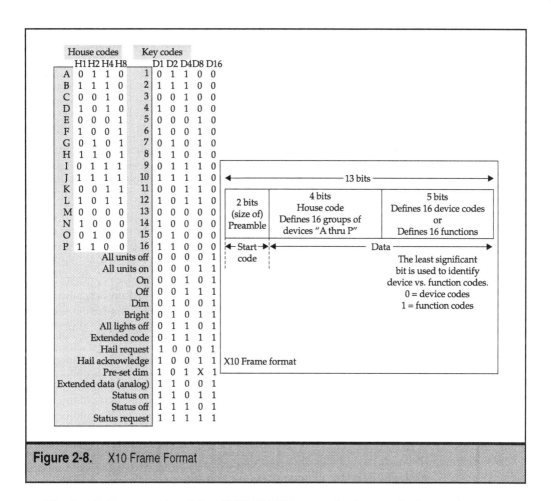

Figure 2-8.   X10 Frame Format

The LonTalk protocol and the ANSI EIA709.1 standard are at the heart of a LonWork network, which is media independent and works on power-line systems among others. In a power-line application, a speed of 5.4 Kbps is achievable and the number of devices and physical extent of the network are dependent on the environment.

The *EIA-709.2—Control Network Powerline Channel Specification*, developed by committees and subcommittees operating under the auspices of the CEMA (Consumer Electronics Manufacturers Association) Engineering Policy Council is based on the LonWorks control networking platform (by Echelon). The EIA-709 standard is divided into three parts:

▼   EIA-709.1 defines a communication protocol for networked control systems in a home.

■   EIA-709.2 addresses transceivers for networking consumer products over existing power lines using narrow-band signaling.

▲   EIA-709.3 provides for free-topology twisted pair media.

EIA 709.2 further defines physical communication over 120V to 240V AC power lines inside and outside of homes. The channel occupies the bandwidth from 125 KHz to 140 KHz and communicates at 5.65 kilobits per second. The standard supports both two- and three-phase electrical configurations and defines a narrow-band signaling technology that meets regulatory requirements for North America and the European Community. This specification originated from Echelon's PLT-22 Powerline Transceiver.

In addition to the EIA-709 standard, an entire suite of specifications—also developed by CEMA—has been designed for home automation called the Consumer Electronic Bus (CEBus). It provides for a standardized communication facility for exchange of control information as data among devices and services in the home. The CEBus set includes the following:

▼ EIA-600.10 Introduction to the CEBus Standard

■ EIA-600.31 Powerline Physical Layer and Medium Specification Describes the CEBus, power-line physical layer, and medium portion of the CEBus system. Its purpose is to present all the information necessary for the development of a power-line physical layer for a CEBus device. The specification occupies bandwidth from 100 KHz to 400 KHz.

■ EIA-600.32 Twisted Pair Physical Layer and Medium Specification

■ EIA-600.33 Coaxial Cable Physical Layer and Medium Specification

■ EIA-600.34 Infrared Physical Layer and Medium Specification

■ EIA-600.35 RF Physical Layer and Medium Specification

■ EIA-600.37 Symbol-Encoding Sublayer

■ EIA-600.38 Powerline/Radio Frequency Symbol Encoding Sublayer

■ EIA-600.41 Description of the Data Link Layer

■ EIA-600.42 Node Medium Access Control Sublayer

■ EIA-600.43 Node Logical Link Control Sublayer

■ EIA-600.45 Node Network Layer Specification

■ EIA-600.46 Node Application Layer Specification

■ EIA-600.81 Common Application Language (CAL) Specification

▲ EIA-600.82 CAL Context Description

In Europe, the European Committee for Electrotechnical Standardization (CENELEC) has adopted the *EN50065-1* standard, which has been in existence since 1990 and covers the frequency range of 3 KHz to 148.5 KHz. The standard segments the available frequency range according to application using generic bands A, B and C, as shown in Figure 2-9.

Band A (3–95 KHz) is defined for utility applications, and Bands B and C (95–148.5 KHz) for home automation. In addition to frequency segmentation, the Standard defines

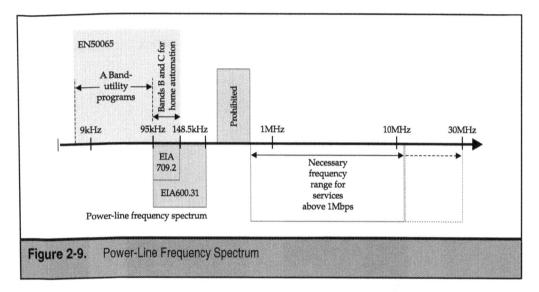

Figure 2-9.   Power-Line Frequency Spectrum

maximum voltage levels and total harmonic distortion (THD) limits at the point of signal injection.

Wireless Networks

In a wireless network (such as the home network shown in Figure 2-10), all devices communicate by using radio frequencies. This makes it the most mobile of the "no new wires" networks. As mentioned previously, phone-line networks are great, but their connection points are limited to the rooms that are equipped with telephone jacks. In many cases, this limits the choice to only a few rooms, and often a room has only a single jack. As a result, you end up either unplugging your phone or installing a splitter jack that still renders the phone useless. In either case, you are tethered to that one location in the room. Power-line networks offer greater mobility—wherever a power outlet exists, so does a network access point. In the typical house, there are three plugs per room on average, yielding a choice of three different network access points in a single location.

With a wireless network, you are not bound to any one location; you are free to roam about the house and still have access to the network—the ultimate in mobility. As with power-line networks, there are several different types of wireless networks ranging in speeds from slow and inexpensive to fast and fairly expensive. The three types of networks covered in this section are these:

▼   WiFi

■   HomeRF (SWAP)

▲   Bluetooth

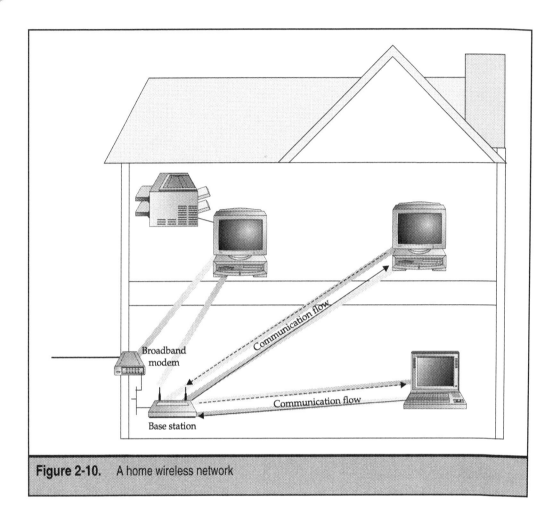

Figure 2-10. A home wireless network

WECA (WiFi) Networks

WECA, short for the Wireless Ethernet Compatibility Alliance, is an alliance of companies seeking to ensure interoperability among IEEE 802.11b High Rate products from multiple manufacturers, and to promote the technology within both the business and consumer markets. The Alliance, formed in 1999, was founded by Cisco Systems, 3Com, Intersil, Agere Systems, Symbol Technologies, and Nokia; it is now supported by over 110 companies. Agilent Interoperability Certification Lab, an independent company, does interoperability testing.

WiFi is short for Wireless Fidelity, just as HiFi is short for High Fidelity for audio equipment. It's a brand mark of certification, or a seal of approval, which confirms that a particular piece of equipment is compliant with the standard on which WiFi is based. WiFi targets the higher-end spectrum of networking and is becoming more pervasive in office buildings, public places such as cafés and airports, and the home. The technology operates at 11 Mbps and is targeted more at offices as an alternative to 10Base-T-wired Ethernet.

WiFi is based on a variation the IEEE 802.11 specification for wireless networks called 802.11b. The original IEEE 802.11 specification addressed two ways of communicating between devices at speeds up to 2 Mbps. The methods are *direct-sequence spread spectrum (DSSS)* and *frequency-hopping spread spectrum (FHSS)*. Both use frequency-shift keying (FSK) modulation techniques and are based on the spread spectrum radio waves in the worldwide 2.4 GHz Industry, Science, and Medicine (ISM) band.

IEEE 802.11   In 1997, the IEEE completed the 802.11 standard for wireless networks (WLAN), and in 1999, it ratified a revision of the 802.11 standard called 802.11 High Rate (802.11HR or 802.11b). The revised version provides a significantly higher speed—11 Mbps—over the original version, which operated at speeds up to 2 Mbps, while maintaining the 802.11 protocol.

The 802.11 standard specified a peak rate of 2 Mbps and offered a choice of two different physical (PHY) layers, DSSS or FHSS. Vendors were left to choose the one that was more appropriate to their specific application. Many chose to go the DSSS route because the 802.11b standard was also based on DSSS, making for a smoother migration from 2 Mbps systems to the 11 Mbps versions.

The difference between 802.11 and 802.11b lies in the PHY layer. 802.11b uses DSSS exclusively because of its potential for higher data rates. The original 802.11 uses both FHSS and DSSS and specifies a peak data rate of 2 Mbps. Other than the PHY layer, 802.11 and 802.11b protocol operation are the same.

Spread Spectrum Modulation Techniques   In spread spectrum modulation, the data is sent in small chunks over a number of discrete channels. By spreading the signal over a wider band of frequencies, some bandwidth is sacrificed in order to gain better signal-to-noise performance. This undermines a basic rule of conserving frequency bandwidth where possible, but the spreading process makes the data signal much less susceptible to electrical noise than do conventional radio modulation techniques, and so it is worth the trade-off. Other transmission and electrical noise, typically narrow in bandwidth, will interfere with only a small portion of the spread spectrum signal, resulting in much less interference and fewer errors when the receiver demodulates the signal. Spread spectrum was first developed for use by the military because its use of wideband signals makes it difficult to detect and more resistant to attempts at jamming.

Frequency-hopping spread spectrum (FHSS), or *frequency-hopping code division multiple access (FH-CDMA)*, modulates the signal onto a carrier signal that hops from frequency to frequency over a wide bandwidth. The act of changing the carrier frequency periodically reduces interference from rogue signals. To properly receive the signal, the receiving station must use the same hopping code as the sender and must be listening to the incoming signal at the right time and correct frequency.

FCC regulations require that 75 or more frequencies be used per transmission channel with a maximum time of 400 ms spent at any single frequency. It is possible for two stations to use the same frequency without interfering with each other if they use different hop codes. By using this modulation technique, speeds up to 2 Mbps can be achieved. The 802.11 standard calls for normal operation at 1 Mbps with an optional 2 Mbps in extremely clean environments.

Direct-sequence spread spectrum (DSSS), also known as *direct sequence code division multiple access (DS-CDMA)*, uses a technique that combines the data with a higher data rate bit sequence called a *chipping code*. A chipping code uses a specific set of bit patterns to represent a 0 or 1 bit of the data stream. As a result, a 0 bit may be sent as a stream of bits such as 01001100010. The redundant chipping code helps the signal resist interference and also enables the original data to be recovered if data bits are damaged during transmission.

The ratio of chips per bit is called the *spreading ratio* or *processing gain,* and the higher the processing gain, the greater the resistance to interference. The FCC has set the minimum linear processing gain at 10; 802.11 specification sets their minimum at 11. With this technique, it is possible to achieve speeds in excess of 2 Mbps, but the 802.11 specifications specify a peak data rate of 2 Mbps with a fallback rate of 1 Mbps in extremely noisy conditions. The 802.11b specification amends only the PHY layer of the architecture, DSSS is used exclusively because of its higher data speed potential, and two data rates were added: 5.5 Mbps and 11 Mbps. Table 2-1 compares the two modulation techniques.

802.11 MAC Layer   The data link layer within 802.11 consists of two sublayers: logical link control (LLC) and media access control (MAC). 802.11 uses the same 802.2 LLC and 48-bit addressing as other 802 LANs, allowing for seamless operation with wired Ethernet via a bridge or *access point*. The MAC remains unique to WLANs, but an access point that con-

	Frequency Hopping	Direct Sequence
Cost	Lower	Higher
Power consumption	Lower	Higher
Tolerance to interference	Higher	Lower
Data rate	Lower potential data rates	Higher potential data rates
Capacity	Highest aggregate capacity	Lower aggregate capacity
Range	Less	More
Operation	1 Mbps but allows for 2 Mbps in very clean environments	2 Mbps and greater, with a fallback rate of 1 Mbps in very noisy conditions. 802.11b achieves speeds of 11 Mbps with fallback rates of 5.5 Mbps, 2 Mbps, and 1 Mbps.

Table 2-1.   IEEE 802.11 Frequency Hopping Versus Direct Sequence Relative Comparison

nects a WLAN channel to a LAN backbone masks any differences. Some of the unique features of the frame format and MAC allow for fast acknowledgment, the handling of hidden stations, power management, and data security.

In standard Ethernet, the CSMA/CD (Carrier Sense Multiple Access with Collision Detection) protocol regulates how each station gains access to the medium and how it behaves once a collision is detected. The 802.11 WLAN standard uses a collision avoidance (CSMA/CA) scheme, as collision detection is not possible in radio systems because of what is known as the "near/far" problem.

In radio communication, the transmission drowns out the stations' ability to hear the collision, so a scheme for avoiding collisions is more appropriate. Here's how CSMA/CA works: A station listens to the airwaves before sending; if the channel appears to be clear, the station transmits. The receiving station issues an ACK once it receives the packet. The sending station assumes a collision if an ACK is not received within a set time frame and resends the frame after waiting for a random period.

The 802.11 frame format relies on a 50 ms interframe gap, and the standard requires a receiving station to send an ACK within 10 ms of receipt, assuming the CRC check is correct. The 10 ms delay ensures that the receiving station is given an opportunity to immediately seize the airwaves before the 50 ms interframe gap expires, at which time any station can transmit.

WiFI Topology   Wireless devices operate in two basic modes: peer-to-peer (without an access point) or client/server (with an access point). In peer-to-peer mode—also called ad hoc mode—two or more wireless devices that have recognized each other can begin communicating within a single cell without the presence of an access point. This mode of operation is called an *ad hoc network* or an *independent basic service set*. Figure 2-11 shows the different components of a WiFI network and how they interact.

A *basic service set (BSS)* is a cell that normally includes one or more wireless devices and an *access point (AP)*. The AP, the bridge to the wired world, supports up to 250 stations, depending on the configuration. It is important to remember that once an AP is present, all communication between wireless devices flows through the AP. The devices cease direct communication with each other—in other words, they quit operating as an independent basic service set or as an ad hoc network—and the method of communication changes from a peer-to-peer mode to a client/server approach.

When the AP is connected to a wired LAN, the wireless network is considered to be operating in *infrastructure mode*. The connection between two APs, or an AP and a wired device, is called the *distribution system (DS)*. The access point, or the antenna that is attached to it, is usually mounted in a position that offers the best coverage—high on a wall or on the ceiling. Multiple APs can be deployed in a way that allows overlapping of cells—overlapping BSSs—to increase coverage and decrease congestion, thereby forming an *extended service set*.

For home use, WiFi offers Ethernet speeds without the wires, but at a price. The access points and transceiver cards tend to be expensive. Unlike phone-line and power-line networks that can be bought in complete packages for networking two computers at a minimum, the components are typically sold separately, which makes it a more daunting proposition for the do-it-yourselfer.

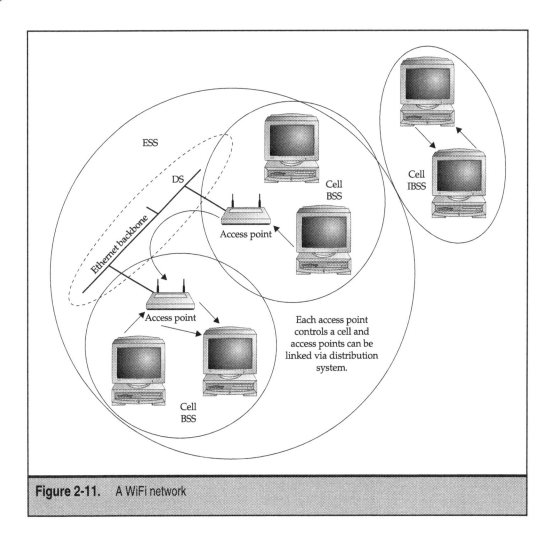

Figure 2-11. A WiFi network

HomeRF Networks

The Home Radio Frequency Working Group, an alliance of companies that initially began with five member companies in 1998 and has now grown to include seventy, developed HomeRF or Home Radio Frequency. The group was formed to establish an open industry specification for wireless digital communication between PCs and consumer electronic devices anywhere in and around the home. In fall 2001, it announced the formation of a HomeRF European Working Group to focus on market differences in that region.

The working group has developed a standard called Shared Wireless Access Protocol (SWAP), which defines a common interface that supports wireless voice and data networking in the home. Designed to carry both voice and data traffic and to interoperate with the Public Switched Telephone Network (PSTN) and the Internet, it operates at 2.4 GHz using frequency hopping at speeds up to 2 Mbps.

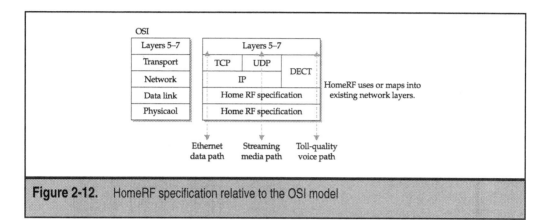

Figure 2-12. HomeRF specification relative to the OSI model

The protocol has built-in support for up to six full-duplex Toll-Quality voice channels. The technology is a combination of the Digital Enhanced Cordless Telephone (DECT) and WLAN technology. Both time-division multiple Access (TDMA) and CSMA/CA are supported. TDMA is used for voice and time-critical applications, while CSMA/CA is used for high-speed data. As is evident from Figure 2-12, the protocol operates at the physical and data link layers of the OSI model and uses or maps into existing network layer protocols.

When only data communication is supported, the network operates in a typical wireless fashion in that all nodes are equal and have equal access to the network. For interactive voice, a connection point (CP) is required that attaches to the PC via a USB port and provides gateway functions to the PSTN network. The CP can also be used to schedule device wake-up and polling.

HomeRF Feature Summary   A summary of the HomeRF features are provided in the following table:

Feature	Description
Uses frequency hopping	50 hops/second
Frequency range	2.4 GHz
Power	100 mW
Data rate	1 Mbps using 2FSK modulation 2 Mbps using 4FSK
Number of stations	127 nodes which may include: A connection point support voice and data services A voice terminal that uses only TDMA A data node that uses CSMA/CA A voice and data node providing both functions

Feature	Description
Voice channels	6 full duplex
Security	Blowfish encryption
Compression	LZRW3-A algorithm
48-bit network ID	Enables multiple networks

HomeRF v2.0   In 2001, the HomeRF work group introduced the HomeRF 2.0 standard, which is meant to be ideal for the broadband, multimedia home environment. HomeRF 2.0 enables Toll-Quality wireless voice with up to eight channels, priority access for streaming media, and 10 Mbps data rates with fallback rates of 5.5 Mbps or lower as conditions warrant. Designed for consumers, HomeRF is touted to be easy to install, secure, reliable, and affordable. HomeRF proponents expect to achieve speeds of up to 25 Mbps by 2002 with full backward compatibility.

In October 2001, Proxim, a networking peripheral maker and a member of the HomeRF Alliance, announced the availability of the HomeRF v2.0 in its latest generation of wireless networking devices. Intel, in contrast, announced in March of the same year that their next-generation Anypoint WLAN product would support the IEEE 802.11b standard instead of HomeRF v2.0. The original equipment supported HomeRF v1.0. The reason given was the surprisingly fast penetration of the IEEE 802.11b standard into homes. The competition in this area is just beginning to heat up, and 2002 should prove rather interesting.

Bluetooth

Bluetooth promises to eliminate the need to interconnect devices with wires through the use of short-distance wireless connections. For instance, the link between a computer and a printer would be accomplished using radio waves. In 1995, the telecommunications and information technology industries commissioned a study to examine the feasibility of a low-cost, low-power radio-based cable replacement, a wireless link that would eliminate the need for communication cables for short distances. The vision was to facilitate a network where small portable devices would be able to communicate in an ad hoc way leading to the integration of the technology into future devices. The study led to the birth of Bluetooth. Today, over 1000 companies belong to the Bluetooth Special Interest Group, and they would all like to see Bluetooth's radio communications take the place of wires for connecting peripherals, telephones, and computers.

Bluetooth communicates on the same frequency that is used by WiFi and is also based on the same IEEE 802.11 standard. A potential misconception is that Bluetooth competes with, or is a replacement for, WiFi. Bluetooth, however, was not developed to be a LAN solution, nor was it meant to transmit large amounts of data. Its specific aim was to provide a dynamic way of handling different mobile devices. As result, we will cover the topic with only enough depth to provide an appreciation of its features and how it differs from WiFi.

Bluetooth Functional Description   As mentioned earlier, Bluetooth communicates on a frequency of 2.45 GHz, the same frequency used by IEEE 802.11, but so do other devices that are already found in the home. Cordless phones, for instance, make use of this frequency band, so a design point must be to ensure that devices do not interfere with each other. A representative Bluetooth network is shown in Figure 2-13.

Bluetooth devices use extremely weak signals of 1 milliwatt (mW)—a cell phone is 3000 times more powerful—which limits the range of the signal to about ten meters but does not prevent the signals from penetrating walls. This is an important aspect of the technology that differentiates it from infrared-based technologies such as IrDA. Infrared requires a direct line of sight and so can control only objects that are unobstructed and are directly within its narrow-range line of sight. Bluetooth, by contrast, is able to control devices in another room.

Even though the transmission power will in most cases be limited to 1mW, power levels of up to 100mW are possible depending on the equipment power class. In general, Bluetooth devices will have a nominal operating range of about 30 to 40 feet. References have been made to Bluetooth as a means for long distance wireless communication; in this application a bluetooth connection would be established between a computer and a cell phone. The cell phone would be used to establish a connection and facilitate communication with a remote device. The Nokia web site provides a very good vision of how they see the technology evolving.

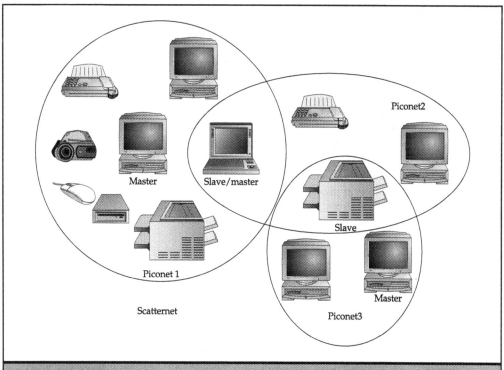

Figure 2-13.   A Bluetooth network

Bluetooth uses spread spectrum frequency hopping, which limits interference (see Table 2-1 for spread spectrum features). The signal hops between 79 different frequencies 1600 times per second, which minimizes the risk of interference, as any interference on a particular frequency will last only as long as the transmission remains on that frequency. Bluetooth systems create a *personal area network (PAN)* or *piconet*. Each device has a built-in high-performance integrated radio transceiver with a unique 48-bit address, 3 bits of which are used for the media access control (MAC) address.

A piconet starts with two devices and may grow up to eight (hence the three bits, which is the minimum number needed to define eight unique addresses). During the normal operation of a piconet, all devices are considered peers. When the piconet is being established, however, one device will act as a master for synchronization purposes and the others act as slaves. The master uses its own clock and hopping sequence to synchronize all the slaves.

Once the piconet is established, the members stay in contact with each other by randomly hopping frequencies in unison to avoid interference, including other piconets that may be operating in the same room. As new Bluetooth-enabled devices enter the range of the piconet, an electronic conversation automatically takes place to determine whether they have data to share or whether one needs to control the other. A slave device in one piconet may be a master device in another, or a slave device may communicate in multiple piconets. When piconets overlap in this fashion, a *scatter net* is formed.

Before a piconet is formed, all devices operate in a STANDBY state and periodically listen for messages every 1.28 seconds on a set of 32 hop frequencies (most countries use 32 frequencies). Any device is able to initiate the formation of a piconet and does so by issuing a PAGE message or an INQUIRY message; in so doing, the initiating device becomes the master device.

The PAGE is used in instances when the target device address is known; when it is not known, an INQUIRY is sent followed by the PAGE. The initial PAGE is sent on 16 different hop frequencies, and if a response is not received, the message will be resent on the remaining 16 frequencies. Devices may operate in a HOLD, SNIFF, or PARK mode. HOLD mode—a power-saving mode—is used if no data is to be transmitted; it is especially useful when connecting several piconets. In PARK mode, a device gives up its MAC address and responds only to broadcast messages. In SNIFF mode, a slave listens to the piconet at a reduced rate at an interval that is programmable.

Bluetooth Feature Summary   A summary of the key features of Bluetooth follows:

▼ Devices in a piconet share a common communication channel with a total capacity of 1 Mbps. About 80 percent of this capacity is available to data; the rest is consumed by protocol overhead for management, handshakes, headers, etc.

■ The available frequency range is 2400–2483.5 MHz, with 79 1 MHz channels (the portion used is actually 2402 MHz to 2480 MHz). Japan uses the frequency range of 2472–2497 MHz with 23 1 Mhz channels.

- A data channel hops randomly 1600 times per second between the 79 (23 for Japan) RF channels, and each channel is divided into 625 microsecond–long time slots. Data in a packet can be up to 2745 bits in length, and a packet can be up to 5 time slots wide.

- A piconet is a personal area network consisting of one master and up to seven slaves. The master transmits in even time slots, slaves in odd time slots.

- Two types of data transfer are used between devices: *Synchronous Connection-Oriented (SCO)* and *Asynchronous Connection-Less (ACL)*.

- There can be up to three SCO links of 64,000 bits per second in a piconet. To avoid timing and collision problems, SCO links use reserved slots set up by the master. Masters can support up to three SCO links with one, two, or three slaves. Other slots not reserved for SCO links can be used for ACL links.

- One master and one slave can have a single ACL link. ACL is either point-to-point (master to one slave) or broadcast to all the slaves. ACL slaves can transmit only when requested by the master.

Wired Networks: Ethernet

Several options are available for wired networks: Ethernet, Arcnet, Token Ring, and FDDI. Of the four, we will focus on Ethernet throughout this section because of its pervasiveness, its cost, and its availability on copper wires. Arcnet is a token-passing bus technology with bus structure that is similar to the IEEE 802.4. It uses RG-62 A/U 93Ω coaxial cable, which is different from the cable installed in your home for cable television. Coaxial cable in general is difficult to work with and is typically not the first choice for wiring a home network.

The Token Ring protocol was originally developed by IBM, which led to a standard version, called IEEE 802.5. A Token Ring network uses 150Ω shielded twisted pair cables or IBM Type-3 unshielded twisted pair cables. Type-3 cable has impedance of 100Ω in contrast to the 150Ω impedance of STP cabling and usually includes four twisted pairs as opposed to STP cabling, which has only two twisted pairs.

The technology uses a token-passing scheme in a logical ring and operates at 4 or 16 Mbps. FDDI (Fiber Distributed Data Interface) also uses a Token Ring protocol on two rings for redundancy. Operating at 100 Mbps, FDDI is a standard for data transmission on fiber optic lines that can extend up to 200 km (124 miles). Both Token Ring and FDDI technologies are not suited to home networking because of cost and complexity.

Ethernet is the most widely installed local area networking technology. The protocol operates on a wide variety of cabling systems from coaxial to copper at either 10 Mbps or 100 Mbps (Fast Ethernet). Another version, called Gigabit Ethernet, operates at an even greater speed, 1000 megabits per second, or 1 Gbps (gigabit or billion bits per second).

Originally, 100 Mbps systems were used for Ethernet backbone systems. The cost of Ethernet components has fallen to the point that there is not much difference in price between a 100 Mbps card and a 10 Mbps card. In fact, the typical card now comes in both

10 Mbps and 100 Mbps, with the capability of automatically sensing the speed of the network to which it is connected.

Most Ethernet networks today are built using UTP (unshielded twisted pair) Category 5 (Cat 5) wiring. This grade of wiring is better than the wiring used in the typical U.S. home for PSTN services, and thus additional wiring would be needed in the home before an Ethernet network could be built. For this reason, the technology does not fall within the "no new wires" camp.

Overview of an Ethernet Network

Figure 2-14 outlines a typical Ethernet configuration. In most applications, a twisted pair Ethernet network requires a *hub* that serves as a concentration point for all devices. Hubs are typically available in increments of four ports for four devices up to twenty-four ports for twenty-four devices. Each device is connected to a port on the hub via a cable that should not exceed 100 meters for 10Base-T with Category 3 wiring. 100Base-T can be unpredictable on Cat 3. Higher-quality cabling such as Category 5 wiring may be able to achieve longer segment lengths, in the order of 150 meters, while still maintaining the signal quality required by the standard. For a home network, it is advisable to err on the side of caution and always use at least Cat 5 wiring with a segment length of no more than 100 meters (328 feet) regardless of whether a 10 Mbps or 100 Mbps solution is employed.

The CSMA/CD Protocol   The CSMA/CD protocol—*Carrier Sense Multiple Access/Collision Detection*—provides for the sharing of a common transmission medium between multi-

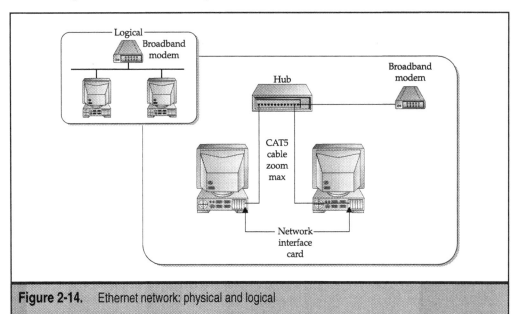

Figure 2-14.   Ethernet network: physical and logical

ple devices. It was designed on the probability that stations competing for time on a shared medium will eventually collide. It overcomes this eventuality by introducing a scheme that detects collisions and provides for an orderly method for resumption of transmission once a collision has been detected.

Before transmitting a frame, the device listens for a period of no activity on the medium. It then broadcasts its data frame onto the medium and is heard by all devices. If another device attempts to send at the same time, a collision condition occurs. The transmitting stations send a jam sequence to ensure that all stations are aware of the collision. The stations wait a random period before attempting the transmission again, a process that is repeated until the transmission is successful. A summary of the process follows:

1. A station with data to transmit monitors for the presence of a transmitting station through a process called *carrier sense*.

2. If a carrier is detected, the station waits before transmitting. The station must wait for a period that is equal to or greater than the interframe gap before transmitting after a carrier ends.

3. The station broadcasts when the medium becomes free of a carrier and a period equal to, or greater than, the interframe gap is reached.

4. The station continually monitors for a collision while it is broadcasting.

5. If a collision is detected, the station ceases its broadcast and sends a 32-bit jam sequence to ensure that all devices on the medium are aware of the collision. If the collision happens very early in the broadcast, the station will continue to broadcast the preamble of the frame before stating the jam sequence.

6. The station waits a random period of time—called the *backoff delay*—that is determined by a random number generator before resuming step 1. The backoff delay increases if multiple collisions occur. Once a station successfully transmits a frame, it resets its collision counter, which readjusts the backoff timer.

Practical Ethernet Tips   Err on the side of caution: always use Category 5e cables and keep the length of the cable less than or equal to 100 meters or 328 feet. Cat 5e cables may seem like overkill for the types of application that you will be using, but this choice positions you for any further improvements in the protocol.

A hub can be attached to another hub. When attaching two hubs in this fashion or when attaching a hub with another device such as a broadband modem, a special port on the hub is used. Normally this port is labeled as "Uplink," with a switch that switches it between operating as a normal Ethernet port or as an "Uplink" port. In the Uplink position, a normal straight-through cable is used to connect the devices, and the switch in effect cross-connects cable leads. If the hub is not equipped with an uplink port, the same thing can be accomplished by linking the two devices with a crossover cable—a cable specially built to cross-connect signal leads.

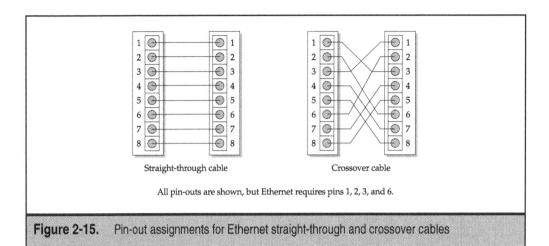

Straight-through cable   Crossover cable

All pin-outs are shown, but Ethernet requires pins 1, 2, 3, and 6.

Figure 2-15. Pin-out assignments for Ethernet straight-through and crossover cables

Two Ethernet devices can be attached to each other without a hub. In this situation, a crossover cable should be used to directly attach the two devices. Figure 2-15 shows an example of the pinout assignments for both a straight-through cable and a crossover cable.

Both Category 5 and Category 5e consist of unshielded twisted pair with 100Ω impedance and electrical characteristics supporting transmissions up to 100 MHz. The main differences are in transmission performance achieved through more twists per inch and better insulation on Cat 5e cables. Category 5e components are most suitable for high-speed applications and outperform Category 5 equipment during high–data transfer scenarios. It is advisable to use Category 5e cabling if possible to ensure an infrastructure that is capable of handling future applications.

A Historical Perspective

Xerox developed the first Ethernet system in the early 1970s. Since then, it has morphed into the standard that we know today and has grown to become the most widely used local area networking technology. It is now supported on numerous different types of transmission media, from coax cable to copper and fiber. The topology too has seen changes: a star topology is now supported in addition to the original bus architecture. Table 2-2 offers an overview of the timeline for the development of Ethernet.

Home Networking at a Glance

As is evident from the different technologies that have been covered, there is a wide choice of options available for building a home network. Each option has its strengths

Year	Event
1976	Metcalfe and Boggs of the Xerox Palo Alto Research Center (PARC)—the developers of the first experimental Ethernet system, which interconnected Xerox Alto computers and laser printers at a data transmission rate of 2.94 Mbps—publish their paper entitled "Ethernet Distributed Packet Switching for Local Computer Networks" in the Communications of the Association for Computing Machinery (ACM) (http://www.acm.org/classics/apr96/).
1979	Digital Equipment Corporation (DEC), Intel, and Xerox join forces to standardize an Ethernet system for public use.
1980	The three companies release version 1.0 of the specification called the "Ethernet Blue Book," or "DIX standard"—for DEC, Intel, and Xerox. It defines the Ethernet standard based on thick coaxial cable, or "thick" Ethernet. The version is based on a 10 Mbps CSMA/CD (Carrier Sense Multiple Access with Collision Detection) protocol.
1982	The first Ethernet controller boards based on the DIX standard become available, and the second and final version of the DIX standard, version 2.0, is released.
1983	The IEEE (Institute of Electrical and Electronic Engineers) releases a standard for Ethernet developed by the 802.3 Working Group called *IEEE 802.3 Carrier Sense Multiple Access with Collision Detection (CSMA/CD) Access Method and Physical Layer Specifications*. The standard changes some elements of the original DIX frame format but is mindful of allowing hardware based on the two standard to interoperate on the LAN.
1985	IEEE 802.3a defines a second version of Ethernet called "thin" Ethernet, "cheapernet," or 10Base-2. In this version, a thinner and cheaper coaxial cable is used and the interconnections are simplified. Both continue to use a CSMA/CD protocol and a bus topology. Another standard defining a 10 Mbps Ethernet over a broadband system is also released. This version of Ethernet is called IEEE 802.3b or 10Broad36.

Table 2-2.   A Historical Perspective of Ethernet

Year	Event
1987	The IEEE 802.3d and IEEE 802.3e standards are released. 802.3d defines the Fiber Optic Inter-Repeater Link (FOIRL), which is used to extend the maximum distance between repeaters of a 10 Mbps Ethernet to 1000 meters. This is accomplished by using two fiber cables. 802.3e defines a version that operates on twisted pair wiring at a speed of 1 Mbps but that never does get much traction in its adoption.
1990	The 10Base-T or IEEE 802.3i version of the standard is released. It is based on simple unshielded twisted pair Category 3 cabling that supports 10 Mbps. This catches the interest of many companies that are interested in Ethernet because of the number of buildings that have already been wired with UTP cable. The introduction of a *hub,* a central point for cable runs, also makes cabling a simpler task. A single cable failure can be easily isolated without affecting the entire network.
1993	The IEEE 802.3j standard for 10Base-F improves on the earlier 802.3d FOIRL standard by increasing the distance from 1000 to 2000 meters.
1995	A ten-fold improvement in Ethernet is defined in IEEE 802.3u, the 100Base-T standard that is commonly known as Fast Ethernet running at 100 Mbps. This standard supports three different media types: 100Base-TX uses two pairs of Category 5 twisted pair cable 100Base-T4 uses four pairs of Category 3 twisted pair cable 100Base-FX uses multimode fiber
1997	Full-duplex Ethernet is introduced in IEEE 802.3x. Instead of sharing a common transmission medium, it operates by using a point-to-point link. The dedicated connection eliminates the need to listen for other devices that may be sharing the link, and so the contention requirement is eliminated. As a result, it is possible to have separate transmit and receive channels (separate wires), which allow for full duplex or the capability to send and receive simultaneously. This feature effectively doubles the data rate, as a device could be sending a 10 Mbps data stream while receiving another at the same time. In previous versions, it could not do that, because a device was allowed only to transmit or to receive at any given instance. Both 10 Mbps and 100 Mbps Ethernet benefit from the full-duplex feature. This year also sees the addition of another media type for 100 Mbps Ethernet called IEEE 802.3y 100Base-T2. This standard defines the operation of 100 Mbps over two pairs of Category 3 balanced cabling.

Table 2-2.   A Historical Perspective of Ethernet *(continued)*

Year	Event
1998	Gigabit Ethernet technology—the 1 Gbps IEEE 802.3z 1000Base-X standard—is introduced, offering a 100-fold improvement over the original specification and a ten-fold improvement over 100 Mbps Ethernet. Additionally, the IEEE 802.3ac standard, which defines the support of Virtual LAN (VLAN), is released. The announced Gigabit Ethernet standard supports three media types: 1000Base-SX operates with a 850 nm laser over multimode fiber 1000Base-LX operates with a 1300 nm laser over single and multimode fiber, and1000Base-CX operates over short-haul copper "twinax" shielded twisted pair (STP) cable
1999	The IEEE802.3ab 1000Base-T standard is released. It defines 1 Gbps operation over four pairs of Category 5 UTP cabling.

Table 2-2.   A Historical Perspective of Ethernet *(continued)*

and weaknesses, and some are more mature than others. Table 2-3 provides a quick summary of some of the various aspects of each technology. It is not an exhaustive list but should be sufficient to give a quick flavor of the benefits of each.

MULTIUNIT SOLUTIONS

The multidwelling, multitenant, and hospitality industry, collectively called *multiunits* throughout the rest of this section, represents a largely untapped growth market for high-speed networking broadband solutions. According to Cahners In-Stat Group, the broadband market for multiunits is expected to grow to $2 billion by 2004 from $137 million in 2000.

Before proceeding, a definition of each term is appropriate. By *multidwelling units*, we mean apartment complexes, condominiums, town houses, university dormitories, or any other facility where multiple families live. The *hospitality industry* represents hotels, airports, hospitals, and convention centers or any facility that provides short-term stays, overnight or in transit. Finally, *multitenant units* are office complexes, commercial and industrial properties, factories, or any other facility where people work or operate a business. The collective term *multiunit* is used to cover them all.

As people become used to high-speed networks and the applications that they facilitate at school and at work, they grow accustomed to the benefits that these networks provide and begin to long for the same level of performance at home. For single-family residences, the options are numerous, but for apartment dwellers and hotel guests, the choices are usually limited to the types of service that are available in the building or on the grounds.

Feature	Phoneline	Powerline	Wireless	Ethernet
Leverage existing infrastructure	Yes	Yes	Yes	No
Leverage Standards	Good	Poor	Medium (quite a few)	Excellent
Quality of Service Support (QoS)	Yes	Emerging	Mixed (some do, some do not)	No
Robustness	Good	Medium	Challenging	Good
Speeds	10 to 100 Mbps (emerging)	bits to 2 Mbps	1 to 11 Mbps	10 to 100 Mbps (1,000 Mbps too expensive for home market)
Privacy of physical medium	Good	Poor	Poor	Good
Cost	Good	Good	Medium	Medium
Maturity	Good	Emerging	Good	Excellent

Table 2-3. A Quick Comparison of Home Networking Solutions

The same is true for some office buildings. While a business may have a high-speed network installed within the confines of each of its offices, access to the rest of the world can be limited by the facilities that exist within the building. An office at the top of the Empire State Building, for instance, has a greater challenge gaining access to broadband facilities than an office that is located in a newer corporate park. The age of the building, the state and the type of wiring, the construction material used, all play a part in deciding the types of new services the building can accommodate.

Factors for Consideration

Multiunit property owners and managers, as well as service providers, recognize the market potential of broadband, but they also understand the hurdles, which are numerous. They recognize its competitive advantage in attracting new guests or tenants and, just as important, if not more so, its potential for an additional revenue stream.

The cost of building the infrastructure to provide these services can be very high, especially with older buildings. Consider an older high-rise building. It may be framed in concrete, asbestos may have been a component in its construction, the wiring ducts between the ground and top floors—which are typically installed in the elevator shaft—may be bulging at the seams with substandard wiring that is good enough for voice but not for a data network. The cost to retrofit a building like this can be extremely high when you consider the cost of labor, that of material, and the potential loss of business that could result if sections of the building have to be closed to accommodate the work. Property managers must take a lot more into consideration in deciding on a solution than would a single-family homeowner.

Whereas a homeowner gravitates toward the simplest and most reliable solution, a multiunit manager must take into consideration the cost and the price/performance of the solution, which includes features of reliability, management, scalability, security, and return on investment. The multiunit manager is usually a lot more cautious and conservative in his choices and will usually gravitate toward a solution that has a proven track record and is backed by a widely accepted standard. The reasoning is based on the need to ensure a cost-effective and stable solution. Leading-edge technology may be too risky, especially when considering the implications of its failing and the potential cost of providing and managing the service. All these factors narrow the playing field of available options.

Possible Solutions

Of the solutions that have been presented for home networks, wireless would appear to be a logical choice. While it is definitely an option, it still has the "image of something new," which leads to hesitation. Many are taking a wait-and-see attitude before taking the plunge. Another consideration with wireless is the method used to link the access points. In a typical application, there is a single broadband connection from the building, or maybe two. All users of the multiunit network share the common broadband connection. To facilitate this, in the case of a wireless network, the APs throughout the building must be connected to a common distribution system, which has a link to the shared broadband facility.

The most effective way of accomplishing this, from a performance viewpoint, is to use a wired LAN to link the APs and the broadband facilities. The problem is, this solution is fine if the building is wired with the correct grade of wiring. What if it is not? We are still faced with the challenges of rewiring. It is possible to forego the use of a wired LAN and have the APs act as repeaters to form a wireless DS. The problem with this approach is that it uses some of the bandwidth that would otherwise be used by other wireless devices and thus degrades performance.

Wireless, however, does have a place in the multiunit market. Many schools are deploying wireless networks because of the mobility that it affords. Wireless-enabled workstations are being installed on carts and are being rolled from classroom to classroom as needed. The benefits are obvious in this application. Instead of having every classroom prewired, the workstations can be deployed wherever they are needed and still be connected to the school network. Having a wireless solution in this environment allows for network connectivity in places that normally would not have any. The ubiquitous nature of the network access allows connectivity in the gym, the cafeteria, or even the halls.

Long-Reach Ethernet

As is evident from Table 2-2, Ethernet has been around and has seen many enhancements since the early 1970s. Consistently throughout its evolution, the technology has always depended on exacting standards for the transmission medium on which it operates. The original standard was limited to coaxial cable, but as technology improved, it became possible to build an Ethernet network using twisted pair copper wiring. This was a tremendous achievement that played a major role in the growth of the technology. The introduction of copper as a viable transmission medium was a major event, but the quality of the cable remained an important factor. Using a grade of wire inferior to the one the standard called for potentially rendered the network useless, especially under heavy loads. So while the use of copper wires was a significant breakthrough, it was not perfect.

In February 2001, Cisco Systems announced the availability of a new line of Ethernet-based networking products capable of delivering high-speed broadband capabilities over voice-grade telephone lines. Based on a new technology called Long-Reach Ethernet (LRE), the products remove the requirement for higher-grade copper cables. The potential of this technology is as vast as it is revolutionary. It significantly broadens the application of Ethernet and addresses some key hurdles in equipping older buildings with broadband capabilities. With its introduction, Ethernet has the potential to expand from a predominantly corporate network application to a more universal access methodology by using the existing unconditioned wiring infrastructure of a building.

LRE allows for speeds of 15 Mbps for a distance of up to 3500 feet, 10 Mbps for a distance of up to 4000 feet, and 5 Mbps for up to 5000 feet. It allows for full broadband-type features by enabling the coexistence of different traffic types on a common medium. Data, voice, video, and streaming applications can all share the same wiring facility.

The Applications   Using our previous example of the Empire State Building, it is easy to imagine the cost benefits of LRE. The existing unconditioned wiring that is most definitely substandard to Cat 3 standards can now be used to provide broadband facilities within the building. For similar older buildings that could be wired only at considerable challenge and expense (if at all), LRE represents a natural choice. The fact that the technology is based on Ethernet—a proven standard with a track record—also helps remove some of the concerns a building manager may have in implementing a new untried technology.

In fact, many higher-end hotels are now offering high-speed terminal access in guest rooms by using LRE. It is just a matter of time before this type of service becomes more generally available in a wider range of hotels. The possibility of an additional revenue stream makes the provision of the service even more attractive to the hotel manager, especially now that the cost of provisioning the service becomes less restrictive as a result of the ability to leverage the existing infrastructure.

For the multidwelling market, the technology allows the apartment manager to realize additional revenue from the growth of high-speed broadband solutions. Where cable or DSL service would once have been the only choices for apartment dwellers (assuming the infrastructure was there to support it), the apartment management is now able to get a piece of the pie by provisioning each apartment with high-speed access—a new service that can now be bundled into a higher rent.

LRE can also be integrated into a wireless solution. The technology can be used as the backbone of the access point interconnectivity as well as access to server-based applications. Another benefit is its ability to overcome the distance limitations of traditional Ethernet. Instead of being limited to 100 meters on each segment, LRE extends the reach by providing improved Ethernet-based performance from distances of 3500 feet to 5000 feet.

Features of the Technology   The Cisco LRE solution—an extension to IEEE 802.3–compliant Ethernet—uses a point-to-point topology, as shown in Figure 2-16. It provides for half- or full-duplex operation with support for POTS, PBX, or ISDN signaling concurrent

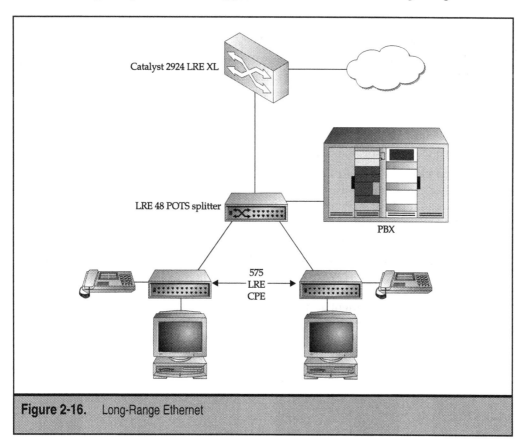

Figure 2-16.   Long-Range Ethernet

with data on standard telephone-grade wiring. Frequency division multiplexing is employed to separate the downstream, upstream, and other signaling services into different channels across a frequency band of 300 Hz to 3.4 KHz over Category 1, 2, or 3 single copper wiring.

The product line consists of four components: the Cisco Catalyst 2900 LRE XL switch, the Cisco 575 CPE Device, the Cisco LRE 48 POTS Splitter, and the Cisco Building Broadband Service Manager (BBSM).

▼ **Cisco Catalyst 2900 LRE XL** Based on the Catalyst 2900 Series XL 3.2 Gbps switch architecture, the switch has 12 or 24 Long-Reach Ethernet ports and four 10/100 Ethernet ports with integrated LRE technology, which means that they do not require an external modem pool. It delivers dedicated bandwidth per port at rates up to 15 Mbps. Each switch can be configured to support the following modes:

■ 5 Mbps symmetrical rate (up to 5000 feet)

■ 10 Mbps symmetrical rate (up to 4000 feet)

■ 15 Mbps symmetrical rate (up to 3500 feet)

The 10/100 Ethernet ports can be used to connect servers, daisy-chain multiple LRE switches, or uplink to other Ethernet switches.

▼ **Cisco 575 CPE Device** This terminates the LRE connection at the user's desktop. This device provides one RJ-45 Ethernet connection and two RJ-11 connectors for plain old telephone service support. The telephone is plugged into one connector; the other is used to connect to the wall.

■ **Cisco LRE 48 POTS Splitter** This is used in applications where telephone services are being multiplexed along with data. It provides connectivity between the LRE switches and a building's PBX. It supports 48 ports for aggregating 575 CPE links and six RJ-21 ports—two each for connectivity to a patch panel, Catalyst 2900 LRE XL switches, and an on-site PBX.

▲ **Cisco Building Broadband Service Manager (BBSM)** This registers and provisions network service according to the guest or tenant profile. It offers bandwidth throttling for efficiency and congestion avoidance; a web portal for local service; content and advertising specific to the guest, tenant, or organization; and integration with hotel billing systems for network usage charge-backs.

To the user logging on to the network, the BBSM presents the user with a welcome screen the first time the browser is launched. The welcome screen can be tailored to give terms and conditions, approve billing, and present any other information management wishes to display.

CORPORATE NETWORKING SOLUTIONS

A lot has been published on Ethernet, Token Ring, and Arcnet—the types of networks usually deployed by businesses or industrial plants. The most pervasive network architecture today is Ethernet, which we have covered in the section on home networks. Token Ring, originally an IBM solution, is still found in many IBM-centric SNA applications. In fact, it remains the only way of null-attaching two SNA networks (SNI), which is a way of interconnecting two SNA networks so that they are totally independent of each other's addressing requirements. Other than specific needs like this, Token Ring networks are becoming less popular, as is Arcnet. So instead of explaining each network type, we will take a different approach in this section and look at some broader networking applications and concerns found within the corporate environment.

Corporate networks are used to connect remote offices as well as to facilitate data interchange and communication with partner companies, suppliers, customers, and employees. Throughout the 1970s and 1980s, corporate networks consisted mainly of direct data links between sites through either private facilities like T1s or a shared infrastructure like X.25. Any solution that required the transport of any traffic type other than data, or very high speeds, pretty much meant a build-your-own approach.

Networks like X.25 supported only data and were not suitable for solutions requiring very high speeds. Building a private leased line solution meant the appropriate consideration for bandwidth allocation for each traffic type, growth requirements of each, and choosing the right topology to get the best price/performance. Security, which was just as important then as it is now, was less of a factor in the design of the network than it is today, mainly because the network components then were largely within the control of the owning company.

As Figure 2-17 shows, the original requirements of a corporate network have not changed; they have only grown more compelling. New ways of doing business, a geographically dispersed workforce, the global reach of the Internet, and the availability of high-speed networking solutions that support multiple traffic types and services—all of these have contributed to the growth of brand-new industries.

Today, many companies are turning to VPN solutions to provide the security and reach required for linking remote workers, offices, and other companies across a readily available infrastructure—the Internet. The corporate network that was once confined to offices where people work now extends into the home. Additionally, areas of the corporate network now need to be made available to the general public, a fact that has made security considerations more prominent than they have ever been.

In this section, we will examine *virtual private networks (VPN)* and some of their security features. We will also discuss other approaches for securing the network.

VPNs

Virtual private networks are becoming commonplace and are being used as a means to extend the corporate network to employees' homes, remote offices, traveling users' laptops,

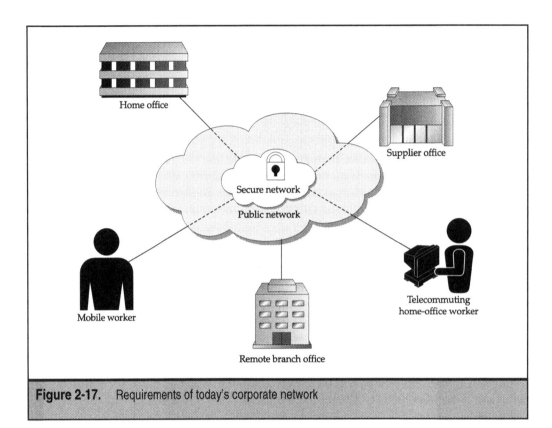

Figure 2-17. Requirements of today's corporate network

and business partners across a common shared network. Once the infrastructure is in place, the benefits of a VPN solution are immediate. It creates a dynamic environment for extending the reach of the network without breaching security. The network boundary becomes elastic, and the topology expands or contracts as appropriate. The VPN provides significant economies of scale with a faster return on investment, as operational costs are lower when compared to operating a private circuit network. The flexibility that it provides also helps increase productivity, as the workforce is "always in the office."

Because the transport is being carved out of a public or shared network, security considerations are extremely important. A good security policy does not define just one process but several processes, each intended as a separate hurdle to be overcome by a potential wrongdoer. In other words, a building that depends entirely on a key card entry system for its security is less secure than one that has a key card entry system, guard dogs inside the building, a locked door to every room, and a password for the file system. The point being, as one hurdle is overcome, another is presented to help frustrate and thwart any attempts to infiltrate the system. This is an important feature of any good VPN solution—multiple security features are required.

To ensure a secure transmission between a home office and the main office, most VPN solutions employ some method of encryption, of which there are many. The basic principle, however, regardless of the technique, is to scramble the data in such a way that only the intended recipient is able to unscramble the content. In some applications, the header information as well as the data in encrypted; in others, only the payload (the data) is encrypted.

IPSec Securing the Data

IPSec provides a way to create a secured environment within a larger unsecure network. It is a set of extensions to the IP protocol family that defines a set of protocols and cryptographic algorithms for creating secure IP traffic sessions between IPSec gateways. Developed by the IETF, IPSec is supplied natively in Windows 2000 Server and Professional Editions, and across Cisco, Lucent, and other networking vendors' product lines. IPSec operates at the network layer, so an application does not need to have any knowledge of IPSec to use it—nor do the protocols that operate at the transport layer like TCP and UDP.

IPSec works in any of the following ways:

▼ Computer-computer or host-host

■ Computer-router or host-network

▲ Router-router or network-network

In essence, IPSec seeks to perform four basic functions through the use of two protocols: Authentication Header (AH) and Encapsulated Security Payload (ESP). The four basic functions and the protocol that delivers each are listed in Table 2-4.

Function	Protocol	Notes
Confidentiality Ensures that only the receiver of the message is able to understand.	ESP	Can be used with or without authentication and integrity
Integrity Ensures that the data does not get changed along the way.	AH, ESP	Does not require confidentiality
Authentication Provides a means by which the receiver knows that the message was originated by the sender.	AH, ESP	Does not require confidentiality
Replay protection Ensures that a transaction can be carried out only once unless it was authorized to be repeated.	AH, ESP	Always requires authentication and integrity

Table 2-4. IPSec Basic Functions

The Authentication Header (AH), which comes after the IP header in the packet structure, creates an envelope that provides integrity, data origin authentication, and protection against replays. This protocol authenticates every packet, which protects against session-stealing programs. Even though the IP header is outside of the AH envelope, AH provides authentication for as much of the IP header as is possible. As a result, it is susceptible to any attempts to manipulate IP headers in transit, so it cannot be used in environments that employ Network Address Translation (NAT), which manipulates these addresses. Several RFCs provide a choice of actual algorithms to use in the AH, but they all conform to the guidelines specified in RFC 2402.

The Encapsulated Security Payload (ESP) protocol includes all of the functions of AH to protect against attacks on encrypted, nonauthenticated data streams. The specification does allow for use of ESP without the AH functions. In applications that employ NAT, ESP can be used in null encryption, which basically means using AH without authentication of the IP header. When used in this way, it is compatible with NAT.

IPSec functions in either transport mode or tunnel mode. The difference between the two basically consists of where the IPSec data is placed in the packet structure. In transport mode, it is inserted after the IP address and before the transport header. In tunnel mode, a normal IP packet is placed into an IPSec envelope, which in turn is placed into another IP envelope. This mode of operation is easily deployable, since all that is required is for the end points to be running the tunneling software. The most common use for tunneling is to support remote and mobile users. Gateways may use both modes: transport mode to protect information that flows between it and another gateway, and tunnel mode to send data to end users.

IPSec Algorithms and Security Keys   For every IPSec connection, there are two *security associations (SA)*, one for each direction. An SA defines the parameters—including authentication algorithms and session keys—that will be used by the communicating end systems for its direction of the flow.

The authentication algorithms used by both AH and ESP are HMAC-MD5 and HMAC-SHA1, which are both key-based authentication algorithms where participants share a secret key. MD5—Message Digest version 5—uses a 128-bit key, and SHA1—Secure Hash Algorithm version 1—a 160-bit key. Because HMAC assumes only the sender and the receiver know the key, both data integrity and origin authentication are assured.

Several encryption algorithms are available, with DES (Data Encryption Algorithm) being the default encryption algorithm used in ESP. In addition to DES, there are Triple DES, Blowfish, CAST-128, RC5, IDEA, and ARCfour. Implementers are left to choose the algorithm specific to their implementation. Most forms of encryption belong in one of two categories:

▼ *Symmetric-key encryption* uses a single key that both the sender and the recipient possess. This single key is used for both encryption and decryption, so the security is dependent on no unauthorized person's gaining the key. Communicating end points must securely exchange the key before they can exchange encrypted data. The larger the key, the greater the strength of the encryption.

▲   *Public-key encryption* uses two keys, a public key and a private key. The public key can be passed openly between the communicating devices or published in a public repository, but the related private key must remain private. Only the private key can decrypt data encrypted with the public key. Data encrypted with the private key can be decrypted only using the public key.

Digital signatures are used in public-key environments to help secure transactions by verifying the identity of the sender. A digital signature is a means by which the sender binds its identity with a particular message, which serves as a unique signature. The process entails combining the information with some secret information held by the sender into a signature or a tag.

VPN Tunneling

Tunneling techniques employ an approach that takes an original packet and encapsulates it into a new packet. This new packet is then transmitted to the destination. At first glance, the new packet looks like any other data packet. The data of the new packet, however, is actually another fully formatted packet with its own header and data. The original packet (tunneled packet) may or may not be encrypted. Some tunneling protocols such as GRE just encapsulate the original packet without making any changes; others, including IPSec, encrypt the original packet before encapsulating it. Once the new packet reaches its destination, the outer packet is removed and the original packet is processed as if it had never been encapsulated.

In tunneling, there are two end points. The first end point exists at the device for which the original packet was intended—this is the true end point of the message. The second end point exists at the device that de-encapsulated the packet, or the tunnel end point. The true end point of the message is not aware of the path that was taken to get the message there, nor is it aware of the fact that the message was encapsulated. In the case of the tunnel end point, both ends must be aware that the original packet is being encapsulated and the type of tunneling method that is employed.

From the perspective of the message flow, the message is sent from the sending device—the true source of the message. It is then encapsulated by the device that encapsulates the message. On the other end, the tunnel end point—the device that does the de-encapulation—de-encapsulates the message and sends it on to the intended receiver—the true end point.

Firewalls

Firewalls represent the first line of defense—hopefully, not the last—and are usually placed at the point where the corporate network touches an unsecure network. In its most basic form, a firewall is just a piece of software that has a list of users or addresses that are allowed into a network. In its more complex form, it is hardware and software working cooperatively to implement multiple rules governing who, what, when, and how. Attacks from malicious users may come from many different sources. An important policy that goes hand in hand with a firewall is to constantly analyze who has accessed, and has attempted to access, the network and a constant review of existing policies as new threats are identified.

Some of the most common methods used to control the flow of traffic in and out of a secure network are:

▼ **Packet filtering**   Packets of data are analyzed against a set of filters that may be a simple list of addresses or a complex "if, then" permutation. Packet filtering allows control over every aspect of the packet, from the length to the type of application that the packet represents.

■ **Proxy service**   Inside users of the Internet, those that are on the secure side, appear to have full unobstructed access to the Internet, but every request is routed through the proxy server. A proxy server has the ability to hide the real address of the user on the inside from the target server on the outside and only present to the unsecure server an address that appears to be from the Internet. Proxy servers can also operate in a more restrictive manner and deny the flow of certain types of data. Video streaming, for instance, may be blocked at the proxy server.

▲ **Stateful inspection**   Traffic flowing into the network is examined to see if it matches the characteristics of the packet that flowed out. As an example, if a device on the inside opens a Telnet session to a target server on the outside using the known Telnet port of 23, the firewall would expect a response with a port number that matches the Telnet session.

Additional Security Hurdles   AAA servers—authentication, authorization, and accounting—can provide further levels of security. Each request for a connection to the network is sent to the AAA server, which performs the following checks:

▼ **Authentication**   Checks to ensure that you are who you say you are. Usually, this is accomplished by verifying your user ID and password pair.

■ **Authorization**   Verifies that you are allowed to perform the functions that you are requesting.

▲ **Accounting**   Monitors your activity for the purposes of creating an audit trail that can be used for accounting, security, and reporting purposes.

SUMMARY

These are just some of the tools and applications that are now being used to extend and protect the corporate network. Multiunit and home networking solutions also face the same security issues that corporate networks face today. The borders between home, multiunit, and corporate networks are becoming less distinct from a topological perspective as they all begin to encroach on and touch each other. The only defining area of demarcation will be along security lines, and that too will be quite fluid. As an example of this fluidity, the home network may at one minute be an extension of a secured corporate environment and in the next, a part of the untamed Internet.

In this environment, the only constant is change.

PART II

Access Networks

In Part 1, we reviewed some of the technologies and trends that are driving the growth of broadband. Regulatory, social, and business issues were discussed to give you a better appreciation of some of the forces that are driving the demand for broadband services. To complete the picture, we also looked at networking solutions employed within the home, multiunit buildings, and corporate offices.

Part 2 addresses the different types of access networks that are being deployed to provide broadband services. The proliferation of different access technologies leaves aside the question, why not standardize on a single technology? This obviously would be the simplest solution, and instead of seven chapters to address these technologies, we would need only one or two. But there are very good reasons why we will probably have different solutions for the foreseeable future.

The history and purpose of the network heavily influence its architecture. For example, television service has always been broadcast, and so it is well suited to a medium that can be shared by multiple viewers, such as cable. Attempting to port the service to a telephone infrastructure that was designed for point-to-point communication is an expensive proposition. Conversely, porting telephone service to a shared medium creates the need for technology to ensure security and privacy. These hurdles will not disappear anytime soon, and until they do, service providers will continue to get the most from their existing investment, which means different access technologies.

CHAPTER 3

Evolution and Technical Foundation

In this chapter, we will look at the evolution of three different kind of networks: telecommunications, data, and broadcast networks. Reviewing the history helps us understand their evolution and the internal and external forces that have acted to change them. Understanding the limitations of each underscores the challenges and investment required for the provision of new services on the existing infrastructure.

To provide a solid technical foundation for the remaining chapters, some of the basic terminology, techniques, and technologies will be reviewed. We will cover the various types of modulation techniques, describe the key challenges to each media type, and review different error correction and noise mitigation schemes.

EVOLUTION AND CONVERGENCE

We are at a point in history where three major areas—telecommunications, computer data networks, and broadcast networks—are rapidly approaching a point of convergence. The question is, will we need one or three different access methodologies?

Telecommunications

Since the perfection of the telegraph in 1844, the telecommunications industry has undergone a major technical and administrative transformation. Samuel F.B. Morse's invention of the telegraph led to distant parts of the U.S. continent being linked by network poles and wires. Thirty-two years later, in 1876, Alexander Graham Bell demonstrated the first telephone. By the time of the Second World War, over 16 million telephones had already begun to change the lives of Americans, and today the telephone is as much a part of our lives as the light bulb.

Throughout most of the twentieth century, the telephone industry had a very simple administrative structure. A single company, called American Telephone & Telegraph, AT&T, provided services. In 1984, AT&T, a regulated monopoly, divested itself of its local operating companies that provided telephone services to various regional markets throughout the U.S. As shown in Figure 3-1, these companies, called Regional Bell Operating Companies (RBOCs), provided services to local markets, while AT&T continued to provide long-distance services in competition with other, smaller companies. This was a result of a court-agreed settlement to an antitrust suit filed by the federal government. Twelve years later, in 1996, in an effort to speed up competition, Congress passed the Telecommunications Act. The basic tenet of the act was to allow long-distance companies, RBOCs, and cable companies to compete in each other's markets.

The rest of the world was also seeing change. Government-owned companies called Post, Telephone, and Telegraph (PTT), which offered telephone service, were also contending with competition as a result of privatization. In many countries, such as the United Kingdom, private carriers now compete against British Telecom, which was once a part of the Post Office and the sole operator of the telephone service. The overall impact to the consumer and businesses is a feverish race to leverage existing investments to provide additional multiple services across the last mile of wiring, the portion that touches our homes and offices.

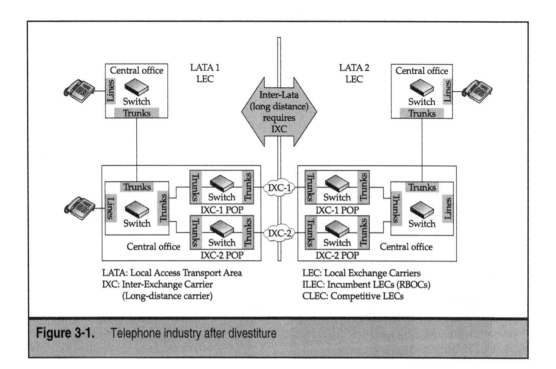

Figure 3-1.   Telephone industry after divestiture

Computer Data Networks

In 1804, Joseph-Marie Jacquard built a loom that weaved intricate patterns by reading holes punched into a small sheet of wood. This was the earliest use of punched cards programmed to control a manufacturing process. Almost 90 years later, using principles of the Jacquard invention, Herman Hollerith developed a punched-card machine that was used to tally the 1890 U.S. census. The counting started in September 1890 and was completed in December of the same year, two years ahead of schedule and at a savings of $5 million.

After merging with another company in 1911, Hollerith's company became the Computer Tabulating Recording Company. He retired from the company in 1921, leaving it in the hands of Thomas J. Watson, who had joined the company in 1918. In 1924, the company was renamed International Business Machines (IBM). Punched cards were used until the early 1970s, but today's ASICs, semiconductors, lasers, and fiber optics make them as obsolete as the abacus.

First developed in the 1960s, time-sharing computers called *hosts* allowed simultaneous access from multiple terminals that were directly attached. Later in the decade, the use of modems facilitated the connection of remote terminals over the telephone system. This was the first glimpse of a computer network, the limitations of which were quickly recognized. What if the connected user wanted to connect to another host or if the host needed to transfer a file to another host? Host-to-host communication quickly became a requirement that gave birth to packet-switching networks and private data networks that used private circuits leased from the telephone company.

Several packet-switching technologies emerged in the late 1960s to early 1970s. Public packet-switched networks such as X.25 were commercially offered to businesses, whereas others had a more specific application. These latter included ARPANet, which was sponsored by the U.S. Department of Defense (DoD). ARPA, short for Advanced Research Projects Agency, was created in response to the Soviet Union's launching of Sputnik in 1957. It had the mission of advancing technology that might be useful to the military. It gave grants to universities and awarded contracts to companies whose ideas looked promising. ARPANet was the predecessor to NSFNet, which is now generally known as the Internet (see Figure 3-2).

Falling prices and improvements in computer processors led to a major shift in trends. Dumb, host-attached terminals were being replaced by intelligent terminals. Today's PC is much more powerful than the earlier hosts that controlled many terminals. As PCs became more powerful, it made sense to interconnect them to share data, and so in the early 1980s, the local area network was born. The resulting landscape was one of hosts, powerful PCs (used to emulate dumb terminals as well as to perform local processing of applications such as word processing), LANs, and a need for a way to interconnect them.

On a parallel growth path, in 1983 the DoD split off a network called MILNet (for military use) from ARPANet, which remained dedicated to research use. ARPA's research on packet switching and an internetworking protocol led to the development of the TCP/IP

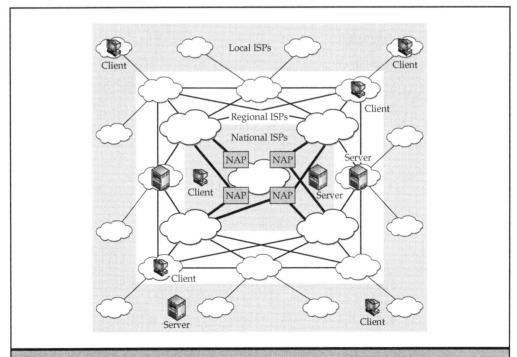

Figure 3-2.   Today's Internet

protocol, a successor to the early Network Control Program (NCP) host-to-host protocol. In 1980, ARPA started converting the machines of its research network, ARPAnet, to use the new TCP/IP, a task that was completed in 1983, when ARPA requested that all computers willing to connect to its ARPANET use TCP/IP.

In 1986, the National Science Foundation established the National Science Foundation Network (NSFnet) to link six supercomputer centers through 56 Kbps lines. NSFnet used TCP/IP at a time when there were strong tendencies toward the rival GOSIP ISO protocols and support for X.25. By 1988, the network was upgraded to a T1 backbone through a contract that was awarded to IBM, MCI Telecommunications Corporation, and Merit Network, Inc. The upgraded network was used to connect seven additional research sites, for a total of 217 interconnected networks. ARPAnet was dissolved in 1990, and responsibility for the Internet was passed to the NSFnet. In 1991, the NSFNet lifted restrictions that had prevented commercial traffic on the Internet, and this proved to be a catalyst for the growth of the network.

Statistics provided by Merit on the use of the NSFnet backbone show a traffic rate of 12 billion packets per month at the beginning of 1992, the year in which the backbone was upgraded to T3 speeds. Over 6000 networks were connected, one third of which were outside of the U.S. The network was also growing at 11 percent per month.

By 1994, traffic had increased to 75 billion packets per month. On April 30, 1995, the NSFNet was officially dissolved, and the traffic was routed through a new OC3 very high speed Backbone Network Service (vBNS), with four network access points administered by MCI. Curiously, although the network had evolved from a series of data networks, the company that was chosen to administer it had its roots not in data technology but in telecommunications.

The selection of TCP/IP for NSFnet resulted in strong acceptance for the protocol worldwide. It had a very large and global installed base, and the protocol began to be deployed in LANs. The commercialization and privatization of the Internet, in 1991 and 1995, respectively, led to its increased use by the private sector. The network was no longer an infrastructure for the provisioning of networking services to the federal government. The improved infrastructure, global reach, and open architecture led to its acceptance as the de facto network for internetworking. This acceptance and reach have led to the development of feature-rich content, which is now driving the requirement to beef up that last mile to our homes and offices.

Broadcast Networks

In 1895, one of Guglielmo Marconi's most important discoveries was the "ground-wave" radio signal, which was created by adding a ground to the transmitter antenna. Grounding the antenna resulted in radio signals that used the earth as a waveguide; in other words, the signal followed the earth's plane, resulting in longer ranges. In contrast, free-space transmission dispersed in all directions, which made the signal weaker a lot faster.

Marconi's success led to a large number of competitors on both sides of the Atlantic, resulting in many important refinements. Reginald Fessenden devised the "continuous wave" theory, demonstrating that radio waves could be used as a carrier for sound. His research led directly to the first long-range transmission of voice, on Christmas Eve 1906 from a station at Brant Rock, Massachusetts. Unlike the telephone, which was quickly adopted for business and home use, it took radio broadcasting another ten years and the First World War before it became commonplace.

In 1915, the first vacuum-tube radio transmitters appeared—a development that would eventually lead to widespread broadcasting. AT&T and its subsidiary Western Electric conducted a series of very successful experiments using the vacuum tubes that increased AT&T's prominence within the radio industry.

When the U.S. entered World War I on April 6, 1917, all broadcasting experiments were halted. All radio stations not needed by the government were closed, and it became illegal (for the duration of the war) for the general population to listen to any radio transmission. A power granted to the president (then President Wilson) through the 1912 Radio Act allowed him to shut down radio stations "in time of war."

After the war, the ban on broadcasting was lifted. Broadcasting blossomed in 1922, and AT&T announced its intent to launch a nationwide broadcasting network. This led to the radio group—a joint venture between General Electric, Westinghouse, and their jointly owned subsidiary, the Radio Corporation of America—forming their own smaller radio network. After four years of competition, the AT&T broadcasting network and the radio group of companies merged and became NBC.

In June 1925, the first synchronized transmission of pictures and sound, using 48 lines and a mechanical system, was achieved by Charles Francis Jenkins, who called it *radiovision*. The ten-minute film was sent from Anacostia Naval Air Station to Washington, D.C., a distance of five miles, where representatives of the Bureau of Standards, the Navy, and the Commerce Department, among others, viewed it. Two years later, the first demonstration of television was presented in front of an audience of members of the American Institute of Electrical Engineers and the Institute of Radio Engineers. The Federal Radio Commission issued the first U.S. television license to Charles Jenkins Laboratories in Washington, D.C., in February 25, 1928, and the first commercial TV license was issued in 1941.

By the late 1940s, radio and television had taken a firm place in the U.S. way of life. But over-the-air broadcast signals were not adequate for sections of the population that lived in mountainous areas. Cable television came into existence in 1948 when two cable systems were built in Pennsylvania and Oregon. Antennas were mounted on the tops of mountains, and homes in the local community were directly wired through coax cables.

By 1962, there were almost 800 cable systems serving 850,000 subscribers, and the focus of cable providers began to expand from the role of a rebroadcaster to that of content provider. Television broadcasters began to see them more as competitors, and in response to concerns, the FCC expanded its jurisdiction and placed restrictions on the cable industry, an action that stagnated the development of cable systems in major markets until the early 1970s.

Industry efforts at the state and federal levels of government saw a gradual lessening of the restrictions, and in 1972 the first pay-TV network, Home Box Office (HBO), was launched. This led to the launch of a national satellite distribution system for cable services. The satellite distribution spurred a new period of growth, and cable subscriptions increased to over 15 million by 1980. In 1984, the industry began complaining that local rate restrictions prevented them from making the necessary investments to compete with satellite broadcasting, which was believed to be imminent. In 1986, the industry was deregulated, leading to a period of soaring cable rates. Subscribers complained in throngs, and in 1993 the Cable Act was passed into law that reregulated cable rates.

The cable industry got a second bite of the apple in 1996 when they were instrumental in crafting some of the provisions of Telecom Act. This time, they cited telephone companies and satellite companies like Direct-TV as possible competition. The act was written to allow local and long-distance telephone and cable companies to compete in each other's markets. Again, AT&T played a prominent role as the corporation began to purchase cable companies with the aim of offering telephone services in addition to regular cable services through a single cable connection. This was the second time in its history that AT&T ventured into the broadcast network market. In December 2001, AT&T announced the planned merger of its broadband division with Comcast Corporation to create the largest cable company in the U.S. Table 3-1 shows the extent to which cable has penetrated the U.S. market.

Cable and over-the-air broadcast networks have always been one-way delivery systems. Both broadcast and cable networks have recognized the potential and limitations of the last mile, the portion of the network that touches our homes and offices. And both are aggressively investing in upgrading their infrastructure in order to provide bidirectional communications in order to offer multiple services across a single connection.

Category	Figure
U.S. television households	105,444,330
Homes passed by cable	98,600,000
Homes passed as a percentage of TV households	96.7 percent
Cable penetration of TV households	69.2 percent
Pay cable units	51,610,000
Cable systems	9,947
National cable networks (video)	281
Annual cable revenue	48,150,000,000
Schools served by cable in the classroom	81,654
Students served by cable in the classroom	43,676,577

Table 3-1.   Penetration of Cable in the U.S. (source: National Cable Television Association)

Convergence

The vision of a common access network for multiple services is not new. The concept of the integrated network was the basis for the development of ISDN. It was hoped that this would be the ubiquitous multimedia technology that would integrate voice, data, and video. Unfortunately, the technology was not embraced at the pace that was originally hoped. One reason for the delay was that two major switch manufacturers, Northern Telecom (now known as Nortel) and AT&T (whose switch business was spun off with Lucent Technologies), chose different ways to implement the ITU standards, ways that did not always interoperate. Subsequent work has led to the development of newer standards, National ISDN-1 and -2, that hide the nuances of each switch from the end user.

Regardless of its hope and promise, ISDN was still a telco solution and so would not have been accepted by the cable industry as a viable solution. Today, the main technologies that are contending for the last mile access include Digital Subscriber Line (DSL—another telco solution), cable, and wireless, having roots in the telecommunications, broadcast cable, or over-the-air industry, respectively. Other forms of access are also emerging; these other access networks will be addressed in subsequent chapters.

Technological advances and regulatory drivers such as the Telecommunications Act of 1996, which opened up competition in local markets, have both contributed to the race to offer an integrated solution. The single most telling statement that testifies to the extent of convergence is the statement that AT&T is the largest cable provider in the U.S. today. This company, which is best known for telephone service, is now a major player in broadcast networking. The merger announced in December 2001 between AT&T and Comcast is a further testament to the state of convergence.

Economic factors have also played a part. A single access for all external services to the home is perceived to be more convenient and affordable than multiple transport services. A common methodology can also be incorporated into the future design of homes, apartments, and office complexes. The final and most important aspect driving the convergence of multiple services across a common access network is the promise of a new revenue stream. The market potential is enormous, and service providers are hoping to get their share of the projected growth.

TECHNICAL FOUNDATIONS

The purpose of the remaining chapters in Part 2 of this book is to examine each access technology in more depth. This section provides a foundation by explaining basic technologies that forms the basis of the different access networks and services. Table 3-2 summarizes features of these access networks and the foundations on which they are built.

Medium	Service	Modulation Technique	Bandwidth	Bit Rate	Distance
Coaxial	Cable downstream	64/256-QAM	750 MHz	3.75 Gbps	< 1 mile before amplifying
Coaxial	Cable upstream	QPSK	200 to 3.2 MHz	400 Kbps to 5 Mbps	< 1 mile before amplifying
Coaxial	Cable upstream	16-QAM	200 to 32 MHz	800 Kbps to 10 Mbps	< 1 mile before amplifying
Phone wire	DSL downstream	DMT, CAP	1 MHz	6 Mbps	3 miles
Phone wire	VDSL	QAM, DMT	30 MHz	51 Mbps	600 ft
Fiber	Fiber	OOK and DWDM	16 frequencies of 2.4 Gbps each	40 Gbps	350 miles
Over-the-air	Broadcast TV	8-VSB	6 MHz	19.39 Mbps	> 50 miles
Over-the-air	LMDS	QPSK	1.1 GHz	2 Gbps	< 3 miles

Table 3-2.   Features of Services and Media

Modulation Techniques

Transmission media, which include wire, fiber, and air, require the use of modulation techniques to place data onto the medium. Multiple modulation techniques are needed because each medium has unique natural characteristics that need to be considered for the efficient transmission of a signal. There are many modulation techniques for each medium, and some techniques are common to different types of media. The choice of a particular technique can easily become a religious debate, as one camp extols the virtue of one over the other. In reality, each access technology will employ its own preferred modulation technique that is optimized for the transmission medium on which it is built. The optimization process takes into consideration the cost of the technique versus the performance trade-offs.

The three basic modulation techniques are *amplitude*, *frequency*, and *phase*, and all three use a carrier signal (see Figure 3-3). Each modulation of the carrier is used to represent a binary 0 or 1 of digital data. Some element of timing is required in all types of modulation, because without the element of timing it would be impossible to interpret consecutive bit patterns that are the same.

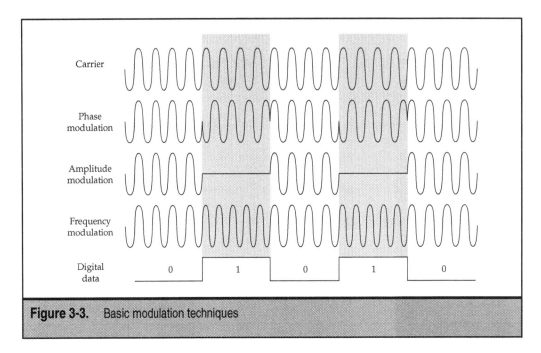

Figure 3-3. Basic modulation techniques

For example, if 1 = on and 0 = off, it would be fairly simple to recognize a pattern of 1010. The receiver would receive on-off-on-off and would know that it translates to 1010. But what happens if the data to be sent is a pattern of 1001? The receiver will see it as on-off-off-on and will interpret it as on-off-on or 101. Because there is no change of state between the two offs, it appears as a continuous off, which will be interpreted as a single 0. To correct this, the receiving system needs a way to know that two of the same bits were transmitted.

In basic modulation techniques, each bit is tied to a set number of cycles of the frequency. So if each bit is represented by three cycles, the receiving station is able to know that consecutive 0s were transmitted, because the signal will have been off for six cycles—two times the normal period required for a bit. Similarly, three consecutive zeros will result in the signal being off for nine cycles.

Amplitude modulation, also known as *amplitude shift keying (ASK)*, modifies the amplitude of the carrier to represent the digital information. A binary 1 is represented by a high carrier amplitude, and a 0, as a zero carrier amplitude. The problem with this approach is that loss of the carrier, whether intentional or not, is interpreted as a 0 bit.

Frequency modulation, also known as *frequency shift keying (FSK)*, manipulates the frequency of the carrier. A higher carrier frequency is used to indicate a 1, and the original carrier frequency, a 0. This technique is a lot more immune to noise, and a loss of connection is never interpreted as data, because the data is represented by the presence of discrete frequencies.

Phase modulation, also known as *phase shift keying (PSK),* modifies the phase of the carrier to signify a 0 or 1 bit, as the name suggests. The original carrier is used to represent a 0 bit, and an inverted or phased version of the same frequency, to represent a 1.

The remainder of this section will cover more complex modulation schemes that are used to gain higher speeds.

OOK: On/Off Keying

This modulation technique is the simplest of the techniques used in fiber optic. It is based on the principle of a light being on or off. A binary 0 is represented by the absence of the light, and a 1, by its presence. The timing element, called the *symbol time,* is based on the duration of time the laser is on or off. The *symbol rate* is the number of symbols per unit of time.

QPSK: Quadrature Phase Shift Keying

As mentioned previously, phase shift keying (PSK) is the simplest way of using a phase shift to represent digital data. It uses two signals phased at 0° and 180°, yielding two possible states. *Quadrature phase shift keying (QPSK)* introduces a second wave that is offset from the original by 90° or one that is 90° out of phase with the other. The digital data is encoded using four (quad)–level differential PSK. This means that for every phase shift, we are able to code two bits, shown in Figure 3-4.

QPSK is widely used in different systems and has been around for a long time. It is the technique that is used in the Bell 212A–compatible modems and V.22, which gives an idea of just how long it has been around. Because it is easy to implement and fairly resistant to noise, QPSK is used for upstream communication from cable subscribers as well as LMDS (Local Multipoint Distribution Service) satellite services.

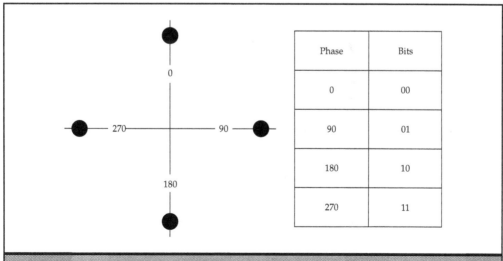

Phase	Bits
0	00
90	01
180	10
270	11

Figure 3-4.   QPSK achieves two bits per phase shift

QAM: Quadrature Amplitude Modulation

Quadrature amplitude modulation (QAM) is a very well understood and widely used technique, common in satellite communications and military applications for many years. It takes QPSK a step further by utilizing amplitude and phase modulation to transmit multiple bits. An unmodified signal has two states, which allow for the transmission of only a 0 or a 1. By increasing the number of states, it becomes possible to transmit more bits. QAM uses two waves that are transmitted at the same frequency but offset by 90°. In other words, they are 90° out of phase and are called a quadrature. Each of the waves is independently amplitude-modulated, hence the name quadrature amplitude modulation. The combination of amplitude and phase results in many more states, making it possible to transmit more bits for every state change.

The constellation pattern depicted in Figure 3-5 depicts the four states of the quadrature at 0, 90, 180, and 270, as well as another four states made possible by adding another amplitude, resulting in eight different states, or 8-QAM. Eight states yield the possibility of representing three bits per state.

There are several different versions of QAM, each using a different number of states; they are represented as *nn*-QAM, where *nn* represents the number of states. The number of bits per state change is k, where $nn=2^k$, and as a result the number of states grow exponentially with the number of bits transmitted. For instance, 64-QAM would be 2^6, resulting in six bits per state, and 16-QAM would be four bits per state. The higher the number of states, the greater the number of bits transmitted. This can be a disadvantage, as transmitting eight bits per state would require 256 possible states, resulting in a very dense constellation pattern requiring very precise equipment to demodulate the signal.

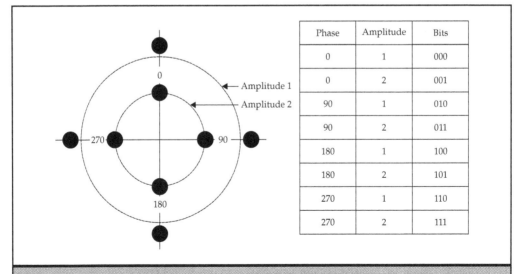

Phase	Amplitude	Bits
0	1	000
0	2	001
90	1	010
90	2	011
180	1	100
180	2	101
270	1	110
270	2	111

Figure 3-5. 8-QAM uses three bits per change

2B1Q: 2Binary 1Quaternary

The *2B1Q* technique is an amplitude modulation line-encoding scheme used in ISDN and HDSL. In this technique, four different voltage (amplitude) levels are used to encode two bits (see Figure 3-6). To encode more bits, more voltage levels would be needed. To encode k kits per state, 2^k voltage levels are needed. In other words, to code four bits per state, 2^4 or 16 voltage levels are needed. The higher the speed, the more difficult it is for the receiving station to discriminate among the different voltage levels.

CAP: Carrierless Amplitude/Phase Modulation

Carrierless amplitude/phase modulation (CAP) is basically the same as QAM, with the difference being the way in which the carrier is used, or not used in this instance. With QAM, analog signals are used, but CAP uses digital filters with equal amplitude characteristics that have a different phase response (Hilbert pair). The signals are then combined, fed into a digital-to-analog converter (DAC), and transmitted.

CAP's advantage over QAM is that implementation is done efficiently in silicon, making it more cost effective. The trade-off, however, is that with QAM the constellation is fixed relative to the carrier, but with CAP the absence of a carrier causes the constellation to rotate. The receiving station is required to employ technology that compensates for the rotation. Even with this requirement, CAP is still more cost effective than QAM.

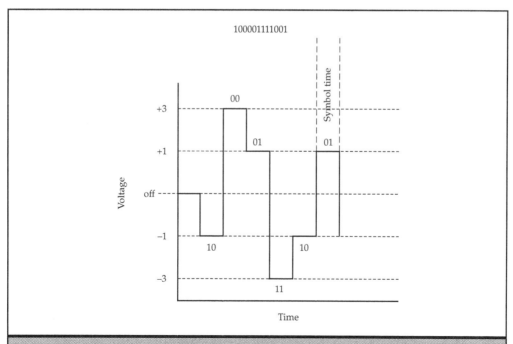

Figure 3-6. 2B1Q uses an amplitude modulation scheme

DMT: Discrete Multitone Modulation

So far, we have discussed modulation techniques that use a single carrier. Multicarrier techniques require considerable digital processing, and for many years they were not commercially feasible. *Digital signal processors (DSPs)* now make it possible to divide the available bandwidth into multiple subchannels, each employing a single-carrier modulation technique like QAM. The result is higher speeds than were previously available for commercial applications.

Discrete Multitone (DMT) modulation's ANSI T1.413 standard specifies 256 subchannels, each with a 4 kHz bandwidth. They can be independently modulated from zero to a maximum of 15 bits/sec/Hz, which allows for up to 60 Kbps per tone. Figure 3-7 illustrates DMT and the use of multiple subchannels.

The best way to conceptualize DMT is to think of it as a technique that uses multiple modems, each representing a single 4 kHz subchannel and each operating concurrently. Each subchannel has its own carrier, which corresponds to the center frequency of the subchannel, and uses QAM modulation. Each subchannel operates independently and can carry up to 15 bits per state depending on the characteristics of the line. If a subchannel is plagued with interference, that subchannel is not used, and because each subchannel operates independently, each can operate at different bit rates depending on the line quality or level of external interference.

An important difference between CAP and DMT is that DMT uses more "smarts" to tailor its signal to the characteristics of the particular channel. Both are designed to make use of the large amount of higher-frequency bandwidth that lies above the voice passband.

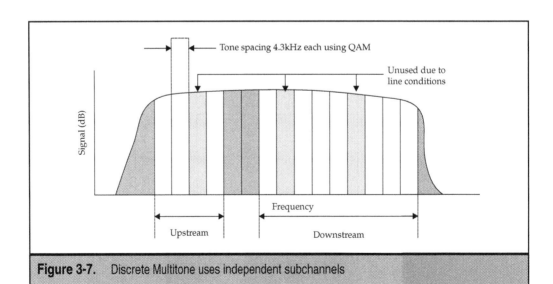

Figure 3-7. Discrete Multitone uses independent subchannels

OFDM: Orthogonal Frequency Division Multiplexing

The difference between OFDM and DMT lies in the way in which the subchannels operate. OFDM subchannels use a common modulation that transfers the same number of bits per second on each subchannel. In contrast, the bit rates of DMT subchannels vary with the quality of the medium. OFDM has been selected for Europe's Digital Audio Broadcast (DAB) standard. DAB will broadcast CD-quality sound and multimedia over the airwaves for automobile and other mobile applications.

Spread Spectrum or CDMA: Code Division Multiple Access

Spread spectrum technology was developed out of research sponsored by the U.S. government for military reasons. During World War II, broadcast signals were susceptible to jamming. This prompted the government to explore ways to make the signal resistant to any attempts at jamming.

Spread spectrum employs a technique that scrambles the signal into different frequencies, a technique that is very similar to the technique used for jamming. In a sense, it is a technique that makes order out of chaos. The technique is especially suitable for wireless communications in cities where a proliferation of wireless signals are used for different purposes. There are two types of spread spectrum: frequency-hopping spread spectrum (FHSS) and direct-sequence spread spectrum (DSSS).

FHSS: Frequency-Hopping Spread Spectrum

Frequency-hopping spread spectrum, sometimes called frequency-hopping code division multiple access (FH-CDMA), modulates the signal onto a carrier signal that hops from frequency to frequency over a wide bandwidth. The act of sending short bursts of data on different frequencies reduces interference from rogue signals (see Figure 3-8).

Both the transmitter and the receiver must agree on the same hopping code in order for the message to be reassembled. To properly receive the signal, the receiving station must use the same hopping code as the sender and must be listening to the incoming signal at the right time and the correct frequency.

FCC regulations require that 75 or more frequencies be used per transmission channel, with a maximum time of 400 ms spent at any single frequency. There are performance trade-offs when using FHSS; for instance, it takes the receiver a finite time to train to each new frequency, time that could have been used for receiving the message had a single frequency been used.

Two stations can use the same frequency without interfering with each other if they use different hop codes.

DSSS: Direct-Sequence Spread Spectrum

Direct-sequence spread spectrum, also known as direct-sequence code division access (DS-CDMA), combines the data with a higher–data rate bit sequence called a *chipping code*. Instead of using a single symbol to indicate a single bit, DSSS operates by using

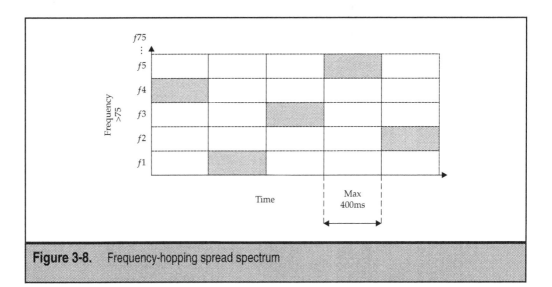

Figure 3-8. Frequency-hopping spread spectrum

multiple symbols to represent a single bit. The multiple symbols, however, are all transmitted in the same window of time it would have taken to transmit a single bit.

To better understand this, we need to do a review of one of the concepts discussed in an earlier section, "Modulation Techniques." There, the following statement was made:

> *In basic modulation techniques, each bit is tied to a set number of cycles of the frequency. So if each bit is represented by three cycles, the receiving station is able to know that consecutive 0s were transmitted, because the signal would have been off for six cycles—two times the normal period required for a bit. Similarly, three consecutive zeros would result in the signal being off for nine cycles.*

The point of the passage is basically that some agreement must be reached between the sender and the receiver on a method for timing when a bit has been received and when another begins. Using the previous example, if the agreed time interval for the transmission of a single bit is three cycles, then in DSSS any number of symbols that is to be used to represent a single bit must be transmitted within three cycles. So if the number of symbols used to make up the pattern for a single bit is 11, all eleven must be transmitted in three cycles. The subinterval of time that each symbol occupies is called a *chip*.

Table 3-3 gives an example of the principle of DSSS.

The redundant chipping code helps the signal resist interference and also enables the original data to be recovered if data bits are damaged during transmission. The ratio of chips per bit is called the *spreading ratio* or *processing gain*; the higher the spreading ratio, the greater the resistance to interference. The FCC has set the minimum spreading ratio at ten.

It should be noted that the sequence of symbols, called the *spreading code,* is unique to each connection. The sequence patterns will be different for any two connections. In other words, the chips used to represent a 0 bit are unique to the connection between a specific transmitter and receiver. Other receiving stations perceive the pattern to be noise.

Action	Coding
Message to be sent	0100
Normally, we would expect	0 = off and 1 = on
But, both the sender and the receiver agree that	0 = off-off-on-off-off-on-on-off-on
and	1 = on-on-on-off-on-off-on-off-off
The transmitted message would then be	off-off-on-off-off-on-on-off-on (0) on-on-on-off-on-off-on-off-off (1) off-off-on-off-off-on-on-off-on (0) off-off-on-off-off-on-on-off-on (0)

Table 3-3.   Example of DSSS Chipping Code

Because of its history and the motivation behind the research for its development, spread spectrum is also a very good noise mitigation technique. The use of multiple frequencies in the case of FHSS makes it less susceptible to interference. The entire spectral range would need to be affected for the entire transmission to be affected. DSSS is even less prone because of its use of a chipping code to represent a single bit. In the event a few bits are affected, the remaining bits may be unique enough to interpret the sender's intent.

Error Detection and Correction Techniques

Error detection and error correction are two entirely different schemes. In *error detection*, the receiver may be aware there is an error but makes no attempt to correct it. Some protocols are designed to know that there is a problem and respond by ignoring the transmission or packet completely. Protocols such as IP use checksums to determine the integrity of a packet; if a packet is deemed corrupt, it is dropped. Other higher-layer protocols are left to clean up any discrepancies. In the case of IP, a protocol like TCP will let the sender know that a packet was not received and the sender will send it again.

Error correction is an attempt to correct an error that is known to have occurred. Attempts to correct an error may be made by the receiving station at the time of occurrence; in other applications, another protocol is left to correct it. Even though IP discards a packet that is in error, the end-to-end transmission does have a measure of error control when a higher-layer protocol like TCP is used. In cases where a protocol like UDP is used to control the end-to-end flow, error conditions are totally ignored and the application is left to worry about it.

Error control schemes that rely on retransmissions have a direct impact on latency. Implicit in a scheme that relies on retransmissions is the requirement to have some type of acknowledgment scheme, which serves only to slow things down. High-speed systems typically use schemes that allow the receiver to detect and correct the error on the fly in order to negate the need for retransmissions. Noise mitigation techniques are also a type of error control.

Parity

The simplest form of error detection is the use of *parity*. The basic tenet of parity checking is to introduce a single redundant bit into a stream to determine the integrity of the stream upon reception. If the stream is 0000111, then an odd parity error detection scheme would insert an extra 0 to ensure that the number of 1 bits in the stream is an odd number, three in this case. The receiving station, having been configured to expect a odd parity scheme, would determine the received stream to be in error if the number of 1 bits received is an even number. The reverse is true. If an even parity scheme were employed, the receiver would expect to receive the transmission with the number of 1 bits being an even number.

This is a simple yet effective way of dealing with errors that affect a single bit. The technique, however, no longer works if two or more bits are erred. Two transposed bits result in a bad packet with the same parity as the one that was originally sent. On receipt, the packet would appear to correct but would in fact be in error. Parity checks are no guarantee for more than two wrong bits.

FEC: Forward Error Correction

Forward error correction schemes fix errors at the time of reception and do not require retransmission of the data. This is especially important on wired media where errors can be introduced through crosstalk, which may cause subsequent retransmitted blocks to be in error again, thereby creating a loop until some other, higher-layer protocol times out the connection.

FEC segments the data into blocks, and redundant bits are added before transmission. These redundant bits are used to detect errors that may be introduced during transmission. The principles of FEC are similar to the principles of parity checking. Examples of FEC schemes are Hamming codes and Reed Solomon (RS) codes.

Hamming Codes *Hamming codes* provide for FEC by using a "block parity" mechanism that can be inexpensively implemented. In general, their use makes it possible to correct one error or detect two errors but not to do both simultaneously. Thus, Hamming codes may be used as an error detection scheme to detect up to two wrong bits *or* to correct a single wrong bit.

The error correction is accomplished by using more than one parity bit, each computed from the bits in the data. The number of parity bits required is defined by the Hamming rule and is a function of the number of bits of information transmitted.

The Hamming rule is expressed by the following equation:

$$d + p + 1 <= 2^p$$

where d = number of data bits and p = number of parity bits. Thus, if the number of parity bits is four, then the rule shows that four parity bits can provide error correction for five to eleven data bits.

The result of adding the computed parity bits to the data bits is called the *Hamming code word*; it is expressed as (n,d), where d is the number of data bits and n is the total number of bits in the code word. It follows then that the number of parity bits is $n - d$.

The best way to understand the benefits and application of FEC techniques is to go through an exercise. The following rules outline the steps to take to determine a Hamming code for any number of data bits. As you review the example in Figure 3-9, try to correlate each step with the rules.

The rules for the exercise are:

1. Mark all bit positions that are powers of two as parity bits: positions 1, 2, 4, 8, 16....

2. The actual data bits occupy the remaining bit positions: 3, 5, 6, 7, 9, 10, 11, 12....

3. Each parity bit calculates the parity for specific bits in the code word using the following logic:
   Parity bit in position 1: check every other bit (1,3,5,7,9...)
   Parity bit in position 2: check 2 bits, skip 2 bits (2–3,6–7,10–11,14–15...)
   Parity bit in position 4: check 4 bits, skip 4 bits (4–7,12–15, 20–23...)
   Parity bit in position 8: check 8 bits, skip 8 bits (8–15,24–31,40–47...), etc.

4. Set the parity to 1 if the total number of ones is odd. Set the parity bit to 0 if the total number of ones is even.

RS: Reed Solomon   *Reed Solomon* is also a block mode scheme, but unlike Hamming codes, RS is able to operate on symbols or groups of bits, so it is able to process much larger blocks and deal with longer burst interference. This coding system provides superior burst-error correcting capability. The principles are the same as for all FEC schemes, in that redundant data is added to the data stream to facilitate error correction. RS schemes also provide for the ability to interleave, further reducing errors. An RS scheme is normally expressed as RS(n,k), where n is the total number of symbols per block and k is the number of symbols that were intended (or the symbols that represent the original data).

An expression of RS(204,188) would mean a Reed Solomon scheme where 188 symbols (or bytes, when used to express MPEG packets) make up the block. The number of error-correcting bytes is 16 (204 – 188), making for a total block size of 204 symbols. The formula $n - k / 2$ is used to determine the number of symbols that can be corrected by the coder. Thus, an RS(188,204) scheme provides burst error protection for any eight bytes in the block (204 – 188 = 16, 16 / 2 = 8).

Interleaving

Interleaving is designed to protect against impulse noise or burst interference. Mixing bits from several data words in a fixed way, or interleaving the bits, helps ensure that no single

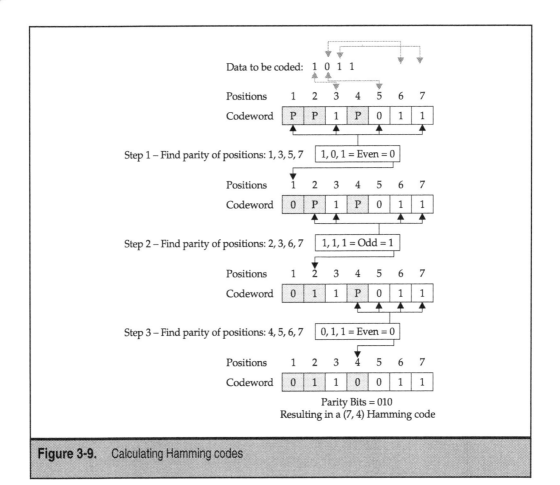

Figure 3-9. Calculating Hamming codes

word is affected by impulse noise. If impulse noise affects the stream, several bits may be affected, but interleaving the bits increases the probability that the bits affected are from different words and not all from the same word. This allows for better reconstruction of the original data.

For example, *x* data words are to be transmitted. The words are placed in an interleave buffer where all the bits are aligned in a matrix. Instead of transmitting the words in a serial fashion, each column of the matrix is transmitted. The receiver collects all the columns and reconstructs the matrix. Now if a few bits were transposed, it would be easy to error-correct using a Hamming code, which is effective in reconstructing a block that has only a single bit error.

The error rate of a transmission can be significantly reduced if a Hamming code is combined with interleaving. A trade-off of interleaving is latency. Both the sending station and the receiving station must place the data into a buffer to perform interleaving and reconstruction of the stream, which introduces serialization delay. The processing time required to determine the additional Hamming code bits, as well as the additional bits generated by the process, also adds delay.

Challenges and Limitations of the Transmission Medium

Different media have different qualities, and each is subject to its own set of unique limitations and challenges. The need to increase the speed and performance of a transmission medium, whether it be wire line, fiber, or wireless, requires constant research on new methods and technologies to overcome the shortcomings of that medium.

Signal Attenuation

The longer a signal travels, the more it attenuates, or the weaker it becomes. *Attenuation* is distance related and is common in both analog and digital signals on all transmission media. The challenge that this presents is that the weaker the signal, the more difficult it is for the receiver to differentiate between the intended signal and noise. This relationship between the signal and noise is expressed as a ratio $x:y$—the lower x is, the less chance there is of differentiating it from noise.

Imagine yourself at a party in a room full of people. You are seated close to a friend and having a discussion. As long as your friend is close by, he is able to distinguish your words and your meaning. Now, let's say your friend decides to get a drink on the other side of room. He finds it increasingly difficult to hear your words as he moves further away. Eventually, you are not understood at all because of all the competing noise in the room. At that time, your only option is to shout, or turn up the power. So too in transmission, a way to get a better signal-to-noise ratio is to boost the power of the signal.

Now imagine the same example, but this time, the room is extremely large with lots of echo, like an auditorium. In this environment, your friend would probably lose track of what you are saying a lot faster because of the reverberant qualities of the room. The difference between this and the previous example is the size of the room. The room size introduced different acoustic qualities that made the challenge of communicating at a distance even greater. This is also the case for the type of medium that is used; different media have different transfer qualities. Different wire grades, for instance, have different levels of attenuation, and fiber has the least attenuation. And as a final consideration, the higher the frequency, the more prone it is to attenuation.

The extent of attenuation or gain is normally expressed in logarithmic units called decibels (dB). The doubling of power or a 50 percent decrease in power is equal to 3 dB gain or loss. So, a 6 dB gain is four times the power and a 6 dB loss is one quarter the original power. When it is necessary to transmit signals over long distances via cable, repeaters are used at different lengths of the cable to boost the signal strength to overcome attenuation. This greatly increases the maximum attainable range of communication.

Power attenuation in decibels is given by the formula:

$$A_p = 10 \log_{10}(P_s / P_d)$$

where:

A_p = Attenuated power,
P_s = Power at source, and
P_d = Power at destination

Attenuation can also be expressed in terms of voltage:

$$A_v = 20 \log_{10}(V_s / V_d)$$

where:
A_v = Attenuated voltage,
V_s = Voltage at source, and
V_d = Voltage at destination

Resistance

Resistance is another characteristic of a transmission medium that affects signal strength. It is a measure of how much a medium obstructs the flow of an electric current. Resistance is measured in ohms and noted by using either the word ohms or the symbol Ω. As high frequencies are transmitted through wire, electricity tries to migrate to the outside of the wire, leaving little conductivity in the middle. This effect, which is sometimes called the *skin effect,* explains why wired media cannot be used for very high frequencies, while over-the-air transmissions are able to achieve ranges up to 30 GHz.

Noise

In addition to the challenges the natural properties of the transmission medium present, engineers also have to be concerned about external impairments. *Noise* comes in many different flavors and from many sources and is a factor in both analog and digital transmissions on all transmission media. It is any unwanted electrical or electromagnetic energy that interferes with a signal.

White Noise *White noise* is always present on the medium. White noise has a frequency spectrum that is uniform and continuous over a specific frequency band. The range of frequencies is generally defined as those that are within the range of human hearing, from 20 Hz to 20 kHz. White noise is not caused by a single source but is the result of a confluence of sources. The cumulative effect is sometimes called Gaussian noise, which is the random fluctuations and oscillations that occur when electricity moves across a communications medium. You can see and hear it when your television is having fuzzy reception.

Impulse Noise or Burst Noise *Impulse* or *burst* noise is the random occurrences of electric or electromagnetic spikes having random amplitude and spectral content. Impulse noise is very short in duration and may affect a wide range of frequencies. Some sources include motors, lightning, and corroded light switches.

RFI: Radio Frequency Interference Both over-the-air transmissions and wired media can experience *RFI*. The nature of this type of interference makes its possible impact on broadcast signals clear. Its impact on wired communication is not as obvious. Wire can act as an antenna when radio frequencies are present. At lower frequencies, this is not as problematic, but the impact becomes more pronounced at higher speeds. DSL and cable modem systems experience RF noise called *RF ingress.*

RF ingress increases with housing density and is especially prevalent in high-rise apartment buildings containing 200–300 units. It negatively affects operation of DSL and cable modems within multidwelling units, and it can interfere with cable modems in single-family homes that share the same return path. The primary source of RFI is AM broadcast and ham radios, and the extent of the interference is proportional to the proximity to the source.

Ham radios are more a problem to the operation of VDSL because their frequency bands overlap, but QAM and DMT modulation techniques have proved to very resilient in this environment. In a DMT system, RF ingress noise is mitigated in part because the channel is partitioned into narrow subchannels. In a worst-case scenario, subchannels that overlap the band of the interfering radio are simply turned off. Because the interference may affect several subchannels, a window can be applied to the received signal before demodulation to reduce the number of affected subchannels. Texas Instruments has developed an additional, proprietary receiver technique that confines RF ingress to only a few subchannels, improving system performance.

Crosstalk

Crosstalk is another form of external impairment. It occurs when a disturbance is caused by the electromagnetic fields of one telecommunications signal affecting a signal in an adjacent circuit. The two conductors do not need to actually touch for crosstalk to occur; they need only be close enough that the radiating signal of one of the wires interferes with the signal traveling on the other. Reference is sometimes made to near-end crosstalk (NEXT) and far-end crosstalk (FEXT); the difference between the two lies in where the crosstalk occurs, as Figure 3-10 demonstrates.

In order to certify a cable as Category 3 or 5, the NEXT test is performed. NEXT is an error condition that can occur when a connector is connected to a cable. The NEXT test is performed from 1 MHz to 16 MHz for Cat 3, and 1 MHz to 100 MHz for Cat 5, in increments of 1 MHz. Since signal loss can change with frequency, a range of frequencies must be used to determine if the cable is acceptable. Crosstalk is more detrimental at higher speeds because higher frequencies are more susceptible. An infrastructure that operates perfectly for a low-frequency application like voice will not necessarily perform when higher frequencies are applied.

Loading Coils

As we discussed previously, transmitting a signal over long distances attenuates or weakens the signal. This effect is especially pronounced in wired media. Wires stretched

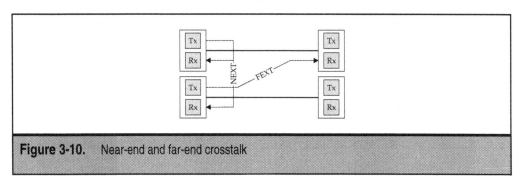

Figure 3-10.   Near-end and far-end crosstalk

over long distances act as capacitors that tend to store the signal rather than aid in its flow. Twisting the pairs helps somewhat by adding an electrical characteristic of *mutual inductance*. Inductance has the effect of offsetting the natural capacitance of the wire, aiding in the transmission of the signal. The benefits derived from mutual inductance are limited, however, and another method of gaining a more pronounced effect of inductance is to add loading coils to the length of the wire.

Acceptable voice quality is limited to cable runs of about 18,000 feet, in most cases. The grade of wire, of course, directly affects the distance, but 18,000 feet has been generally accepted as the limit. If the wire is too long, voice becomes more muffled because attenuation has a more pronounced effect on higher frequencies. Loading coils are used to extend the wire beyond the limit mentioned, and they have the effect of tuning the circuit for voice transmission, which has a frequency range of 300 to 3000 Hz.

By adding loading coils, shown in Figure 3-11, the attenuation rate remains flat up to and beyond 3 kHz depending on the loading architecture used. Loading coils, which look like round doughnuts around which the wires are wrapped, are placed at different distances along the length of the line. In practice, load coils are placed at either 3000 feet, 4500 feet, or 6000 feet. No xDSL services can be supported on a loop with loading coils.

Bridged Taps

Most illustrations of a telephone connection show a pair of copper wires running from your home directly to the central office. This is a good logical depiction of the connection but is far from accurate from the perspective of the physical engineering. The cable from the central office may extend a mile beyond your house. The wire that enters your house is often a separate cable that runs from your *demarc*—the point at which the telephone service meets the house wiring—to a junction box by the street, where it is tapped into a spare cable pair. The pair is then flagged as used to ensure that another customer does not use it further down the street. This approach allows the phone company to reuse the pair anywhere along the street when you move.

These taps, called bridged taps, are nuisances because they have the potential to introduce noise impairments for high-speed applications. People move in and out, and engineers install taps as a matter of course without thought for future types of service. After

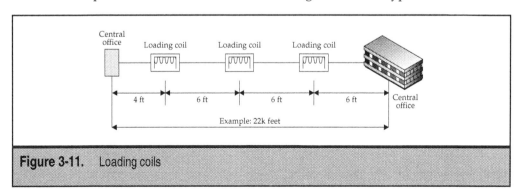

Figure 3-11. Loading coils

all, bridge taps do not affect voice calls, and voice service is what the wires were meant to provide. As a result, the service provider usually does not have adequate records of bridge taps, so any interference they may cause in any future service must be addressed through some type of noise mitigating technique.

SUMMARY

In this chapter, we have touched on some of the more fundamental aspects and terminology of analog and digital communication. The information provided in this chapter forms the basis for further discussion on the different access technologies. As we progress through the remaining chapters, reference will be made to the techniques, terminology, and technologies described in this chapter. These references will be made in passing without further explanation. As we encounter a new aspect of a particular technique as it applies to the specific access technology being discussed, we will spend some time explaining the new aspect. Any new technique not already covered will be discussed at the time it is mentioned.

CHAPTER 4

DSL Access Networks

T he last mile of the network, as it is commonly called, represents the portion of the service provider's network that touches our homes and businesses. Throughout the remainder of this book, we will refer to this portion of the network as the last mile or the access network and, within this chapter, the local loop. This portion of the infrastructure presents the greatest challenge for the provision of integrated digital services because it evolved from networks that were built for specific analog applications. The access network throughout its evolution was tuned to the specific requirements of the applications for which it was built. This tuning resulted in an improved service but ironically, more often than not, made the network inefficient at delivering an integrated digital service without a significant investment to either undo the tuning, deploy solutions that compensate for the limitations caused by the tuning, or build entirely new networks.

Obviously, service providers have a vested interest in leveraging their existing investments, thereby rendering the third option the least attractive. It is a lot more tenable to develop solutions that would compensate for the limitations of the network or require the least retrofitting possible. We consider ISDN to be the first in the line of technologies and services called DSL or *Digital Subscriber Line*. ISDN held the promise of the *Integrated Services Digital Network* and was developed to address telephony services and lower-speed data applications. It was the vision of the ITU, that ISDN would be the catalyst for the total end-to-end digitization of the *Public Switched Telephone Network (PSTN)*—from end user to end user and end device to end device.

The lessons learned from the development and deployment of ISDN, plus the advances in *digital signal processors (DSPs)* led to the development of a new type of technology and service, *Digital Subscriber Line (DSL)*. Many of the improvements and advanced features of DSL are a direct result of the experience gained with ISDN. There are many different types of DSL technologies, each having its own distinct features. Two collective terms are used to describe this generic family of technologies—xDSL and DSL. We have chosen to use the term DSL. As we discuss the different DSL variants, we will sometimes make reference to its chronology. In this context, the contemporary definition of DSL will be used and basic-rate ISDN will not be factored into the chronology of the service, even though we consider it to be part of the family.

DSL service is being launched in most of the major countries of the world and is also beginning to appear in many less-developed ones. Worldwide, the residential market represents the larger installed base, but the business market generates the greater portion of the revenue. The growth of DSL is believed to be a direct result of an increased reliance on the Internet as a tool that facilitates communications, business, and entertainment.

THE STATE OF AFFAIRS

Since the signing of the Telecommunications Act of 1996, any company that is certified to do so by the state in which they operate, may provide the services of a LEC or *local exchange carrier*. Since that time, the number of LECs has grown to the point that the term LEC has been further refined to differentiate between the RBOCs that originally provided service and the new emerging competing companies. The original Bell companies are called ILECs or *incumbent local exchange carriers,* and the competitors, CLECs or *competitive local exchange carriers.*

The passage of the act granted the following provisions to CLECs:

▼   They must have access to UNEs or *unbundled network elements* through a collocation arrangement. The act defined the UNE as any facility or equipment used in the provision of a telecommunications service, and the features and functions that are provided by them. To a CLEC, the most important piece of the UNE is the local loop, as it is the portion of the network that touches the customer. With access to the local loop, CLECs are given the means by which they can offer competing services without the need to lay new copper.

▲   Any service offered by an ILEC to the public must be available to CLECs at a wholesale rate, thereby laying the groundwork for the resale of services. This provision was crucial to the business case of many CLECs, as it negated the need to invest in expensive equipment and allowed small *Internet service providers (ISPs)* the ability to offer services like DSL.

The act paved the way for increased competition. ILECs found themselves in competition with their own services, a fact that has led to competitive rates and services. By 1999, there were over 1500 CLECs operating in the U.S. Inadequate business plans, a lack of venture capital, and a hostile economic environment in 2000 and 2001 contributed to the demise of many, some being forced into bankruptcy. Those that survived grew stronger, and many new ones are still emerging. The Telecommunications Act opened up competition, and now both ILECs and CLECs are in competition to offer value-added services such as DSL.

PSTN Architecture and Optimization

The PSTN has been around for over 100 years and has been optimized to transmit speech through the use of a very narrow range of frequencies. The audible frequency range of the human ear is about 20 Hz to 20 KHz, but the technology used in the PSTN is designed to transmit a much narrower range, 300 Hz to 3.3 KHz. This range was found to be the best compromise between quality and economics, the range in which speech can be economically transported and still be generally understood without a lot of degradation. In reality, there is no fixed range for speech, because everyone has a different voice range, and every syllable has its own set of harmonic frequencies that accompany it. The brain compensates for the missing frequencies, and so the limited range that was chosen was found to be the most efficient range that provides an acceptable quality. The PSTN can be broken down into three sections:

▼   **The local loop**   The portion of the network that connects the subscriber to the telephone company central office through the use of a pair of unshielded twisted copper wires

- ■ **The central office (CO)** Provides the switching and signaling functions of the network
- ▲ **Trunks** Connect the switches across which many voice connections travel

For the purposes of our discussion, we will not dwell on the architecture of the network other than to give a passing understanding of the major components that facilitate the operation of the network. We will, however, examine in more detail the elements of the network that present the greatest challenges, and those that aid in the delivery of new integrated digital solutions.

Limiting the voice passband, as shown in Figure 4-1, helped reduce the cost and complexity of building the network. On the links between switches, multiple voice channels were multiplexed across trunks. The frequency ranges on either side of the voice passband were used as guard bands to ensure accurate demultiplexing.

Challenges of the Local Loop

The telephone network has its roots in, and was engineered for, analog voice calls. The network in its most basic form has a much wider frequency range than is required for what it is intended to do; in fact, it has the capability to deliver millions of bits per second. At its limits, phone wire can transmit at 1 MHz for about three miles, or 30 MHz for a distance of about 600 feet, but because of the application for which it was engineered, analog voice, it was optimized for a frequency range up to about 4000 Hz. As a result, the data

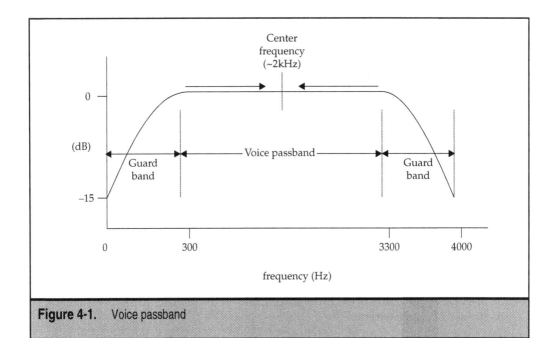

Figure 4-1.   Voice passband

carrying capabilities were limited to a data rate around 64 Kbps, which is a lot lower than what is required for high-speed digital integrated applications.

In 1928, Henry Nyquist published a paper called "Certain Topics in Telegraph Transmission Theory," in which he postulated that a sample of twice the highest signal frequency rate is needed in order to reconstruct the original signal. Applying this theorem to a 4000 Hz passband, it would take 8000 samples to capture a 4 KHz signal. Assuming an 8-bit sample, it would take 64 Kbps (8 bits * 8000) to communicate the digitally encoded signal in real time over a circuit. This calculation formed the basis on which T1-carrier circuits were designed, since they were designed to carry voice that was digitized. The 4 KHz (voice channel) is sampled and multiplexed using time division multiplexing into 24 channels, as shown in Table 4-1. But that is enough for now on T1-carrier circuits, as we will revisit the topic later, during our discussion on HDSL.

The point is, a perfectly good network was totally compromised to provide a basic analog voice service. Or that would be the contemporary opinion, but the fact is, who would have known way back then that the infrastructure would be a contender for the provision of digital voice, video, and data? What would have been the frame of reference for understanding the speeds now required for these new services, for after all, a present-day handheld computer could compete with the most powerful computer in existence back then. In fact, several factors in the engineering of the local loop make it a less-than-favorable environment for the delivery of high-speed integrated digital services.

Mixed Gauges

Phone wire consists of two thin unshielded copper wires that are twisted together to minimize crosstalk. The original construct of the local loop was a single wire using ground return, one that used a common return path. Later, two parallel wires were used, an arrangement that improved voice quality but was subject to signal loss. Eventually, the pair was twisted, decreasing the amount of crosstalk and helping, to a degree, to decrease attenuation through a process of mutual inductance, which offset the capacitance of the wire.

Description	Applying the Calculation
For DS-1 transmissions, each frame contains 8 bits per channel. There are 24 channels, which requires 1 "framing bit."	24 channels * 8 bits per channel = 192 + 1 framing bit = 193 bits per frame
Applying Nyquist theorem.	193 bits per frame * 8000 samples = 1,544,000 bps (the size of a T1-carrier circuit)

Table 4-1. Nyquist Theorem and T1-Carrier Circuits

The local loop is defined by two parameters—the loop length and the wire gauge. These two values determine the frequency response of the local loop. Table 4-2 provides information on common wire gauges of copper wire: the higher the number, the thinner the wire gets and the less distance served.

The connection between the subscriber and the central office is made up of many wire segments of about 500 feet each, spliced together. In the past, a common practice was to mix a 26- and 24-gauge wires or 26- and 19-gauge wires. The thinnest wire (26 gauge) would be used in the first 500 feet from the CO, where it would be spliced to a 24-gauge or the thicker 19-gauge wire. The thicker-gauge wires were used for subscribers that were farther away from the CO because the thicker wires covered greater distances.

For analog voice, this works fine, but mixing different gauges introduces the propensity for "echoing" caused by the transmission being reflected as well as transmitted when the electrical characteristics of the two wires change. The actual splice is also a source of potential problems. If the joint oxidizes, the circuit develops high resistance, causing further problems that were not factored into the design process.

A set of rules called the *resistance design rules* was established to ensure proper quality of service. The rules governed the distribution of twisted wire pairs from the CO to the subscriber; they are summarized as follows:

▼ Loop resistance is not to exceed 1500 ohms.

■ Inductive loading is to be used whenever the sum of all cable lengths, including bridged taps, exceeds 15,000 feet.

■ For loaded cables, 88 mH loading coils are placed 3000 feet from the CO and thereafter at intervals of 6000 feet.

■ For loaded cables, the total amount of cable, including bridged taps, in the section beyond the loading coil furthest from the CO should be between 3000 feet and 12,000 feet.

▲ There are to be no bridged taps between loading coils and no loaded bridged taps.

Gauge	Diameter in Inches (mm)	Feet per Ohm	Ohms per 1000 Feet	Maximum Current in Amperes
16	0.051 (1.29)	249	4.0160	0.8530
18	0.0403 (1.02)	156.6	6.3850	0.5390
20	0.0320 (0.813)	98.5	10.1500	0.2690
22	0.0253 (0.635)	62.0	16.1400	0.2120
24	0.0201 (0.511)	39.0	25.6700	0.1330
26	0.0159 (0.406)	24.5	40.8100	0.0853

Table 4-2. Wire Gauges

A revised version of the resistance rules are defined in the Bellcore document SR–TSV–002275. The revised rules require loop resistance to be 1,300 ohms at 18,000 feet and 1,500 ohms on loops that extend between 18,000–24,000 feet. The point to remember here is that the rules focused on the maintaining a consistent resistance across the loop at different distances. No reference was made to mixed gauges. Later we will introduce another set of rules, called CSA rules, that focused more on containing mixed gauges, and that eventually replaced these rules, not by mandate but by default.

Loading Coils and Bridge Taps

To counteract the natural capacitance of copper wires over long distances, loading coils were used to add extra inductance. Capacitance is the natural tendency of a wire to store an electric current instead of letting it flow. In an electrical circuit, capacitance has the effect of changing the phase of the current relative to the voltage. As a result, the maximum current is not achieved at the same time as the maximum voltage, an effect that limits the power delivered.

Inductance, whose unit of measurement is a *millihenry (mH)*, also changes the phase, but in the opposite direction of the effect of capacitance. The application of induction aids in delivery of power or transference of the signal. As is the case with most things, too much of a good thing can also be bad. Loaded circuits need balancing because too much loading will induce a hum on the loop and too little has an attenuating effect on the voice passband. Guidelines were developed to ensure a balanced loop, and three different loading schemes were commonly deployed to apply different levels of inductance, as shown in Table 4-3.

The distance from the CO to the first coil is half the normal spacing of the scheme being used. So in an H-88 scheme, which is the most common scheme used, the distance from the CO to the first coil is always 3,000 feet.

The net effect of a loading coil is a more consistent and flatter signal loss profile for frequencies in the range of 3.3–5 KHz, as shown in Figure 4-2. Attenuation increases rapidly above the 3.3–5 KHz range, depending on the loading scheme and wire gauge employed. For analog voice, this is not a problem because the passband is confined to 300 Hz–3.3 KHz. The negative side of this approach for contemporary digital applications is the rate of attenuation beyond the 3.3–5 KHz range. Attenuation increases more rapidly for the higher frequencies than it would have in an unloaded circuit, and these are the very frequencies required by higher speeds.

Scheme	Loading	Spacing Distance
H-88	88 mH	6,000 feet
D-66	66 mH	4,500 feet
B-44	44 mH	3,000 feet

Table 4-3.   Common Loading Schemes

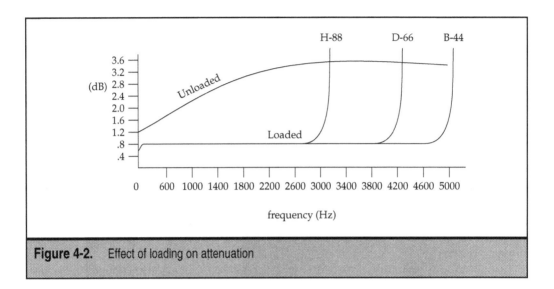

Figure 4-2. Effect of loading on attenuation

There was a time when a single loop was shared between subscribers (in what was called a party line), and bridged taps were the means by which each party tapped into the main loop (see Figure 4-3). The taps were still used on single subscriber loops because of the convenience of tapping into a main cable run from the CO, instead of having to run different cable lengths for different subscribers. Taps cause echoes that adversely affect the timing of the circuit, and those that are left unterminated pose the additional problem of signal loss.

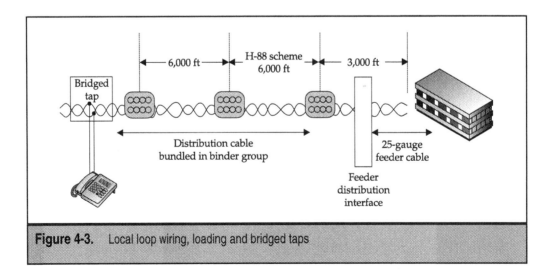

Figure 4-3. Local loop wiring, loading and bridged taps

Approaches to Digitization

The advances made and the experience gained in the development and deployment of digital trunk technology led to the birth of a new form of loop technology called a *digital loop carrier (DLC)*. The basic concept was to use T1 carrier technology to extend the reach of the local loop. A T1 carrier was used to digitize a number of copper pairs, usually ten pairs, from the CO to a central point in a neighborhood called the *carrier serving area (CSA)*. Eight of the ten pairs were used to provide four T1s or 96 channels of digitized voice, and the remaining two, for management.

The approach provided two immediate benefits: it extended the reach of the local loop, and it increased loop capacity through a more efficient use of existing copper pairs. Instead of using 96 copper pairs for 96 voice channels, the new approach used ten pairs, which is a significant saving. The act of regaining the use of pairs through more efficient use of technology is called *pair gain*; the pair gain in this instance is almost 10:1. Another benefit of the new type of loop was the fact that it was more feature-rich than the old analog carrier system and was capable of deploying some of the newer services such as digital data service (DDS). Finally, it was more reliable, easier to install, and easier to maintain.

CSA design guidelines, also called CSA rules, were established to define the distribution of twisted wire pairs from digital loop carrier (DLC) systems (see Figure 4-4). These rules eventually supplanted the resistance design rules and also became the rules that were applied to straight copper runs. The CSA rules are defined as follows:

▼ Only nonloaded cable should be used.

■ If multiple gauges are to be used, they should be restricted to two gauges only.

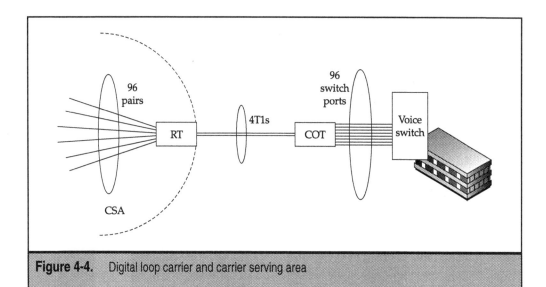

Figure 4-4. Digital loop carrier and carrier serving area

- Total bridged-tap length may not exceed 2500 feet, and no single bridged tap may exceed 2000 feet.

- The amount of 26 AWG cable may not exceed a total length of 9000 feet, including bridged taps.

- For single-gauge or multigauge cables containing only 19 and 22, or 19 and 24 AWG cable, the total cable length may not exceed 12,000 feet, including bridged taps.

▲ The total cable length including bridged taps of a multigauge cable that contains 26-gauge wire may not exceed 9000 feet.

The CSA concept led to a more manageable local loop arrangement that extended the reach of the CO. Subscribers that were too far from the CO could now be serviced by DLCs through some type of CSA architecture. A typical CSA served 200 to 600 residential units and 500 to 1000 access lines; loops were limited to 9000 feet (26 AWG) or 12,000 feet (24 AWG) of unloaded twisted pair wiring.

A device called a *central office terminal (COT)* was installed in the CO and was linked to a *remote terminal (RT)* that was placed in the field close to the subscribers to be served. The loops from the subscribers would run as far as the RT. In a specific example of a CSA architecture, a COT was used to interface to 96 switch ports. Four T1s were used to connect the RT to the COT, and 96 subscriber loops were terminated into the RT. The 96 loops were multiplexed on each of the 96 channels of the 4 T1s (each T1 has 24 channels and 24×4 = 96). This configuration served 96 subscriber lines in a very efficient manner.

The net result of this approach was that the reach of the CO was extended through the use of a digital loop carrier without violating the CSA rules. The 9000 feet or 12,000 feet rule applied to the loops that terminated at the RT. The link between the COT and RT was governed by the limitations of the technology used to link them. So if T1 carriers were used, the rules governing, or limitations of, T1 carriers applied.

DAML: Digital Added Main Line   As the Internet gained in popularity, there was an increase in demand for second telephone lines into homes. This rush on the telephone company for new lines created a demand for wire pairs that was not projected in previous growth models. The risk of running out of pairs became a real possibility, especially in densely populated areas, so new ways of leveraging the existing copper base had to be found. To solve the dilemma, some carriers began introducing a type of local loop technology called *Digital Added Main Line (DAML)*, sometimes called a pair-gain device because it enables the provider to provision an additional line without using an extra loop.

This type of loop allows a single copper pair to be used to provide two voice channels. It uses the BRI (ANSI T1.601) standard that is used in ISDN. BRI transports a total of 160 Kbps of symmetric digital data over loops up to 18,000 feet long. BRI consists of three channels: two B-channels of 64 Kbps each and one 16 Kbps D-channel, along with another 16 Kbps for framing and line control. In DAML applications, BRI transceivers are used to enable a single loop to transport two voice channels. The D-channel is ignored, yielding a 2B+0 service, which is not compatible with ISDN switches.

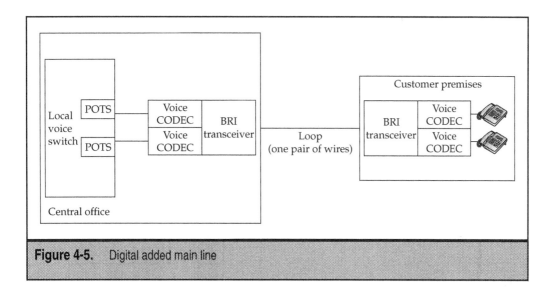

Figure 4-5. Digital added main line

As shown in Figure 4-5, the service uses two CODECs (voice coder/decoders) on both ends of the connection to convert the B-channels (which are digital) into an analog telephone interface, thereby providing the traditional voice telephony interface to the CO switch and telephone equipment at the customer premises.

THE DSL FAMILY

The advances made in *digital signal processing (DSP)* have made DSL technology, as we know it today, possible. The limits of technology are constantly being tested, and as a result, caution should be exercised when making any reference to speeds or distance. DSL began at a modest data rate of 144 Kbps for basic rate ISDN, evolving to 8 Mbps for ADSL and now up to 52 Mbps for its newest addition, VDSL. This rapid evolution is an indication of the technological miles we have traveled since the early days of digital T1 circuits. It is also the reason why any reference we make to speeds and distances should be read with caution, as it is only a matter of time before they too are improved.

DSL technology falls into two different camps, those that provide *symmetric* data rates (the same speed upstream as downstream) and those that use different speeds downstream and upstream, called *asymmetric* (see Figure 4-6). The variants are further described in Table 4-4. Upstream communication is always from the perspective of the user to the service provider, and downstream is from the service provider to the user. At first glance, asymmetric data rates may appear to be a problem or inferior, but in actuality they closely match the way data typically flows between a user and a server. In most applications, the bulk of the data flows from the server to the user, with communications from the user to the server consisting of short commands. For example, take a typical

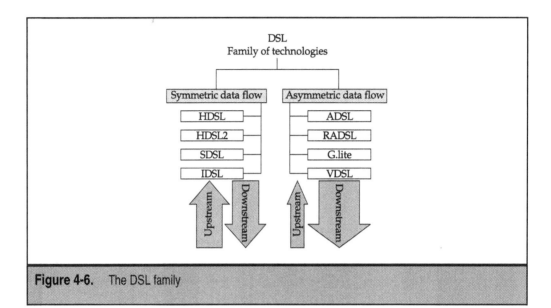

Figure 4-6. The DSL family

DSL Variation	Symmetry	Downstream Rate	Upstream Rate	POTS Support	Wire Pairs
HDSL: High Bit-Rate DSL	Symmetric	1.544 or 2.048 Mbps	Same	No	2–3
HDSL2: High Bit-Rate DSL #2 or SHDSL: Single-Pair DSL	Symmetric	1.544 or 2.048 Mbps	Same	No	1
SDSL: Symmetric DSL	Symmetric	768 Kbps to 2.0 Mbps	Same	No	1
IDSL: ISDN DSL	Symmetric	ISDN rates: 144 Kbps total	Same	No	1
ADSL: Asymmetric DSL	Asymmetric	1.5–8 Mbps	16–640 Kbps	Yes	1
RADSL: Rate-Adaptive DSL	Asymmetric	1–8 Mbps	16–640 Kbps	Yes	1
G.lite: "G"—ITU series "Lite" type of ADSL	Asymmetric	Up to 1.5 Mbps	Up to 512 Kbps	Yes Splitterless	1
VDSL: Very High Data-Rate DSL	Asymmetric	13–52 Mbps	1.5–6.0 Mbps	Yes	1

Table 4-4. The Major DSL Technologies

browser session. Whenever the user requests a page from Yahoo, the request takes the form of a short packet, which results in a lot of data, words, and images being presented on the screen. It takes only a twist of the wrist to get a gush of water flowing from a tap.

An asymmetric solution, therefore, is not necessarily bad from a user perspective unless an application like FTP *(File Transfer Protocol)* is being used. In a file transfer, the user could be uploading a file (as opposed to downloading from the server), which would be a problem with an asymmetric data rate. When uploading, most of the data flows upward, with short commands flowing back to the user. In this type of application, asymmetric data rates are a problem because the bandwidth allocation is opposite to what the user requires. Other than applications such as FTP, a basic principle in selecting a DSL service is: if you are a user, an asymmetric technology will be fine most of the time; if, however, you are hosting a server, a symmetric solution should be employed.

All the different types of DSL use *framed transport*. Framed transport means the link sends a series of frames with no pause in between. If there is no data to be sent, the frames are still sent but are populated with a bit pattern that is understood to mean no data present. The concept is similar to a conveyor belt with attached boxes. If there is something to be sent, it is placed in a box. If there is nothing to be sent, the conveyor belt still transports the empty boxes. This is different from the techniques used in packet-based transport, where data is sent only when data is available to be sent.

Power Spectral Density: A Common Issue

The proliferation of DSL technologies makes it almost impossible to avoid mixing the different technologies in the same binder group, distribution frames, connector panels, or other components that handle multiple transmission loops. Crosstalk is an ever-present challenge and must be contained to acceptable levels in order to provide a reliable service.

Power spectral density and transmit power are two important aspects in the design of all DSL technologies. Power spectral density impacts the compatibility of different DSL technologies that share the same binder group. The transmit power of a DSL system is usually set to provide a satisfactory *signal-to-noise ratio (SNR)* under worst-case conditions. This margin is set to a 6 dB SNR, which means that the DSL will provide a 10^7 *bit error rate (BER)* when the crosstalk signal power is 6 dB greater than the defined "worst-case" crosstalk model.

The power spectral density of each service is shaped to minimize interference with that of other signals. The power levels are set, and are fixed, at a compromise level to operate over the CSA to meet "worst case" scenarios. For typical conditions, the 6 dB assures that DSLs will operate at a BER that will provide a reliable service even when the transmission environment is worse than normal. The 6 dB value originated in work done on basic rate ISDN. The value provides for cable aging, splicing, imperfect transceivers, and noise introduced by the wires in the CO and customer premises, among other variables. The bottom line is, it is the best trade-off between a reliable service and the use of the technology across the longest loops.

HDSL: High Bit-Rate DSL

High Bit-Rate DSL, also called High Data-Rate DSL, was the first DSL technology to be put into operation. Developed by Bellcore in the late 1980s, it was intended to be an economical solution to the growing corporate demand for T1 carriers. The first version of the technology was placed into service in March 1992, and since then most major telephone companies around the world have adopted the technology. HDSL operates at symmetric speeds of 1.544 Mbps and 2.048 Mbps, the same as T1/E1, and is becoming the preferred option over T1/E1 service. To understand the reasons, we need a quick primer in T1/E1 systems.

T1 and E1 Transmission Systems

Originally designed for the provision of trunk service between central offices, T1/E1 over time became popular as a way to provide high-speed digital connections to customer premises (see Figure 4-7). The lines became very popular and were used for different types of applications like linking PBXs or for connections to data networks. The new use subjected it to the perils of the local loop, which was hostile to anything digital.

T1/E1 operates over telephone wires but was expensive to deploy and operate, because special engineering on the loop was required. First, the loop had to be conditioned and loading coils and bridged taps had to be removed; second, the line had to be spliced, and repeaters inserted, at every 3000–6000 feet. The repeaters were used to regenerate the signals that weakened over long distances. The technology at the time was not as advanced as today, so the capability to recover a weakened signal required repeaters placed at much shorter distances than would now be required. Each repeater was specially engineered for the circuit on which it was installed, as it was critical to ensure that each section of the line remained within specifications so as to limit signal loss. Each repeater then had to be installed in a special hardened case that had to be mounted on a pole or in a manhole.

The power levels required for T1/E1 service created crosstalk for other services, and so special care had to taken to ensure that T1/E1 lines did not interfere with other services traveling on other copper pairs close by. This meant that T1/E1 had to be separated into

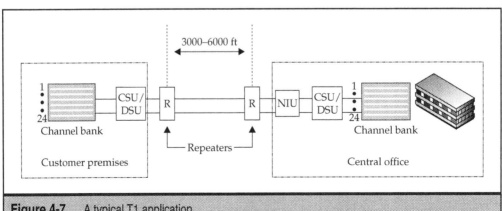

Figure 4-7. A typical T1 application

separate binder groups. The repeaters, too, were a source of crosstalk—one reason for the hardened case—and special consideration had to be given to where it was being placed. A customer premises location 18,000 feet away from the CO would require at least six repeaters for a T1/E1 line (three for each copper pair), each having its own hardened case, which was more expensive than the repeater itself.

There were also power considerations. Repeaters were normally line-powered from the CO, requiring a special power supply at the central office. The resistance in the loop created an additional challenge, as it wasted a lot of the power being sent to the repeater. Additionally, if a repeater failed, it required a special service trip for repairs. The engineering requirements of T1/E1 services led to very long lead times, measured in weeks and sometime months, for the delivery of the service.

In spite of its challenges, T1/E1 service saw much success and was popular for digital access to the PSTN. It was this success that created the desire to find other, more cost-effective, solutions. The technology increased the capacity of two pairs from 2 analog conversations to 24 digital conversations, a 12:1 ratio. The popularity of the service and the high cost of providing T1/E1 led to the early and speedy adoption of HDSL. Because of its comparable speeds and its lower provisioning and operational costs, HDSL became a natural choice and eventually emerged as the solution of choice wherever possible.

HDSL as a Solution to T1/E1 Service

The single most important aspect of HDSL that makes it an attractive option over T1 and E1 is the elimination of the midspan repeater. The repeater, together with the case in which it was installed, was a major contributor to the cost of provisioning T1/E1 service, so any technology that was not dependent on these elements was a winner from the start. In fact, HDSL was originally designed to overcome the limitations of T1/E1 and was originally called a "repeaterless or nonrepeated" T1/E1 replacement technology. Other differences between the two services are summed up in Table 4-5.

T1 transmission requires a repeater every 6000 feet to regenerate the digital signal. HDSL overcomes the need for repeaters by using a line code adapted from ISDN BRI DSL (IDSL) called 2B1Q (two binary, one quaternary). Using this line code, the effective range of transmission was doubled from 6000 to 12,000 feet without the need for repeaters, with an added benefit of reducing crosstalk. Each transceiver was capable of full-duplex operation at 784 Kbps (representing twelve 64 Kbps channels plus the normal 16 Kbps D-channel). When duplexed onto two separate pairs, HDSL was able to attain T1 speeds of 1.544 Mbps in each direction. A third transceiver and wire pair were required to attain E1 speeds. Today's HDSL now operates on two pairs with transceivers that are capable of running at 1.168 Mbps. Other versions of HDSL that use a single pair are also in existence today.

We have stated that HDSL has become the preferred solution over T1/E1 services, yet the term is still not as popular as T1 or E1. Today, people still refer to T1 and E1 service when reference is made to digital circuits of 1.544 Mbps or 2.048 Mpbs. The reason for this is more perception than reality—to the customer the service appears the same, but the technology that provides it has undergone significant changes that are not immediately obvious (see Figure 4-8).

Feature	T1/E1	HDSL
Speeds	1.544 Mbps and 2.048 Mbps	1.544 Mbps and 2.048 Mbps
Repeaters	Every 3000 to 6000 ft	12,000 ft on 24-gauge wire or 9000 ft on 26-gauge wire
Loading coils	Must be removed	Must be removed
Bridged taps	Must be removed	Do not need to be removed
Binder groups	Separate groups	No need for separate groups
Pairs	Two pairs	Two pairs (earlier versions used three pairs for E1 speeds)
Line code	Bipolar AMI	2B1Q (some earlier versions used CAP and DMT)
Transmission scheme	Dual simplex—each pair carries data in one direction or the other	Dual duplex—both pairs carry data in both directions
Annual maintenance	Higher (more components such as repeaters)	Lower (less components to fail)

Table 4-5. A Comparison of T1/E1 and HDSL

In the central office, T1 circuits bypass circuit switches and are routed through a DACS (digital access and cross-connect system) to the trunk network. Instead of a T1 line to the DACS, a special device called an *HDSL termination unit–central office (HTU-C)* is used. On the other end of the connection, a companion device called an *HDSL termination unit–remote (HTU-R)* is installed at the customer premises. The link between the HTU-C

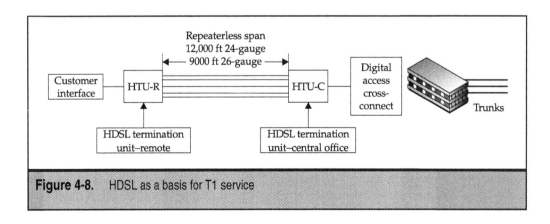

Figure 4-8. HDSL as a basis for T1 service

and the HTU-R can be a distance of up to 12,000 feet depending on the wire gauge in use and must conform to the CSA rules. All loading coils must have been removed, but bridged taps may be left in place.

For E1 applications, the HDSL functionality is built into the *line terminating unit (LTU)* or *network terminating unit (NTU)* along with a module to map the HDSL frames into E1 frames. To the user, the service appears as an E1 service even though HDSL is being used as the basis for the transport of the frames (see Figure 4-9). In fact, the user is presented with an E1 interface and so may never know what is under the covers. The HDSL transceiver that is built into the LTU or NTU is equipped to operate in one of the following modes, each using 2B1Q line coding:

▼ Three pairs of wires with each pair running in full-duplex mode at 784 Kbps on each pair

■ Two pairs with each pair running in full-duplex mode at 1.168 Mbps on each pair, or

▲ A single pair running in full duplex at 2.320 Mbps

The links between the LTU and the NTU must also conform to CSA rules.

The benefits of HDSL over T1/E1 services are numerous, especially to the service provider. The service eliminates the need for repeaters by doubling the distance limits of traditional T1, and it presents less crosstalk. The length of the circuit can also be improved either by using a lower-gauge wire or by using repeaters that are made for HDSL circuits. The use of repeaters allows an HDSL line to double its length to 24,000 feet with a single repeater or to attain 36,000 feet with two repeaters. The repeaters in an HDSL line are much more efficient and powerful than those used for T1 circuits. Originally in two-repeater HDSL systems, the repeater closest to the CO was powered from the CO, and the one closest to the customer was powered from the customer premises. Improvements in repeater technology and reductions in transceiver power consumption now allow both repeaters to be powered from the CO.

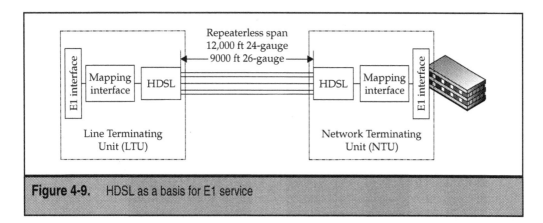

Figure 4-9.   HDSL as a basis for E1 service

HDSL Applications

HDSL is not ideal for residential broadband services for two reasons. First, it uses two copper pairs, which is not cost effective for the service provider, especially in light of the fact that other DSL solutions available require only a single pair. Second, it cannot coexist with voice services on the same copper pairs.

Some common uses of HDSL include:

▼ Internet access from server farms. The symmetrical data rates allow for the same speed in both directions, so the user accessing a server benefits from the same data rate regardless of whether or not he is doing an upload or a download of a file.

■ Linking RT to COT in a CSA architecture, thereby extending the reach of the central office.

■ Connecting PBX and packet-based data networks to a public network.

■ Campus solutions. Many University campuses, for example, have an extensive copper network throughout the campus. HDSL is used to get T1 speeds between buildings.

■ Connecting wireless-based stations into landline networks.

▲ Any corporate solution that requires the use of a T1 or E1 circuit.

The Drive to HDSL2 or SHDSL

The success of HDSL is undeniable, but despite being an economical solution for T1/E1 services, it still has the drawback of requiring two wire pairs. As stated earlier, the popularity of the Internet has created an increased demand for circuits, and so copper has become a valuable commodity for carriers. While the price of raw copper may be low, the cost of installing new copper facilities is higher than ever. The drive to conserve copper makes HDSL less attractive than it would be if it required only a single pair yet offered the same performance benefits.

Another important aspect of HDSL that needed improvement lay in the details in the implementation that were defined within the specifications. Or maybe we should say the lack of details that were defined. Only the essentials of the technology were addressed in the specifications, and as a result, matters of interoperability became a concern because of the way vendors chose to implement certain features. Proprietary versions were also being developed that attempted to address the requirement for a single-pair solution. Without doubt there was a need for a version that would address these concerns. It is this quest that led to the development of HDSL2, the new and improved second-generation version of HDSL.

HDSL2, as it is called by ANSI, is also known as SHDSL to the ITU; originally known as G.shdsl, the standard is now approved and defined in ITU-T G.991.2.

Even though HDSL is the most widely deployed DSL, it is not governed by a standard and interoperability was a major concern. The pending HDSL2 standard as defined by the ANSI T1E1.4 committee hopes to address this question by ensuring that enough technical details are incorporated to let vendors deploy products that are interoperable. To further facilitate this goal, the industry-led consortium of leading telecommunications product and chip vendors will work with the University of New Hampshire to test the interoperability of new products.

The specification for HDSL2 is now complete, and products are now available on the market. HDSL2 provides full T-1 service using a single copper pair on loops that meet the Carrier Service Area deployment guidelines. The economic benefits to service providers are considerable because the use of a single pair instead of two creates the potential to generate twice the revenue stream without the need to upgrade the cable plant. The line coding was changed from 2B1Q to a pulse code modulation scheme called 16-TCPAM with Spectral Shaping. Changing the line coding makes HDSL2 more resistant than HDSL to noise resulting from coexistence with other DSL services.

When the line first comes up, the remote ends do an initial handshake to determine the features that are supported. The handshake uses a 2-PAM scheme. Once the connection has been established, a 16-PAM scheme is used that encodes four bits of data per state, and trellis coding is used for error detection, hence the 16-TCPAM. Of the four bits that are used per state change, one bit is used for the trellis coding FEC.

HDSL2 Power Controls Impact on Spectral Compatibility

The HDSL specification includes the option of variable power, which has implications that could affect the design of other DSL technologies. The variability of the power in HDSL2 allows the power to adjust to loop conditions, thereby limiting crosstalk in nearby loops.

There are obvious ways in which one could address noise. The interaction between services can be reduced by unbundling and segregating different cables; this, however, limits flexibility and may exhaust capacity in cable ducts. The length of the loop could be decreased, and that would most certainly help, in that the longer the loop, the higher the noise level. This approach, however, decreases the reach of the CO and results in more expense to reach the areas that would have otherwise been covered had longer cable runs been permitted. Noise can also be mitigated using lower speeds, but this defeats the objective of providing high-speed service. Varying the power is also a solution in mitigating noise. Varying the power according to conditions has a great appeal, as it is adaptive to the environment.

HDSL2 power control operates at start-up and during run time. Both ends of an HDSL2 connection adjust power according to the BER and SNR on the cable. During start-up, if the BER and SNR are good, the power may be lowered to a power level that is adequate to keep the same BER and SNR. This is a key feature, as it takes into consideration other services that may be operating in the immediate area and adjusts the power downward to lessen the possibility of generating interference, even though the level at which it was operating was fine in terms of its own connection. During run time, the power may be adjusted if the BER rises above the threshold set for DSL operation, which is 10^7.

In order to accommodate the tracking ability of the receiver and to prevent oscillation due to interaction with neighboring variable-power loops, the rate of power level change has been limited to 0.01 db per minute. At this rate, it would take five hours to do a 3 dB-power change. Once the initial adjustments are made at start-up, the power may not change during the life of the circuit. If, however, the run-time feature is enabled, the levels will adjust as necessary.

A preliminary estimate suggests that up to 80 percent of HDSL2 circuits could operate at reduced power, which could reduce crosstalk in adjacent loops by 50 to 75 percent. The implications of this feature for ADSL, which has rate-adaptive capabilities, may be significant when the loop traverses the same binder group as an HDSL2 loop. The variability of the power and the reduction in crosstalk could negate the need for ADSL to adjust its upstream data rate because of noise that would have been caused by the HDSL2 loop had the power not been adjusted.

Feature	HDSL	HDSL2
Line speed	1.544 Mbps or 2.048 Mbps	1.544 Mbps or 2.048 Mbps
Pairs	2	1
Bits per baud	2 bits/baud	3 bits/baud plus 1 for FEC
Baud rate (state changes)	2@392 KBauds/second	517.3 KBauds/second
Line code	2B1Q	16-TCPAM (2-PAM for handshake)
Power	Fixed	Variable according to line conditions
Attached standards	No	Yes—focus of ANSI T1E1.4 committee
Maximum single Bridged Tap length	2,000 feet	2,000 feet
Maximum total Bridged Tap length	2,500 feet	2,500 feet
Loading coils	None	None

SDSL: Symmetric DSL

Symmetric DSL, or single-line DSL as it is sometimes called, is distinct from HDSL in that it uses a single pair. The technology, however, also has a lot in common with HDSL in that it started out as a technology based on the same chip sets as HDSL. Using multiple copper pairs to provide residential digital services is not the ideal solution to a service provider. If one is to digitize the loop to the home, it makes more sense to use the existing copper than to waste an additional pair.

SDSL was intended to solve this problem. Instead of using two transceivers and two copper pairs, SDSL uses a single transceiver and a single copper pair to provide fractional

T1 services. In many cases, customers require higher data speeds but not a full T1 service, and this is where SDSL fits in. Advances in technology now make it possible to gain higher data rates than the 784 Kbps of the original HDSL transceivers. Different vendors now have solutions that extend the range of the SDSL modem to 1.5 and 2 Mbps.

Table 4-6 lists the data rates and speeds that are achievable using the technology. The rates given are conservative and do not address specific vendor models that achieve higher rates. The fate of SDSL is unknown with the advent of HDSL2. Both technologies have a lot in common in terms of speeds and the number of loops. HDSL, however, has the added feature of variable power, which is very desirable. SDSL, in contrast, tends to boost power levels to gain greater distances, with undesirable consequences.

IDSL: ISDN DSL

Between 1982 and 1988, ANSI (the American National Standards Institute) developed standards that defined ISDN DSL (IDSL). The technology functions in much the same way as basic rate interface ISDN in that it uses the same 2B+D, which provides for a data capacity of 144 Kbps. In ISDN, the two B-channels are circuit switched, each capable of carrying voice or data in both directions. The D-Channel carries control signals and customer call data in a packet-switched mode and operates at 16 Kbps. Remaining throughput is absorbed by operational, administrative, maintenance, and provisioning (OAM&P) channels operating at 16 Kbps.

IDSL runs on a single pair of wires at a maximum distance of 18 kft (about 3.4 miles or 5.4 km). In traditional ISDN applications, the ISDN link requires a connection to a voice switch in the central office. IDSL eliminates this requirement and the entire connection is provided by DSL equipment; for this reason, the technology is sometimes called "BRI without the switch" or "switchless BRI." Some versions of IDSL allow for the full use of the 144 Kbps bandwidth or full 2B+D operation, while others allow only 2B operation or 128 Kbps.

Data Rate	Distance
128 Kbps	22,000 feet
256 Kbps	21,500 feet
384 Kbps	14,500 feet
768 Kbps	13,000 feet
1024 Kbps	11,500 feet

Table 4-6.   SDSL Speeds and Reaches

A variation on IDSL exists that is based on the primary rate interface ISDN model. This version of the technology achieves higher speeds through the bonding of the B channels.

ADSL: Asymmetric DSL

ADSL was developed out of the work done by Joseph Lechleider of Bellcore. A distinguishing feature of ADSL is its ability to transport plain old telephone service along with broadband services. This was achieved by using a guard band to separate the voice-band and broadband frequencies. The broadband frequencies are used for digital services including voice, video, and data.

One of the motivating factors for the development of ADSL was the desire of telephone carriers to compete with cable service providers in the delivery of Video on Demand services. This was a major challenge because of loop conditions already discussed, and the optimization of the network for voice made the infrastructure more challenging to provide high data-rate services. In the early 1990s, Bell Atlantic conducted the first VoD trials in New Jersey. The lessons learned from the early trials indicated that a downstream rate of about 1.5 Mbps was adequate for the delivery of MPEG-1 video streams. And upstream rates up to 64 Kbps were more than adequate to allow users to issue commands like start, stop, pause, rewind, and fast forward to the video server. Speed requirements of this and other common applications are summarized in Table 4-7.

The equipment and operational cost for the delivery of VoD service priced the service out of reach for many consumers, and so the ramp-up of the service never got the traction that was hoped for. This left the carriers and cable companies in search of new markets for the technology. The Internet phenomenon, which was taking off around that time, proved to be one such market. The asymmetric data streaming profile of the ADSL

Application	Downstream	Upstream
Video on Demand (VoD)	1.5–3 Mbps	64 Kbps
Near VoD	1.5–3 Mbps	64 Kbps
Computer gaming	1.5 Mbps	64 Kbps
Video games	64 Kbps to 2.8 Mbps	64 Kbps
Video conferencing	384 Kbps to 1.5 Mbps	384 Kbps to 1.5 Mbps
Broadcast TV	6–8 Mbps	64 Kbps
Internet access	64 Kbps–1.5 Mbps	>10% of downstream
Remote LAN access	64 Kbps–1.5 Mbps	>10% of downstream
Distance learning	64 Kbps–1.5 Mbps	64 to 384 Kbps
POTS	4 KHz	4 KHz
ISDN	160 Kbps	160 Kbps

Table 4-7. Common Applications and Their Speed Requirements

technology matched the data flow profile of web browsing. The downstream data rate was significantly higher than the upstream rate. To address the new market, the following changes were made:

▼ The ratio of the downstream rate to the upstream rate was optimized for TCP/IP traffic. Using a 10:1 ratio, the downstream rate was increased to between 6–8 Mbps, and the upstream to 640 Kbps.

■ Rate adaptation was included to allow the two modems on an ADSL link to adjust their rates according to line conditions.

▲ ADSL was marketed as an "always on" solution like cable.

The line encoding technique used in ADSL modem is DMT, but this was not chosen without much debate, as there was a large installed base of earlier modems that used CAP. Modems using CAP far outnumbered those using DMT, but various standards groups in the U.S. and Europe leaned toward DMT. Line encoding debates tend to be religious, as each side has very convincing arguments for its position. In the end, however, DMT was adopted as the line code of choice.

In 1994, a consortium of companies decided to form the ADSL Forum (now called the DSL Forum), which by charter decided to avoid being embroiled in line coding debates and focus on the architecture of the technology. A lot is owed to this group for the advancement of open specifications, documentation, and service availability. A product of this group, the System Reference Model (an architectural reference model), will be covered in the next section.

Having its roots in VoD, ADSL had elements that made it an expensive option if it was to be used predominantly as an Internet access technology. In 1998, another group was formed, called the UAWG (the Universal ADSL Working Group), whose intention was to produce a mass-market version of the technology with a lower cost basis. As a result, a version of ADSL called ADSL lite was developed that has a maximum downstream rate of 1.5 Mbps. To differentiate between the two, the original version was called full-rate ADSL.

The rationale for the rate reduction was that the backbone of the Internet at the time was a limiting factor, and having a higher data rate at the edges was unnecessary. The rate reduction made the technology more affordable to a larger number of people. It should be noted that the (A)DSL Forum and UAWG are not in competition with each other; ADSL has remained focused on the full-featured version of ADSL, while UAWG has focused on the lite(r) version. The work of UAWG has resulted in a standardized version of the technology called G.992.2, commonly known as G.lite.

ADSL Architectural Reference Model

The DSL Forum, formerly called the ADSL Forum, is a consortium of over 400 companies from various networking, service provider, and equipment industries. Established in 1994, it provides input to international standards bodies and seeks to develop technical guidelines for architecture, interfaces, and protocols for networks that incorporate DSL transceivers.

The best place to start when discussing the subject of ADSL is to review the original system reference model that was published by the forum. Figure 4-10 depicts the original

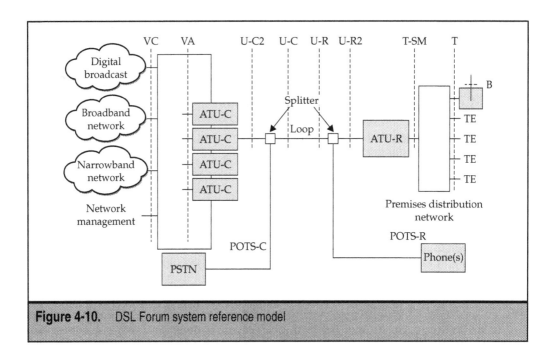

Figure 4-10. DSL Forum system reference model

model as published by the group, and Table 4-8 defines its elements. In it are elements that demonstrate the intended broad reach of the model. Of note is its slant on consumer-type

Element	Description
ATU-C	ADSL transmission unit at the CO end. The ATU-C may be integrated within an access node.
ATU-R	ADSL transmission unit at the customer premises end. The ATU-R may be integrated within a service module (SM).
Access node	Concentration point for broadband and narrowband data. The access node may be located at a central office or a remote site. Also, a remote access node may subtend from a central access node.
B	Auxiliary data input (such as a satellite feed) to service module (such as a set-top box).
Broadcast	Broadband data input in simplex mode (typically broadcast video).

Table 4-8. Definition of DSL Forum System Reference Model

Element	Description
Broadband network	Switching system for data rates above 1.5/2.0 Mbps.
Loop	Twisted pair copper telephone line. Loops may differ in distance, diameter, age, and transmission characteristics depending on the network.
Narrowband network	Switching system for data rates at or below 1.5/2.0 Mbps.
POTS	Plain Old Telephone Service.
POTS-C	Interface between PSTN and POTS splitter at network end or CO.
POTS-R	Interface between phones and POTS splitter at premises end.
PDN	Premises distribution network: System for connecting ATU-R to service modules. May be point-to-point or multipoint; may be passive wiring or an active network. Multipoint may be a bus or a star.
PSTN	Public Switched Telephone Network.
SM	Service module: Performs terminal adaptation functions. Examples are set-top boxes, PC interfaces, and LAN routers.
Splitter	A filters that separates high-frequency (ADSL) and low-frequency (POTS) signals at the network end and the premises end. The splitter may be integrated into the ATU, physically separated from the ATU, or divided between high pass and low pass,with the low-pass function physically separated from the ATU. The provision of POTS splitters and POTS-related functions is optional.
T-SM	Interface between ATU-R and premises distribution network. May be same as T when network is point-to-point passive wiring. An ATU-R may have more than one type of T-SM interface implemented (e.g., a T1/E1 connection and an Ethernet connection). The T-SM interface may be integrated within a service module.
T	Interface between premises distribution network and service modules. May be same as T-SM when network is point-to-point passive wiring. Note that the T interface may disappear at the physical level when ATU-R is integrated within a service module.

Table 4-8.   Definition of DSL Forum System Reference Model *(continued)*

Element	Description
U-C	Interface between loop and POTS splitter on the network side. The need to define both ends of the loop interface separately arises because of the asymmetry of the signals on the line.
U-C$_2$	Interface between POTS splitter and ATU-C. Note that at present, ANSI T1.413 does not define such an interface and separating the POTS splitter from the ATU-C presents some technical difficulties in standardizing the interface.
U-R	Interface between loop and POTS splitter on the premises side.
U-R$_2$	Interface between POTS splitter and ATU-R. Note that at present, ANSI T1.413 does not define such an interface and separating the POTS splitter from the ATU-R presents some technical difficulties in standardizing the interface.
V$_A$	Logical interface between ATU-C and access node. As this interface will often be within circuits on a common board, the ADSL Forum does not consider physical V$_A$ interfaces. The V interface may contain STM or ATM transfer modes, or both. In the primitive case of point-to-point connection between a switch port and an ATU-C (that is, a case without concentration or multiplexing), then the V$_A$ and V$_C$ interfaces become identical (alternatively, the V$_A$ interface disappears).
V$_C$	Interface between access node and network. May have multiple physical connections (as shown in Figure 4-10), although may also carry all signals across a single physical connection. A digital carrier facility (e.g., a SONET or SDH extension) may be interposed at the V$_C$ interface when the access node and ATU-Cs are located at a remote site. Interface to the PSTN may be a universal tip-ring interface or a multiplexed telephony interface such as specified in Bellcore TR-08 or TR-303. The broadband segment of the V$_C$ interface may be STM switching, ATM switching, or private line–type connections.

Table 4-8.   Definition of DSL Forum System Reference Model *(continued)*

applications: digital broadcasting, broadband and narrowband technologies, and regular voice telephony, all applications in which a consumer may have an interest. Another interesting point is that at the time of the development of the model, some of these applications were more a concept than reality. Today, they are all reality to varying degrees, and ADSL is one of the more standardized versions of DSL. This goes to show the benefit of having bodies like these that have a common goal in furthering a technology.

The reference model may appear to be complex, as it depicts the multiple capabilities of the technology and the interfaces that are required to provide them (the dotted lines represent the interfaces). The basic elements are, however, fairly easy to grasp, and Figure 4-11 provides a simplified view of the technology. The basic components of an ADSL connection consist of a DSL modem and a splitter. The copper loop between the customer premises and the CO terminates in a splitter that separates the ADSL traffic from regular POTS traffic. On the customer premises side of the connection, the splitter takes the single copper loop and splits it into two pairs. A single pair connects to the telephone handset (or the wiring that leads to the different RJ-11 jacks throughout the house). The second pair connects to the DSL modem, which is called an *ADSL termination unit-remote (ATU-R)*. This is the device that is used to provide the DSL service.

On the CO side of the connection, the same thing happens: the splitter in the CO sends one wire pair to the voice switch for normal telephone service and the other to a DSL modem called the *ADSL termination unit-central (ATU-C)*. In the CO, multiple ATU-Cs may be aggregated into a single box that multiplex the ADSL service; this device is called a DSLAM, or *DSL access multiplexer*.

ATU-C and ATU-R   These components, sometimes called the DSL modems, are concerned with the physical layer and are responsible for frequency allocation through frequency division and rate adaptation. The POTS service occupies the frequency range up to about 4000 Hz. The bandwidth for digital services is divided using one of two methods, FDM or echo cancellation.

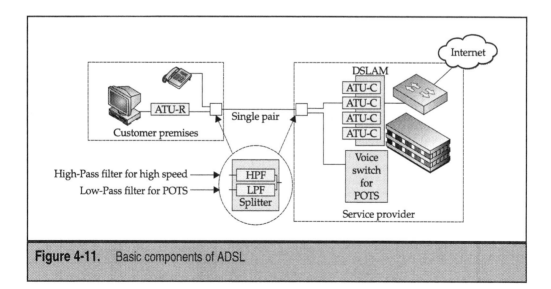

Figure 4-11.   Basic components of ADSL

FDM splits the remaining spectrum into upstream and downstream channels, as shown in Figure 4-12. The upstream channel occupies the range from 25 to 200 KHz, and the downstream, from 250 KHz to 1 MHz. An issue with this approach is that the upstream and downstream channels cannot overlap, which reduces the available bandwidth available for modulation. Its use of higher frequencies also results in a shorter reach.

Echo cancellation, a technique used in V.32 and V.34 modems, allows the overlapping of the downstream and upstream channels, resulting in an extended frequency range available to both channels. Figure 4-13 shows an example of echo cancellation: the downstream bandwidth totally overlaps that of the upstream; echo cancellation is applied in the area where they overlap.

Echo cancellation uses a technique whereby the sender transmits and remembers its transmission. The transmission from the other end is a combination of a new transmission and an echo of the last. When this transmission is received, the receiving station is able to retrieve the new transmission because it recognizes the echo as an echo of its last transmission. This works well when there is only a single echo, but bridged taps may create multiple echoes that makes echo cancellation less effective. Echo cancellation techniques carry a higher cost than FDM, which makes them less favorable. As a result, FDM is typically used, as it keeps the cost of the ATU-R down.

Another function of the ATU-C and ATU-R is rate adaptation, the function by which the modems determine the bit rate at which they will transmit. Through a process of negotiation, both sides agree on the best bit rate given the condition of the line. The agreed bit rate remains constant throughout the life of the connection.

DSLAM The access node in the DSL Forum reference model is called the DSL Access Multiplexer, or DSLAM (see Figure 4-14). Its primary function, in addition to management functions, is to host a set of ATU-C interfaces and multiplex/demultiplex the various ATU-Cs to the relevant transport networks. DSLAMs may be located in the central office, at a remote terminal, or on the customer premises. For multitenant and multidwelling applications, the DSLAM may be placed on location at the customer site.

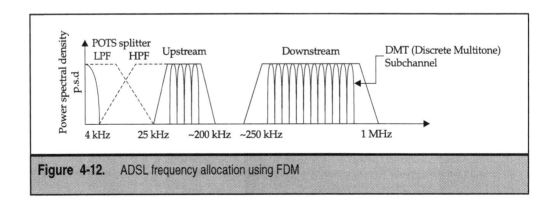

Figure 4-12. ADSL frequency allocation using FDM

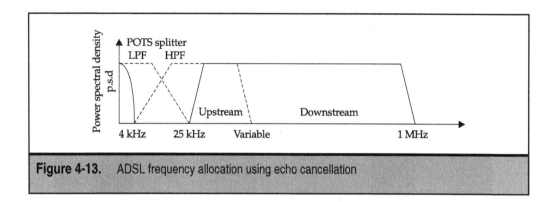

Figure 4-13.   ADSL frequency allocation using echo cancellation

The DSLAM is not tied to any particular DSL technology but instead embodies the different functions that the equipment vendor chooses to include. It handles the various ATU-C data streams through TDM or another version of multiplexing called *statistical time division multiplexing*, which makes use of any available bandwidth that is not being used. TDM, in contrast, allocates the bandwidth in time slots such that a channel that is not busy is still allocated its share of the bandwidth.

The role of the DSLAM is ever expanding as different manufacturers add such features as priority tagging, traffic shaping, and IP routing capabilities. These features add intelligence to the box and decrease the need for additional equipment like routers and hubs. Thus, tomorrow's DSLAM will look a lot different from the ones deployed today.

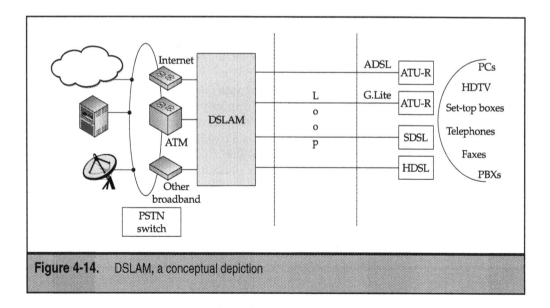

Figure 4-14.   DSLAM, a conceptual depiction

ADSL Bearer Channels and Transport Classes

The bit stream in an ADSL frame can be divided into seven bearer channels operating concurrently, as listed in Table 4-9. The bearers fall into two categories: simplex mode, which means the data flow in one direction only, and duplex mode, where the data flow in both directions. Up to four channels operate in simplex mode; these bearer channels are given the designations of AS0, AS1, AS2, and AS3 and operate independently. The remaining bearers operate in duplex mode and are called LS0, LS1, and LS2. All bearer channels are logical, so bits from all channels are multiplexed across the same physical link as shown in Figure 4-15.

Bearer	Support
AS0	Mandatory support. Unidirectional data rates from 32 Kbps to 6.144 Mbps in multiples of 32 Kbps. The actual upper bound of the channel is the maximum carrying capacity of the link and channel conditions, so AS0 may in fact support higher data rates. ANSI T1.413i2 supports higher data rates, but modem vendors are required to support at least an upper bound of 6.144 Mbps for AS0 bearer channel.
AS1	Optional support. Unidirectional data rates from 32 Kbps to 4.608 Mbps in multiples of 32 Kbps in North America (4.096 Mbps in Europe and elsewhere).
AS2	Optional support. Unidirectional data rates from 32 Kbps to 3.072 Mbps in multiples of 32 Kbps in North America (2.048 Mbps in Europe and elsewhere).
AS3	Optional support. Unidirectional data rates from 32 Kbps to 1.536 Mbps in multiples of 32 Kbps in North America (not applicable elsewhere).
LS0	Mandatory support. Bidirectional data rates of 16 Kbps and the range from 32 Kbps to 640 Kbps in multiples of 32 Kbps. LS0 support of 16 Kbps is an exception to the 32 Kbps multiple rule and applies only to STM mode of operation. It is as a result of the need to support a mandatory control channel called the "C" channel, which is used to transport signaling messaging for selection of services and call setup.
LS1	Optional support. Bidirectional data rates in the range from 32 Kbps to 640 Kbps in multiples of 32 Kbps.
LS2	Optional support. Bidirectional data rates in the range from 32 Kbps to 640 Kbps in multiples of 32 Kbps.

Table 4-9. ADSL Bearer Channel Summary

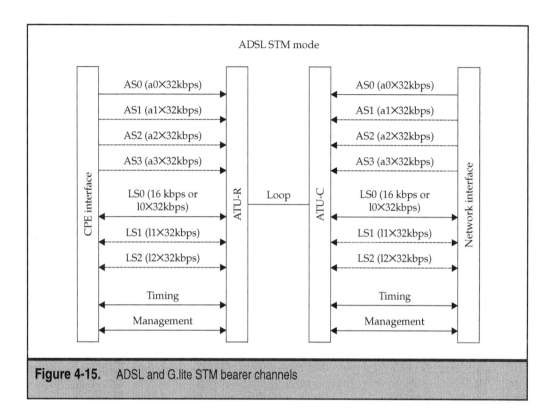

ADSL STM mode

AS0 (a0×32kbps)
AS1 (a1×32kbps)
AS2 (a2×32kbps)
AS3 (a3×32kbps)
LS0 (16 kbps or 10×32kbps)
LS1 (l1×32kbps)
LS2 (l2×32kbps)
Timing
Management

CPE interface   ATU-R   Loop   ATU-C   Network interface

Figure 4-15.   ADSL and G.lite STM bearer channels

In addition to the bearer channels, ADSL provides for an overhead channel that is used to transport bits that maintain synchronization. The overhead channel may also have additional capacity to carry extra bits from bearer channels. The amount of additional capacity available to the overhead channel is a function of the configuration options that contribute to the overhead. Each bearer can be configured to carry bits in multiples of 32 Kbps, which is derived from the granularity of DMT. Any bit rates that are not a multiple of 32 Kbps must use the shared overhead channel of the ADSL frame. So for a bit rate of 100 Kbps, 96 Kbps would be transported in the bearer and the additional 4 Kbps, in the overhead area.

The ATU-R and ATU-C use an *embedded operations channel (EOC)* to communicate in-service and out-of-service maintenance information. The channel is also used to retrieve performance-monitoring parameters. The EOC is transported along with the user data bits as part of the ADSL superframe.

Modes of Operation   The ITU-T specifications for ADSL define two ways of sending data inside ADSL frames. The first is to use *Asynchronous Transfer Mode (ATM)* cells to transport the data. ATM cells are fixed 53-byte cells with a 5-byte header. The data to be transmitted is segmented to fit into the 48 bytes of payload. The cells are then transmitted across *virtual channels*, which consist of a *virtual path identifier (VPI)* and a *virtual channel identifier (VCI)*. Each cell is multiplexed to the link with the VPI and VCI combination being used to identify a specific virtual channel across which the cell travels. In this mode, only AS0 and LS0 are required, as shown in Figure 4-16.

The second way of sending information inside an ADSL frame, as defined by the ITU-T, is by *Synchronous Transfer Mode (STM)*. In this mode, bits are sent in streams without the ADSL equipment being aware of any structure to the stream. All that it is aware of is the fact that bits are flowing in the channels. As a result, no logic can be applied to a stream—the stream cannot be multiplexed, the stream is not recognizable as being in any packet structure—the stream is just a flow of bits. The main reason for having additional bearer channels beyond AS0 and LS0 in ADSL is to allow STM ADSL equipment to multiplex additional data streams such that, for instance, if AS0 is busy with a video stream, AS1 can be used for an audio stream or for downloading a file. Figure 4-15 illustrates the use of the various bearer channels for STM mode of operation.

In addition to the modes of operation defined by the ITU-T, the DSL Forum defined a third mode of operation, called *packet mode* or *frame mode*. The reasoning behind this mode is that most ADSL links are used for Internet traffic that use IP packets. The Internet cares about packets and not about unstructured bit streams like STM or cells like ATM.

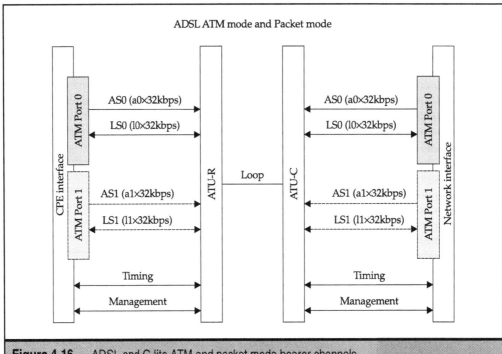

Figure 4-16. ADSL and G.lite ATM and packet mode bearer channels

Furthermore, in ATM operation, each packet would need to be chopped into small cells, adding additional overhead that was deemed inefficient for IP traffic. In packet mode, as with ATM mode, only AS0 and LS0 are required for ADSL (see Figure 4-16).

On a final note regarding STM and ATM mode, STM requires multiple AS bearer channels to multiplex multiple bit streams, and LS0 is used to transfer data upstream to the ATU-C. ATM, in contrast, requires only AS0 for its operation, as multiplexing is inherent to the protocol (which is also true for IP packets and packet mode). Multiplexing in ATM is achieved through the combined inclusion of a VPI and a VCI into the header of each cell. The specifications, however, allow for two ATM ports, with the second optional port utilizing a second AS1 optional bearer channel. The reason for this has to do with the higher-layer applications that may use different data formats for different applications. Take, for instance, a video stream, which may be carried by ATM port 0 and utilizes the AS0 bearer channel, allowing ATM port 1 to be used for Internet traffic that has a totally different format, one based on IP packets. Remember, the STM and ATM modes were a specification of the ITU-T, which intended these two modes to be used for all traffic types, including IP packets. The DSL Forum created the packet mode specifically for Internet access.

Transport Classes   Support for bearer channel AS0 is mandatory in all modes of operation. The maximum number of subchannels that can be active at the same time, and the maximum number of bearer channels that can be transported at the same time, depend on the transport class. The support of a transport class is dependent on the capacity of the ADSL link and the configuration of the subchannels. Transport classes for North American implementations are numbered 1 through 4 and use 1.536 Mbps as the base line multiple.

▼ **Transport Class 1**   This class is mandatory and is intended for the shortest loops. It can carry the following optional configurations, all of which total 6.144 Mbps:

■ One 6.144 Mbps simplex bearer channel (full AS0)

■ One 4.608 Mbps simplex bearer channel and one 1.536 Mbps simplex bearer channel

■ Two 3.072 Mbps simplex bearer channels

■ One 3.072 Mbps simplex bearer channel and two 1.536 Mbps simplex bearer channels

■ Four 1.536 Mbps simplex bearer channels

■ **Transport Class 2**   This class is optional and is intended for medium loops. It can carry the following optional configurations, all of which total 4.608 Mbps:

■ One 4.608 Mbps simplex bearer channel

■ One 3.072 Mbps simplex bearer channel and one 1.536 Mbps simplex bearer channel

■ Three simplex bearer channels at 1.536 Mbps each

- **Transport Class 3** This is an optional class also intended for medium loops. It can carry the following optional configurations, all of which total 3.072 Mbps:
  - One 3.072 Mbps simplex bearer channel
  - Two 1.536 Mbps simplex bearer channels
- ▲ **Transport Class 4** This is a mandatory class intended for long loops.
  - One 1.536 Mbps simplex bearer channel using AS0

Transport classes have also been defined for Europe and other countries that use E1-carrier. These classes use 2.048 Mbps as the baseline for multiple downstream data rates.

- ▼ **Transport Class 2M-1** The following configurations are allowed, each totaling 6.144 Mbps:
  - One 6.144 Mbps simplex bearer channel
  - One 4.096 Mbps simplex bearer channel and the other at 2.048 Mbps
  - Three simplex bearer channels at 2.048 Mbps each
- **Transport Class 2M-2** The following configurations are allowed, each totaling 4.096 Mbps:
  - One 4.096 Mbps simplex bearer channel
  - Two 2.048 Mbps simplex bearer channels
- ▲ **Transport Class 2M-3** The following configurations are allowed, each totaling 2.048 Mbps:
  - One 2.048 Mbps simplex bearer channel using AS0

ADSL Superframe

ADSL uses DMT per the standards, but there are also early implementations that use CAP. Regardless of the technique used, ADSL puts all the bits into what are called *superframes*. The superframe is made up of a sequence of 68 ADSL frames, some of which perform specific functions such as error control or link management. Frame 0, for instance, carries error control information, while frames 2, 34, and 35 carry *indicator bits (ibs)* that are used to manage the link. A synchronization frame that does not carry any information always follows the superframe.

One ADSL frame is sent every 250 microseconds; therefore, it takes 17 milliseconds to transmit a superframe. The ADSL frame is further broken into two parts of 125 micro-seconds each for fast and slower data (see Figure 4-17). ADSL tries to keep latency to a minimum for data that is delay sensitive. This type of data, called fast data, is taken from the fast buffer of the equipment and is placed in the first half of the frame. The second half is used to transport interleaved data from an interleaved buffer. This part of the frame is mainly intended for high-speed Internet access, and the data is interleaved to make it as impervious to noise as is possible. Only the interleaved data path is required in ADSL; the implementation of the fast data path is optional.

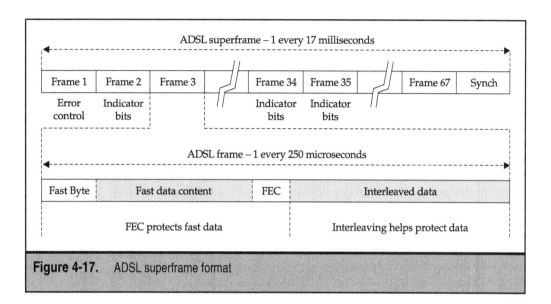

Figure 4-17. ADSL superframe format

G.Lite, G.992.2, or ADSL lite

The splitter's main function in ADSL was to allow the continued use of the customer's telephone across the same wire pair. This caused some concern, as it was an additional piece of equipment to go bad. The expense of support was real, as the following factors had to be considered:

▼ A technician had to make a service call if the splitter failed. This meant the added expense of sending a truck to the customer premises and coordination of schedules.

■ The cost of the splitter was typically priced into the solution, adding an additional capital cost.

■ The splitter introduced the possibility of the technician having to get involved in internal wiring of the customer premises.

▲ Missed appointments could lead to customer dissatisfaction and potential termination of service.

Clearly, it would benefit the service provider if the service could be delivered without the need for a splitter and still be able to provide basic telephone services (see Figure 4-18). This requirement was incorporated into G.Lite, and, apart from the reduction of the downstream speed to 1.5 Mbps for reasons already discussed, this was one of the main differences between ADSL and G.Lite (others are shown in Table 4-10).

In the G.Lite solution, the ATU-R plugs directly into the existing telephone wiring at the customer premises. A splitter is still used at the local exchange to separate the high-speed

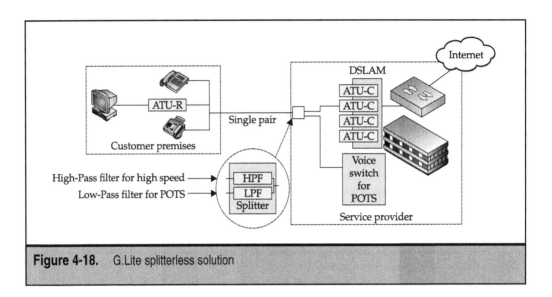

Figure 4-18. G.Lite splitterless solution

data from the normal telephone traffic. This simplified version of ADSL makes it an attractive solution for residential broadband. The downstream speed is adequate for casual Internet use—by users who typically use the connection for e-mail, chat, and browsing—and the cost of the service makes it more affordable.

	ADSL	G.Lite
ITU specification	ITU G.992.1 and ANSI T1.413	ITU G.992.2
Other names	Full-rate ADSL G.dmt	Splitterless ADSL lite
POTS	Uses a splitter to support	Supports without the use of a splitter in the home
Fast data	Supports dual latency for fast data and interleaved data	Supports only interleaved data
Sub-bands	256	128
Speeds	6–8 Mbps downstream Up to 640 Kbps	1.5 Mbps downstream Up to 512 Kbps
Difficulty to install	Requires service tech for splitter and wiring	Does not require technician's help

Table 4-10. Differences Between ADSL and G.Lite

RADSL: Rate-Adaptive DSL

The architecture of RADSL, the maximum speed, and distances supported are all essentially the same as ADSL. The differences between the two lie mainly in the earlier versions of ADSL that needed to be balanced to the conditions of the line. Technicians on both ends of the connection had to fix the speed of the link to match the conditions that existed on the line at the time of the installation. Any variation in line conditions after installation was not addressed.

RADSL addresses this shortcoming, but so does modern ADSL equipment. RADSL and current ADSL technology automatically adapt to changing line conditions each time the link becomes active. RADSL theoretically has an additional feature that is not inherent in modern ADSL: It has the capability to adapt to changing line conditions on the fly for both the upstream and downstream channels (which may be a problem to some types of applications).

In reality, the RADSL products that are deployed do not make allowance for this feature. ATM uses two variables to enforce QoS: the peak cell rate (maximum number of cells transmitted per unit time) and the average cell rate, both of which are dependent on the data rate of the link. How then is ATM to know the data rate, much less enforce QoS, if the data rate is not known until after the link becomes active? Work is ongoing to find an effective way for RADSL to communicate to ATM the data rate of the link at the time of start-up and any changes during operation—a key issue, if ATM is to enforce any quality of service across the link.

DMT modulation techniques are rate adaptive in nature, so modern DSL modems inherently have the capability. The earlier CAP versions of ADSL modems present a greater challenge, as additional circuitry is needed to enable the feature, which adds additional overhead to the operation of the modem. As a result, rate adaptation is almost exclusively found in DMT versions of ADSL modems.

VDSL: Very High Data Rate DSL

Fiber all the way to the home (FTTH) is still prohibitively expensive, as it entails rewiring the entire neighborhood with fiber. An alternative is to use a combination of fiber cables feeding neighborhood *optical network units (ONUs)* and leverage the existing copper loops to the home or business. VDSL is a technology that gets us closer to that dream.

VDSL depends on very short runs over copper loops of up to 6000 feet in order to maximize the available frequency range of the wire, with the remaining loop to the local exchange being served by fiber. The ONU serves as a central distribution point where the fiber from the local exchange terminates and the many VDSL copper loops aggregate. The DSL Forum refers to this arrangement as *fiber to the neighborhood (FTTN)* and extends the concept to include *fiber to the basement (FTTB)* for high-rise buildings with vertical drops and *fiber to the curb (FTTC)* for short drops. The use of FTTB and FTTC gives a good indication of the target market that the technology hopes to address: places with a high concentration of people, such as MDUs and MTUs.

Different documentation uses different terminology to describe the way VDSL is delivered. The common component of all descriptions is the fiber link back to the local exchange and short copper runs (see Figure 4-19). In addition to those previously mentioned, *Fiber to the cabinet (FTTCab)* is also used to describe the scenario where the VDSL loop terminates in a cabinet close to the homes served, with a fiber feedback to the backbone network. *Fiber to the exchange (FTTEx)* has been used to describe those situations where the CO or local exchange serves subscribers in the immediate vicinity.

VDSL is being touted as a full-service access network that addresses the full range of services from POTS and ISDN to linking high-speed LANs. The growing demand for high-bandwidth multimedia solutions and the proliferation of fiber has created a fertile environment for the idea of the technology to seed and grow. The fundamental basis of its application competes, head to head, with FTTH solutions. As was discussed in Chapter 2, solutions for multidwelling and multitennant units are somewhat limited, but VDSL is positioned to be a viable solution for this space, as it leverages the use of existing telephone wires thereby creating hope for buildings like the Empire State Building.

VDSL is asymmetric with downstream speeds that range from 13 Mbps to 52 Mbps across copper loops ranging from 1000 feet to 4000 feet. The upstream rates range from 1.5 to 6 Mbps. This is probably an appropriate place to make reference to the different organizations that are working on the specifications, as different modes of operation, speeds, and reaches have been proposed and are currently under review.

Standards and Operational Issues

Many standards and special interest groups have been actively working on VDSL specifications. The ANSI T1E1.4 working group began studying VDSL requirements in 1996

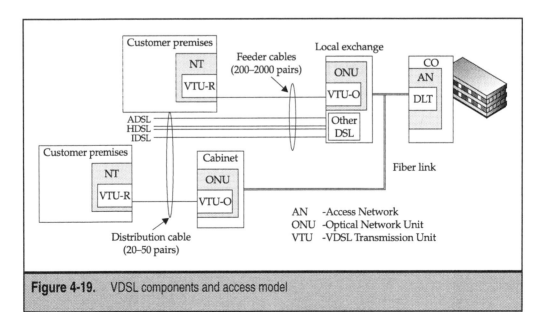

Figure 4-19. VDSL components and access model

and has been focused on U.S.-based specifications. The specification has been divided into three parts: the overall functional requirements and specifications, the transceiver specification using a single-carrier modulation technique, and the technical specification of the modem using a multicarrier modulation line code. Table 4-11 lists the speeds and reaches that have been proposed by ANSI.

The *European Telecommunications Standards Institute (ETSI)* has focused on the specifications for Europe and has divided the specification into two parts. The first part addresses the functional requirement of the technology, and the second, the specification of the modems and line codes employed. They have proposed the services listed in Table 4-12.

The ITU has been focusing on issues for international interoperability. The ITU G.993.1 standard defines recommendations for VDSL and is roughly equivalent to the ETSI TM6 Part 1 VDSL document. It includes worldwide frequency plans that use up to 12 MHz of the spectrum. It also defines asymmetric and symmetric services that can be deployed from central offices or from fiber-fed cabinets located near the customer premises.

The remaining special interest groups include the DSL Forum, the ATM Forum, the *Digital Audio-Visual Council (DAVIC)*, and the *Full Service Access Network Group (FSAN)*.

Loop Length	Service Type	Downstream	Upstream	Reach
Short	Asymmetric	52 Mbps	6.4 Mbps	1,000 feet
	Asymmetric	38.2 Mbps	4.3 Mbps	
	Asymmetric	34 Mbps	4.3 Mbps	
Medium	Asymmetric	26 Mbps	3.2 Mbps	3,000 feet
	Asymmetric	19 Mbps	2.3 Mbps	
Long	Asymmetric	13 Mbps	1.6 Mbps	4,500 feet
	Asymmetric	6.5 Mbps	0.8 Mbps	6,000 feet
		6.5 Mbps	1.6 Mbps	
Short	Symmetric	34 Mbps	34 Mbps	1,000 feet
	Symmetric	26 Mbps	26 Mbps	
Medium	Symmetric	19 Mbps	19 Mbps	3,000 feet
	Symmetric	13 Mbps	13 Mbps	
Long	Symmetric	6.5 Mbps	6.5 Mbps	4,500 feet
	Symmetric	4.3 Mbps	4.3 Mbps	
	Symmetric	2.3 Mbps	2.3 Mbps	

Table 4-11. ANSI Data Rates and Reaches

Service Type	Type	Downstream	Upstream	Best/Worst Reach
Asymmetric	A1	6.4 Mbps	2.048 Mbps	5,875 / 2,765 feet
Asymmetric	A2	8.576 Mbps	2.048 Mbps	5,546 / 2,588 feet
Asymmetric	A3	14.464 Mbps	3.072 Mbps	4,408 / 2,391 feet
Asymmetric	A4	23.168 Mbps	4.096 Mbps	3,264 / 1,486 feet
Symmetric	S1	6.4 Mbps	6.4 Mbps	4,736 / 2,873 feet
Symmetric	S2	8.576 Mbps	8.576 Mbps	4,244 / 2,690 feet
Symmetric	S3	14.464 Mbps	14.464 Mbps	2,771 / 1,886 feet
Symmetric	S4	23.168 Mbps	23.168 Mbps	1,302 / 856 feet
Symmetric	S5	28.288 Mbps	28.288 Mbps	977 / 695 feet

Table 4-12.   ETSI Data Rates and Reaches

The DSL Forum, in keeping with its charter, addresses network, protocol, and architectural aspects of VDSL for all prospective applications, leaving lower-layer functions like line coding to ANSI and ETSI and higher-layer application functions to DAVIC and ATM Forum.

The ATM Forum has defined a 51.84 Mbps interface for private network UNIs and a corresponding transmission technology. It has also addressed premises distribution and delivery of ATM across different access technologies.

DAVIC is focused more on the video aspects and other broadband applications of the technology.

The focus of FSAN, a group consisting of the major telephone companies around the world, has been on the development of a passive optical network (PON) specification and defining common telco requirements for VDSL. The specifications covering PONs have resulted in ITU-T recommendation G.983, and the group has deferred to ANSI and ETSI for recommendations on VDSL specifications.

As a point of reference, Figure 4-20 provides a speed and distance comparison of the ADSL and VDSL.

VDSL Issues   The fact that VDSL uses frequency ranges beyond those employed by other DSL technologies creates a new set of issues and challenges. An impairment that was once tolerated by another DSL, ADSL for instance, may be a major issue for VDSL. The following are just some of the issues and requirements that have surfaced that need to be addressed to ensure its proper operation:

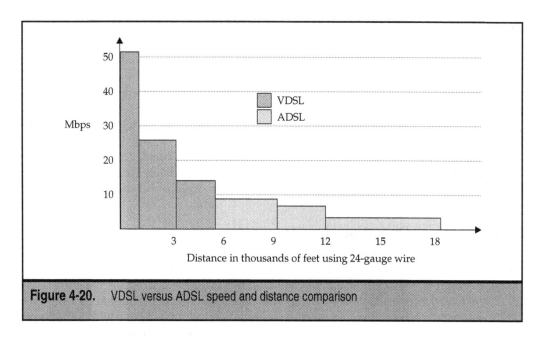

Figure 4-20. VDSL versus ADSL speed and distance comparison

Problems of reliability and robustness in the presence of external interference and impairments are heightened. Bridged taps, for instance, could be tolerated by ADSL but may be disruptive to VDSL operation.

▼ VDSL should be able to coexist and have spectral compatibility with other DSLs.

■ VDSL must strike a balance between achieving the power levels needed for the technology and not creating interference with other services that use the same frequency band. This aspect has wide implications, as industries such as AM broadcasters and civil aviation could be affected by VDSL signals. Special consideration must be given to the deployment of VDSL technology in communities close to airports. The potential interference creates safety concerns.

▲ The state of the cable is also of concern. Tests have shown that VDSL emissions increase when signals are run across untwisted wires. At issue is the question of how to effectively optimize the power requirements to lower emissions if the state of the cable infrastructure is unknown.

Frequency Allocation

VDSL, which was originally called VADSL, was later renamed to VDSL because the specifications allow for both symmetric and asymmetric modes of operation. Its high speed is achieved by using shorter loops and a wider spectral range from 200 KHz to 12 Mhz. The use of higher bands of frequencies creates the need to effectively manage the range and power levels to avoid interference with other services that occupy the same frequency bands.

The allocation and management of the frequency bands have global implications. The bands used by emergency services in the U.S., for instance, may be different in other countries. So VDSL must effectively manage use of the frequency range in order for it to coexist with different services in different parts of the world. Table 4-13 lists, and Figure 4-21 shows graphically, the proposed frequency plans for VDSL that have been accepted and approved for different regions of the world.

VDSL has been known to create problems for ham radio users. The transmit power spectral density must be reduced (or notched) to below –80 dBm/Hz in one or more of these bands, as listed in Table 4-14. When EMI effects are weak, notching of these bands is not required.

VDSL Modes of Operation

VDSL defines five end-to-end transport modes that are combinations of STM, ATM, and packet modes of operation. STM *(synchronous transfer mode)* operation is basically a time division multiplexing scheme where each channel is given a fixed bandwidth. ATM *(asynchronous transfer mode)* takes a more statistical approach to where the channels are multiplexed and offers some amount of discrimination in terms of bandwidth allocation and service quality. The packet mode of operation assembles the data into variable-length packets and sends them on a single channel of the maximum bandwidth.

The combinations for the five different transport modes follow:

▼ STM mode that extends from the user terminal equipment all the way across the copper and fiber links

■ STM mode that extends from the user terminal equipment across the copper link to the ONU, with ATM from the ONU across the fiber

■ ATM mode from the user terminal equipment across both the copper and the fiber

■ Packet mode from the user terminal equipment across both the copper and the fiber

▲ Packet mode that extends from the user terminal equipment across the copper link to the ONU, with ATM from the ONU across the fiber

Frequency Plan	Status	Region
Plan 998	Approved for ANSI T1	North America and Japan
Plans 997, 998	Approved for ETSI	Europe
Plans 997, 998, and Fx (flexible)	Accepted in ITU-T	

Table 4-13. VDSL Frequency Plans and Statuses

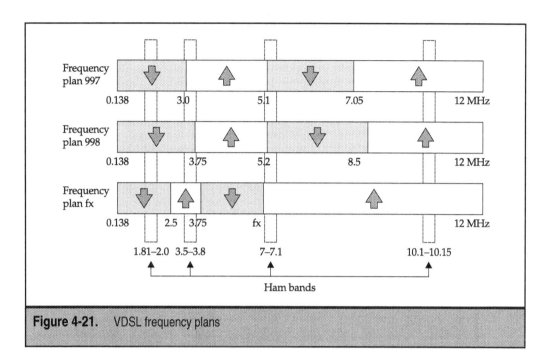

Figure 4-21. VDSL frequency plans

Like ADSL, VDSL has the capability to operate in dual-latency mode to accommodate both fast and slower data (the terms fast and slow are being used relatively to differentiate between delay-sensitive data and data that is less affected by latency issues). In single-latency mode, it operates in either fast mode or slow mode. The mode of operation is determined at the time of start-up.

Band Start MHz	Band Stop Europe MHz	Band Stop U.S. MHz
1.810	2.000	2.000
3.500	3.800	4.000
7.00	7.100	7.300
10.100	10.150	10.150
14.000	14.350	14.350
18.068	18.168	18.168
21.000	21.450	21.450
24.890	24.990	24.990
28.000	29.100	29.700

Table 4-14. International Amateur Radio Bands

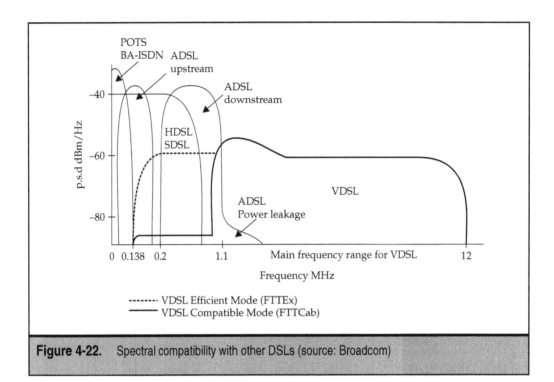

Figure 4-22.   Spectral compatibility with other DSLs (source: Broadcom)

Several modulation techniques have been proposed, with the list of potential candidates being narrowed to CAP/QAM for a *single-carrier modulation (SCM)* line code and DMT, which is a *multicarrier modulation (MCM)* technique. Both the ANSI and ETSI working groups are working on the specifications for transceiver line coding, and the decision, at least initially, will probably be to support both SCM and MCM line codes. The proposed error correction method is standard FEC, Reed Solomon with up to eight correctable octets.

Common to both the ETSI and ANSI specifications for the different line coding is the use of *frequency division duplexing (FDD)*. It is a well understood, mature, and cost-effective solution and can be used to easily mix different services (symmetric/asymmetric, high data rate/low data rate). It does not require central synchronization, so different vendors are not limited to a common timing. With this technique, the available bandwidth is carved into multiple frequency bands for the upstream and downstream transmissions, with each band having its own carrier (see Figure 4-21).

VDSL-Enabled Services

The DSL Forum has identified two service sets for residential and small to medium-sized enterprises (SME), as shown in Tables 4-15 and 4-16. The approximate service rates are given along with the need for a guaranteed service rate. The service set for residential use has a high video content and makes use of asymmetric data rates. The video rates assume

Service Type Asymmetric	Downstream Rate (Mbps)	Upstream Rate (Mbps)	Service Guarantee
Video Services (Assumes 3 simultaneous channels)			
Switched video broadcast	>15 Mbps	<0.2 Mbps	Yes
VoD	>15 Mbsp	<0.2 Mbps	Yes
nVoD	>15 Mbps	<0.2 Mbps	Yes
Audio Services			
Hifi Audio on Demand	>1 Mbps	<0.1 Mbps	Yes
Online radio	>1 Mbps	<0.1 Mbps	Yes
Internet/Intranet			
Download multimedia	>10 Mbps	<0.1 Mbps	Best effort
Download applications	>10 Mbps	<0.1 Mbps	Best effort
Virtual reality gaming	>10 Mbps	<1 Mbps	Best effort
Online shopping	>10 Mbps	<0.1 Mbps	Best effort
Web site hosting	>0.4 Mbps	>2 Mbps	Best effort
Derived Voice			
VoDSL w/ATM <4 channels)	<0.32 Mbps	<0.32 Mbps	Yes
VoIP <4 channels)	<0.32 Mbps	<0.32 Mbps	Best effort

Table 4-15.   VDSL Residential Service Set

Service Type Symmetric	Downstream Rate (Mbps)	Upstream Rate (Mbps)	Service Guarantee
Office Communications			
Derived voice (>16 channel PCM @ 64 Kbps)	<2 Mbps	<2 Mbps	Yes
High-quality video conferencing	<8 Mbps	<8 Mbps	Yes
Internet/Intranet			
Large file transfer	>10 Mbps	>10 Mbps	Best effort

Table 4-16.   VDSL Small-Medium Enterprises Service Set

Service Type Symmetric	Downstream Rate (Mbps)	Upstream Rate (Mbps)	Service Guarantee
Application download	>10 Mbps	<2 Mbps	Best effort
Access virtual reality web sites	>10 Mbps	<2 Mbps	Best effort
Media hosting (Webcast)	<2 Mbps	>10 Mbps	Best effort
Web site hosting	<2 Mbps	>10 Mbps	Best effort
Remote learning applications	>10 Mbps	<2 Mbps	Best effort

Table 4-16. VDSL Small-Medium Enterprises Service Set (continued)

three simultaneous MPEG-2 audio and video channels of at least 5 Mbps per channel. The SME service set assumes that businesses will require equal data rates in both directions, so symmetric data rates are proposed. All service sets' data rates are approximate and are used to give an example of the application of the different service speeds of VDSL.

SUMMARY

ADSL is the most documented and widely available version of the DSL family. Many service providers' DSL offerings begin and end with ADSL. The required splitter for the support of POTS and the high data rates do make it less cost effective for the casual user who may be just looking for a speedy connection to browse the Net. For this class of user, ADSL lite or G.lite is more appropriate. Its high-end speed is more than suitable for this type of use, and the elimination of the splitter makes it a lot easier to work with.

On the business side, HDSL is a proven technology. It is the oldest of the DSL solutions and has been successful in its deployment. Chances are very good that any new T1/E1 service that is installed today is being enabled by the technology. New features such as the variable power capability that is now available in its second-generation version, HDSL2, may eventually be seen in all the other DSL families. The elimination of the requirement for a second loop goes a long way toward realizing pair gain and the conservation of copper pairs, an issue that is becoming more important each day as carriers attempt to keep up with demand for second phone lines for homes.

VDSL is the newest of the DSL families and the one that hopes to facilitate the complete convergence of narrowband and broadband services. It embodies the promise of a full-service access network—a single access technology that delivers services from POTS to e-commerce to campus LAN interconnections. Its target market, MDUs and MTUs, should see substantial growth over the next few years. These trends will benefit the technology and, as it matures, will hopefully realize its promise of being a full-service access network.

CHAPTER 5

Cable Access Networks

To cable operators, the efforts to extract and use the available bandwidth of the local loop are foreign. After all, cable networks evolved from the need to share access to television programming that was broadcast over a wide frequency range. The very cable on which it was built was selected because of its wide spectral capability. The fact that it was shared simplified the topology and made it ideal for broadcast applications. Discrete point-to-point cable runs were not needed between each subscriber and the cable distribution plant, as was the case with the local loop. All in all, cable networks had an advantage over telephone companies and other service providers in delivering broadband services.

There were, however, some obstacles that had be overcome. CATV networks were built to deliver broadcast television in one direction only, so the network had to be upgraded to facilitate two-way communication before it could be used for broadband services. A frequency band had long been set aside for a reverse path; the hurdle, therefore, was to find a way to use it. The equipment used in the cable plant was designed for one-way communication. Amplifiers were used to amplify the forward signals only, and anything flowing in the opposite direction was considered noise.

The topology and access methodology also presented challenges. The fact that cable networks used a shared medium presented security concerns, so the data from multiple subscribers had to have some level of security applied. Finally, the application of CATV networks made it more prevalent in residential than business use. So even though CATV networks had an edge over telco solutions for the delivery of broadband services, they were not necessarily the solution of choice for corporations. This is an undesirable position to be in, because as the data for DSL usage show, worldwide residential installations outnumber business installs, but it is businesses that generate the greater revenue for DSL service.

The potential of the cable networks was not lost on telephone companies, and it was not long after the Telecommunications Act of 1996 that companies like AT&T began purchasing cable assets. Today, AT&T is the largest cable provider in the U.S., a statement that would not have been likely prior to 1996 and one that sums up how fast things are converging. In December 2001, AT&T announced plans to merge with another cable company called Comcast Corporation. The new company will have a combined user base of 22.5 million users—14 million from AT&T, 8.5 million from Comcast at the time of the announcement—and will be called AT&T Comcast. In contrast, the closest rival, Time Warner Communications, had a user base of 12 million during the same period.

While telephone companies were upgrading the local loop to provide broadband services, AT&T was spending billions on upgrading the cable network to provide phone services. As part of the announced merger, both AT&T and Comcast executives promised to make a faster push into cable telephony, taking the necessary steps if cable is to be a single access network to integrated digital services that include new digital video services, digital networks, high-definition television, high-speed Internet access, and telephony.

Cable providers throughout the history of the medium have been tenacious and have carved out an industry in spite of many regulatory and competitive challenges. They have always operated in the shadow of the telecommunications and broadcasting industries, yet the service they have built is ubiquitous, as is evident from the data presented in Table 5-1. As Figure 5-1 shows, the industry has enjoyed a period of rapid growth for new digital services from 1998 to the present. A significant investment has been made in upgrading and expanding the existing infrastructure, which positions cable as a viable competitor to DSL solutions.

FUNDAMENTALS OF CABLE NETWORKS: COAXIAL DISTRIBUTION NETWORK

Cable TV was originally designed for communities in mountainous areas that received poor television reception. The system made use of a powerful antenna that was placed where the signal was strongest, such as at the top of a mountain. The use of a single shared antenna was called Community Antenna Television, which is where the term CATV originates. A cable was run from the antenna to the community, where subscribers could gain access to the shared signal. To be effective, the solution had to be compatible with the same television equipment that was used for receiving over-the-air broadcast, so the over-the-air radio frequency spectrum was re-created within the sealed environment of the cable.

The original systems had a capacity of about 400 MHz. Modern systems that make use of fiber have a much broader frequency range. Modern-day systems have also replaced the shared antenna with a shared satellite dish. The spectrum is divided into two ranges of frequencies for upstream and downstream communication. *Upstream* describes the flow of data from the user to toward the network; it uses the range of frequencies 5–42 MHz.

Homes passed by cable	98,600,000
Homes passed as a percentage of TV households	96.7%
Cable systems	9,947
Annual cable revenue	$48,150,000,000
Digital cable subscribers*	13,700,000
Cable modem subscribers*	6,400,000
Home passed by cable modems	70,000,000
Cable-delivered telephone subscribers*	1,500,000
Industry expenditure on construction/upgrade	$14,290,000,000
* As of November 2001	

Table 5-1.   U.S. Cable 2001 Statistics (source NCTA)

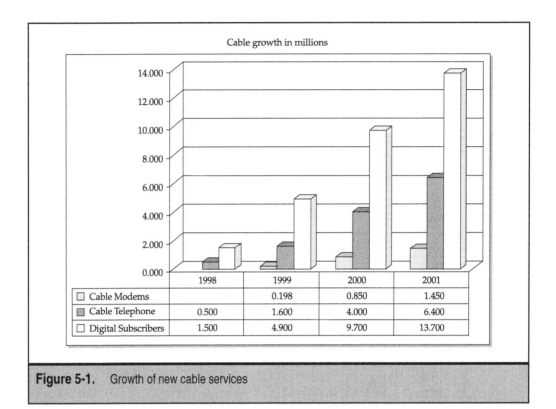

Figure 5-1. Growth of new cable services

Downstream—network to user—uses the range from 50 MHz upward. The bandwidth was divided into channels of 6 MHz each, so the available number of channels in the original system was about 55 channels. More modern systems have a much broader spectrum that more than doubles the capacity of older systems.

The design of the original cable systems utilized only the downstream channels because there was no equipment in the subscribers' homes to send data toward the network. This was the case until the early 1990s, when the cable systems were upgraded to deliver new applications other than TV services.

Cable Architecture

A basic cable system is composed of four major components: (1) the headend; (2) the trunk line; (3) feeder and drop cables; and (4) a terminal equipment or network terminating unit that represents the subscriber. It is as simple as that. An operator of a cable system is called a *cable operator*. Companies that operate multiple cable systems are called *multiple systems operators (MSOs)*, and the systems owned may extend into multiple states. Obviously, there are better economies of scale if the systems owned are contiguous, so MSOs tend to expand by acquiring or merging with cable operators that are adjacent to systems that they own. Examples of MSOs are AT&T Comcast, Time Warner, and Cox.

The distribution of the cables in a cable system has a tree and branch structure. The headend forms the root; the trunk, feeder, and drop cables, the branches; and the subscribers, the leaves. The trunk connects directly to the headend with feeder cables branching from it through the use of splitters. The feeder cable extends the system for miles from the headend and deep into local neighborhoods. Drop cables connect subscribers by tapping into a feeder cable or the trunk.

The center of the cable system is the headend. It is the brains behind the operation, where broadcast and satellite-delivered signals are received and readied for distribution. The headend can receive and process programming in various formats, analog (AM and FM) or digital. It also hosts the equipment used to descramble incoming feeds from satellites and broadcast networks. These feeds are assigned channels that match the channel on which they were originally transmitted before being transmitted over the cable lines. Finally, the headend also has the capability to encrypt signals for security purposes and play local content such as advertising and locally originated programming.

Trunk lines are high-capacity lines that carry signals from the headend to feeder cables that serve the local communities. Trunk lines are usually untapped, which means that drop cables that service subscribers are usually not directly tapped into a trunk. The diagram in Figure 5-2 shows drop cables connected in this way for illustrative purposes and because there are no hard rules that say it cannot be done. The feeder cable is a lesser-quality cable than those used for trunks; it is used to extend the system into local neighborhoods. A quarter-inch drop cable runs from a tap on the feeder cable to the subscriber's premises. The drop cable terminates at a terminal unit.

The distance from the headend to the last subscriber can be as much as 50 miles, and attenuation (or line loss) increases with distance as well as increasing frequency. To maintain signal strength, amplifiers are placed about every 3,000 feet. This poses a tricky problem, though, as higher frequencies attenuate a lot faster than lower frequencies. The power levels required for boosting the frequencies in the higher end of the spectrum might be too great for the channels that occupy the lower frequencies, causing distortion. To rectify this, equalization is used to balance the power levels across the entire band closer to the end points.

There is no way to dictate to subscribers the perfect placement for their homes, which sometimes are situated in very inconvenient spots, like the very end of a cable segment served by an amplifier. To accommodate such users, the cable provider may choose to turn up the power of the serving amplifier. The customer at the end of the segment may be perfectly served by this, but the boosted power levels might adversely affect another subscriber closer to the amplifier. To address this problem, a device called a *pad* is used to induce attenuation. The pad is placed on the drop cable of the subscriber affected by the increased power levels and serves to reduce power to that subscriber.

Hybrid Fiber Coax: The Upgrade Path

The cable system described serves to highlight the components used to deliver cable service, and it also pretty much describes the majority of CATV systems up to the 1990s. By the 1990s, cable operators began to see *direct broadcast satellite (DBS)* service as a serious

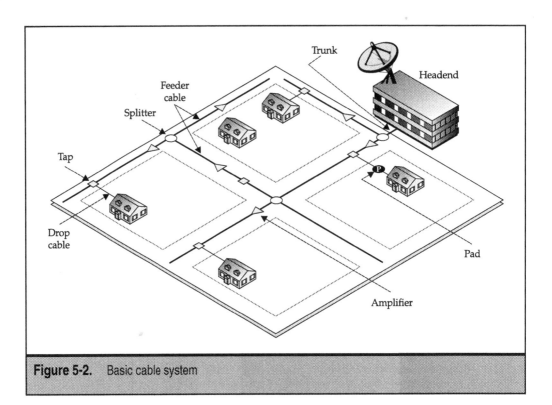

Figure 5-2. Basic cable system

threat and, in a defensive move, began the process of improving the cable system. In addition to needing to remain competitive with DBS services, which had double the channel capacity, the cable plant in general was ready for a major overhaul. The problems associated with wire lines and amplifiers meant high maintenance costs.

The problems of wire attenuation over long distances and the need to amplify the signal every 3,000 feet or so were major headaches. The failure of a single amplifier could degrade or disrupt service to hundreds of subscribers. Take, for instance, the failure of a single amplifier close to the headend—subscribers downstream for as far as the cable extends (which could be 30 or 40 miles away) could be affected.

The quality of the final signal was also a problem. It was a lot like the telephone game that kids play where the first kid whispers something to the second, who in turn whispers what he or she heard to the next kid in line, and so on. Eventually, the last kid tells all the other kids what he or she heard. More often than not, the final version is a variation on the original. The use of multiple amplifiers created similar results. Consider a cable run of 30 miles or more: there could be as many as 40 amplifiers, with each successive amplifier regenerating the version of the signal it was supplied. The end signal was always somewhat distorted from the original. And if that were not enough, the challenge of balancing the power levels across the spectrum was a difficult task, especially when so many amplifiers were involved.

Another very strong incentive to upgrade was the Telecommunications Act of 1996. The carrot of new market opportunities justified the expense to many. And finally, the phenomenon of the Internet and the promise of new service opportunities and revenue streams were enough to convince those who still had doubts.

To overcome the problems of the cable plant, a hybrid version of the network called a *hybrid fiber coax (HFC)* cable evolved, an illustration of which is presented in Figure 5-3. In this new scheme, fiber optic cables were used to replace the coaxial trunks and were run from the headend to a *fiber distribution node*. Coaxial feeder cables were attached to the fiber node distribution node, which converted the analog optical signals into electronics for downstream transmission along the coax feeder cables.

The HFC design transformed a single cable system into a series of "smaller cable systems" with individual serving areas of as few as 200 to 500 subscribers, with the larger areas serving about 500 to 2,500 subscribers. The fiber connection back to the headend meant a significant increase in available bandwidth, signal reliability, and Quality of Service. The fiber also meant a reduction in emissions and so less interference for the existing coax cables. The segmenting of the system into smaller groups also meant that a problem in one neighborhood system would not affect another.

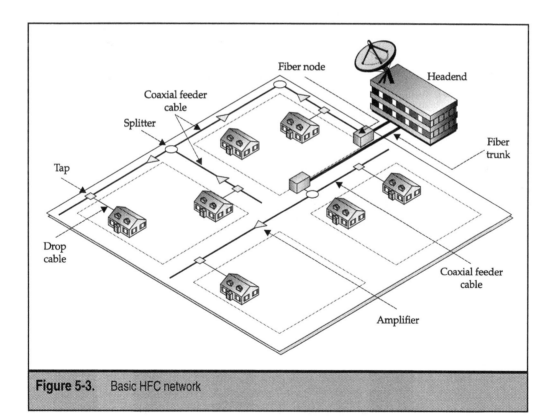

Figure 5-3.   Basic HFC network

In an HFC system, amplifiers are needed only on the coaxial branches, so the number of amplifiers was significantly reduced. The reduction of amplifiers also simplified the task of equalizing the power levels and reduced maintenance, resulting in lower operational costs.

The use of fiber accomplished several things: it significantly reduced a number of transmission issues, it simplified the design, it created smaller and more manageable coax segments, it increased channel capacity by providing a broader range of frequencies, and it set the stage for upstream communication. Cable systems had been predominantly one-way broadcast systems even though the bandwidth was available for upstream communication. If new two-way services like Internet access or VoD were to be rolled out, the network had to be able to facilitate communication along the return path.

Two-Way Upgrade: Return Path Schemes and Challenges

The introduction of the HFC network addressed a major hurdle and was the first step toward the support of new services. The system, however, had to support the flow from the subscriber to the headend if new services like Interactive TV and Internet access were to be offered. The flow of data away from the subscriber and toward the headend is normally called the *return, reverse,* or *upstream path.*

The frequency range from 5 MHz to 42 MHz had been reserved for this purpose, but the network was never equipped to provide a return path. There were a few systems that did have limited return capabilities, but the majority did not. As a side note, in the 1970s Warner Communications had embarked on an interactive TV trial in a small Ohio town. December 1, 1977, saw the first commercial interactive TV service, which offered 30 channels for normal broadcast TV, another ten for pay TV, and the final ten for interactive programming. The project, though popular, was scrapped because of the costs associated with delivering the service. Further details of the project can be found at http://www.media-visions.com/itv-qube.html.

To build an upstream channel, reverse amplifiers were needed on the coax segments to amplify the return signals originating from the subscriber. The HFC upgrade helped in this regard because of the reduction in the amount of coaxial cable used. The fact that lower channels were to be used for the upstream channels meant less attenuation, so fewer amplifiers were needed than was required for the forward path. Finally, the fiber node also needed a laser transmitter to transmit the signal back to the headend.

Cable operators put the cost of upgrading a system for two-way communication at about $750 to $1,500 per mile. Many operators chose to do it at the same time that they upgraded to an HFC network. By taking this approach, they realized a savings of $15 to $16 per home passed, with the majority of the savings being in labor costs. The component costs were: amplifiers for coax segments, $300 to $800; laser transmitters at fiber nodes, about $1,000; and receivers at the headend, 500 to $1,000.

Using a Telephone Return   Not all cable system operators chose to upgrade the network to provide a return path. Some operators took a more cautious wait and see approach on upgrading the system, while others wanted to get a head start on offering new services. Regardless of the reason, some operators opted to use the telephone network to provide the link back to the headend. Cable modems were equipped with the necessary interface to utilize either an *in-band* return path (using the cable network) or an *out-of-band* return path (using a telephone).

Telephone paths limit the available bandwidth for upstream communication. While this may not be a problem if the user is doing basic functions such as browsing the Internet or reading mail, it becomes a bigger issue if they begin uploading files. The upstream channel always has less capacity than the downstream channel regardless of which return path scheme is used, but the cable return is always a lot faster that the telephone solution. The limits on the telephone are anywhere up to 56 Kbps on a clean connection. The lower limit on the cable return path is in the region of 500 Kbps.

Advanced HFC Architectures

The architectures of cable systems and HFC networks are evolving to allow for the integration of multiple two-way services. The new architecture, shown in Figure 5-4, incorporates the features of a well-designed data network, including redundancy, fiber backbones, and interconnectivity of hubs for reliability. The design of the network looks like any other robust network, and this is necessary if the HFC network is to be used as a reliable transport for mission-critical data applications. If the industry wants to attract corporate accounts in addition to its large residential customer base, a reliable and robust infrastructure is needed along with a strong operations and support organization.

Many MSOs are deploying regional hubs using fiber rings to extend the reach of the HFC network and to enable sharing of a headend or multiple headends. The architecture allows multiple operators to share the cost and benefits of a fiber ring. The added benefit of hubs is that they allow sharing headend equipment between companies. As the digital era unfolds and more applications and services are launched, the capital required to build a headend and outfit it with necessary equipment becomes increasingly high. Being able to share the servers, compression, and ad insertion equipment or maybe even a telecommunications switch becomes increasingly attractive as the equipment list and the processing requirements grow.

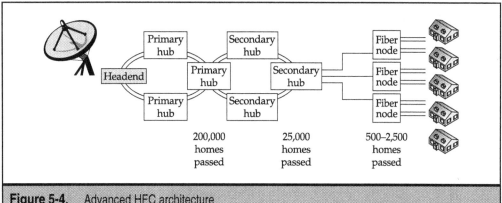

Figure 5-4.   Advanced HFC architecture

NEW SERVICES FOR THE NEW HFC NETWORK

The upgrade of the cable system to HFC was in part a defensive move to stave off competition from direct broadcast satellite and new telco initiatives. A second motivation was to be able to offer new, enhanced services. With the upgrade complete and the capacity in place, new incremental applications could be supported with minimal additional costs. Several applications were targeted, high-speed data services, basic telephone services, and digital video services being a few.

Normally, basic telephone service would not be categorized as a broadband service. It is, however, an important basic service on which we all rely. For this reason, it must be considered in any discussion on integrated digital services. For this reason, we have covered it enough to provide an appreciation of developments in this area.

The remainder of this chapter explores cable's new applications with the aim of explaining the challenges and the ways in which they are being addressed.

Data Services over Cable

The roots of the cable industry were in the one-way distribution of video. As a result, 94 percent of the bandwidth was dedicated to transmissions in a single direction. This meant only 6 percent of the bandwidth was available to be shared by as many as 2,500 users for enhanced services (each fiber node typically served 500 to 2,500 subscribers). This made the upstream path a critical resource that had to be used wisely.

One possible solution was to limit the use of the upstream channels to applications that would not exhaust their capabilities. This, however, was not a viable solution if cable was to compete with other services, such as DSL. Another solution was to upgrade the infrastructure in order to decrease the number of subscribers served by each fiber node, thereby increasing the available bandwidth per subscriber. The downside of this approach was the cost of the upgrade. The industry had already invested billions in the upgrade of the network, and investors would probably not be too happy to hear about more expenditure. A return on the investment made so far would be much more palatable.

A third way to address the problem would be to create an extremely efficient way to utilize the upstream path. Several initiatives were started to do just this, and the sections to follow address some of the approaches that were undertaken and the results that were achieved. Figure 5-5, provided as a reference for the rest of this chapter, illustrates the different channel bandwidth assignments in the U.S., Europe, and Japan.

Standards and Special Interest Groups for Data over Cable

Since 1994, many standards organizations and special interest groups have been working to provide data over cable specifications. The groups that have been active in this area include the IEEE 802 committees, the Multimedia Cable Network Systems partnership (MCNS), the Digital Audio Video Council (DAVIC), the Digital Video Broadcast project (DVB), the Internet Engineering Task Force (IETF), the ATM Forum, and the Society for Cable Telecommunications Engineers (SCTE).

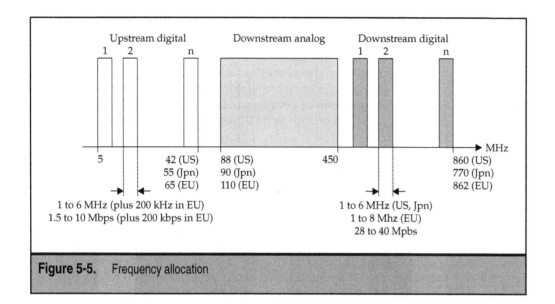

Figure 5-5. Frequency allocation

There are three standards today for cable modems: MCNS/DOCSIS, DVB/DAVIC, and IEEE 802.14. MCNS/DOCSIS is the most widely deployed cable modem technology in North America. DVB/DAVIC is the leading contender in Europe.

IEEE 802 Committees   The IEEE 802 committees are responsible for developing LAN/MAN protocol standards. In 1994, the IEEE 802.14 Cable Media Access Control (MAC) and Physical (PHY) Protocol Working Group was formed to develop an international standard for cable modem equipment. The working group had hoped to publish their specification by the following year but was late in doing so. Two services were defined for an HFC infrastructure with a single downstream channel and multiple aggregated upstream channels. The 802.2 Logical Link Control (LLC) supports the transfer of frames (a mandatory requirement for all IEEE 802 standards) and an ATM service that conforms to ITU-T recommendations.

The IEEE committees have been focused on the next generation of cable modems to address higher speeds. They have teamed with Broadcom and Terayon on the development of a new advanced high-speed physical layer called Hi-Phy.

MCNS   The delay in the publication of the IEEE standards prompted the formation of the *Multimedia Cable Network System Partners, Ltd. (MCNS)*. In January 1996, Comcast, Cox, Time Warner, and TCI Communications (which was later bought by AT&T) formed the MCNS alliance to jump start the process to get a working common specification for cable equipment. They released a draft standard called the *Data Over Cable Service Interface Specification*, DOCSIS, in 1997.

DOCSIS covers the operational elements necessary for the delivery of data services to end users, including service provisioning, security, data interfaces, and radio frequency interfaces (RFI). In the absence of a standard and under growing pressure to capitalize on new market opportunities, many vendors began deploying equipment based on proprietary technology or the draft specification. CableLabs (Cable Television Laboratories, Inc., the research and development arm of the North American cable industry), in an effort to speed the process and to ensure interoperability of different vendor equipment, began a formal certification program in 1998 for the DOCSIS 1.0 specifications. In 1998, the International Telecommunication Union (ITU) accepted DOCSIS as a cable modem standard, called ITU J.112.

CableLabs manages the certification process for DOCSIS modems. The modems that pass the certification tests are given a "CableLabs Certified" seal of approval. All certified modems meet the standard for interoperability regardless of the manufacturer. Certified versions of cable modems became commercially available in 1999.

DVB/DAVIC   A key player behind the introduction of worldwide digital services has been the *Digital Video Broadcasting Project (DVB)*. Using principles of openness and interoperability, the organization has developed standards for all aspects of digital broadcasting since 1993. The DVB Return Channel on Cable Networks (DVB-RCC) specification, also known as DVB version 2 (DVBv2) or ETSI ES 300 800, defines two-way communication over HFC networks. The DVB-RCC specification takes advantage of standardization efforts produced as part of the DAVIC project and as a result is also known as the DVB/DAVIC specification. The ITU has adopted the standard in ITU J112-Annex A.

The DVB/DAVIC specification was originally designed, as a European standard, for digital cable set-top boxes. The specifications have been extended to apply to cable modems and are now known as DVB/DAVIC EuroModem, which is a serious challenger to DOCSIS in Europe.

The DVB project's main focus is the delivery of digital TV over satellite, terrestrial, and cable systems. The initial recommendations for cable networks did not cater to bidirectional communication, but new applications like interactive TV were dependent on a return path for user interaction. DVB went in search of a specification that would meet its requirements and would address a wider range of applications, including high-speed Internet access. The DAVIC 1.2 set of specifications was adopted, and both groups joined forces to seek the adoption of the specifications for a wider range of applications, such as data over cable.

DAVIC was established in 1994 but ceased operation after five years in accordance with its statutes. It remains active only through its web site (http://www.davic.org). Its goal was to promote the success, and maximize interoperability, of interactive digital audio-visual applications and services by promulgating specifications of open interfaces and protocols.

Their DAVIC 1.0 specification was the first in a series of five specifications that define a set of tools for the support of applications such as VoD, nVoD, and teleshopping. The subsequent series either refined existing specifications or added new ones and were backward compatible. The final specification, 1.5, addressed the requirements for

IP-based networks and cable modems. The following are some of the documents that were published in the 1.5 specification:

▼   The DAVIC Intranet specification, which describes the architecture and protocols for an IP-based network intended for digital audiovisual services

■   The DAVIC Cable Modem tools (DVB/DAVIC Standard)

▲   The TV Anytime service definition, describing the contours for a service based on non-real-time use of audiovisual programs, by use of local mass storage

Now that DAVIC has ceased operation, the DVB project has assumed total responsibility for the DVB/DAVIC specifications. Several ETSI documents address the DVB/DAVIC standards; the ones that address HFC networks are ETSI ES 300 429, which addresses downstream transmissions, and ETSI ES 300 800, which addresses the return channels. Because the challenge of data over cable is in the return path, the latter document is usually referenced for data services.

IETF   The *Internet Engineering Task Force (IETF)* formed a working group to produce a framework and requirements for *IP over Cable Data Network (IPCDN)*. They have identified standard MIBs for MCNS/DOCSIS and DVB/DAVIC cable modems and headend-based equipment, in the U.S. and Europe. They also function as a forum for Internet-related issues on data over cable systems, and they advise on the most appropriate body to address these issues if they are deemed not to be within scope.

The ATM Forum   The ATM Forum's *Residential Broadband Working Group (RBWG)* was formed to promote the deployment of ATM over residential network infrastructure. The group's focus is on the delivery of ATM to the home (ATTH) and ATM within the home. The RBWG was instrumental in the IEEE 802.14 decision to use an ATM-based data protocol.

SCTE   The *Society of Cable Telecommunications Engineers* was formed in 1969 to promote the sharing of technical information and knowledge for cable TV communications. In 1995, it became an accredited standards-developing organization for the American National Standards Institute (ANSI). The Data Standards Subcommittee's role is to define standards for equipment interoperability. By charter, it coordinates its effort with the IEEE 802.14, DAVIC, and CableLabs. The SCTE was successful in submitting the DOCSIS specifications for adoption by the ITU.

DOCSIS 1.0 Standard

The basic components of the DOCSIS specification are a cable modem, a *cable modem termination system (CMTS),* and the HFC network. The standard consists of 12 specifications that describe all aspects of DOCSIS from the physical layer to encryption and management (see Figure 5-6). It uses well-known, tried and proven technologies like TDM MPEG for the downstream channels, an 802.2 data link layer for IP, and TCP and UDP for transport. The only new areas are the MAC layer and the minislots that are used for upstream communications. QoS is not supported in the DOCSIS 1.0 version but is available in the

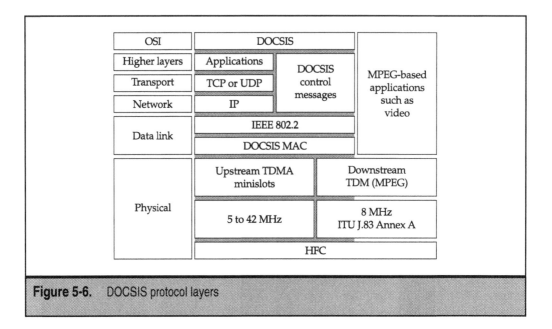

Figure 5-6. DOCSIS protocol layers

1.1 specification. Class of Service (CoS) is supported through the use of 14-bit session IDs (SIDs). The use of CoS allows multiple levels of service to be offered through bandwidth allocation. DOCSIS 1.1, which has many enhanced features, is covered in the section "Cable Telephony," which follows.

In 1999, a group of companies along with the University of Gent (Belgium) developed, tested, and launched EuroDOCSIS, which is based on the DOCSIS specifications. The new specification upped the frequency band of the downstream channel from 6 MHz to 8 MHz and that of the upstream channel from 5–42 MHz to 5–65 MHz in line with European requirements.

CMTS and Cable Modems   In order to enable the cable network for two-way data services, one of the channels in the frequency range of 91–857 MHz is typically allocated to downstream communication and another channel in the range of 5–42 MHz, for upstream communication.

Each downstream channel has a band of 6 MHz and is capable of 27 Mbps throughput when 64-QAM line encoding is used. The capacity can be boosted to 36 Mbps by using 256-QAM. The upstream capacity is dependent on the amount of spectrum allocated for upstream communication. It ranges from 500 Kbps to 10 Mbps using 16-QAM or QPSK. The entire bandwidth is shared by all the subscribers (typically 500 to 2,500) that are connected to a given cable system.

The architecture of DOCSIS consists of the *cable modem termination system (CMTS)*, installed in the headend, the subscriber cable modem, and the HFC wiring infrastructure, as shown in Figure 5-7. Downstream communication flows from point-to-multipoint,

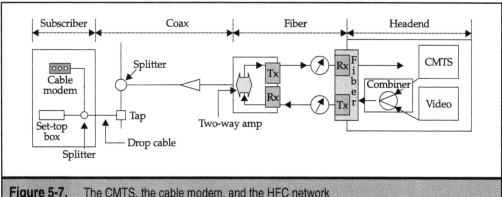

Figure 5-7.   The CMTS, the cable modem, and the HFC network

and upstream, multipoint-to-point. It should be noted that all cable modem traffic traverses the CMTS, so a packet destined from one cable modem to another cable modem in the same cable system is done via the CMTS.

Cable modems translate Ethernet packets into radio frequencies for transmission on to the allocated upstream channel. If the data is being sent from the headend, the CMTS uses the 6 MHz channel that was defined for downstream communication to broadcast the data to every subscriber modem that monitors that channel. On the return path, the CMTS receives the broadcasts of subscriber modems on separate channels from the one used for downstream communication. Figure 5-8 provides a comparative snapshot of the protocol stacks of the CMTS and cable modem as well as the interfaces that are used.

The DOCSIS 1.0 architecture provides for one downstream channel for all subscriber modems and several different upstream channels. This allows the total upstream bandwidth to be increased or decreased by the number of channels used.

The subscriber modem, which has an interface to the HFC network and a 10/100 Ethernet interface to the subscriber LAN or computer, takes the radio signals that it receives from the cable system and translates them into packets for transmission on its 10/100Base-T Ethernet interface. Similarly, any packets received from the 10/100Base-T interface are modulated onto the upstream channels to which they have been assigned.

The MAC downstream protocol used between the CMTS and a cable modem uses a continuous stream of 188-byte MPEG-2 frames. The frame has a 4-byte header that identifies the packet as Data over Cable. Both Data over Cable and Video, or another service, can be multiplexed onto the same MPEG multiprogram transport stream.

The upstream channel, in contrast to the continuous stream used for downstream communication, uses TDMA burst frames. The upstream channel is divided into minislots of 6.25 microseconds each. Each minislot is assigned one of the following classifications:

▼ Contention

■ Contention/data

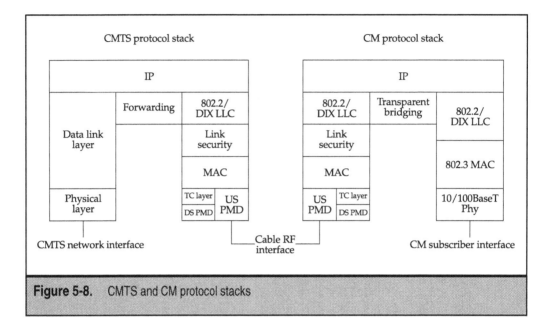

Figure 5-8. CMTS and CM protocol stacks

■ Reserved data

■ Initial maintenance

▲ Station maintenance

The classification of each minislot is described through the use of MAP MAC management messages that are periodically transmitted from the CMTS downstream to all cable modems. The actual description is carried in *information elements (IEs)* within the MAP frame; each IE can describe one or more minislots. A single MAP message can carry as many as 240 IEs and describe up to 4,096 minislots. By using this approach, each cable modem is able to identify the slots that are allocated to it and those that can be used on a contention basis to issue requests for bandwidth. A station that has data to transmit goes through the request–collision resolution–grant–transmission (RCGT) cycle that is shown in Figure 5-9. For each packet that is to be transmitted, a request for the appropriate bandwidth is made to the CMTS either by placing the request in a contention slot or by piggybacking a data transmission.

If a request for bandwidth is to be made by using a contention slot, the cable modem does not use the first contention slot it sees, because this increases the chance of a collision. Instead, it uses a back-off algorithm (similar to Ethernet) to determine the number of contention slots to skip before attempting to use one for transmitting the request. If, however, it was already granted bandwidth for transmission, it may choose to place the next bandwidth request in the same frame as the data.

Chapter 5:   Cable Access Networks

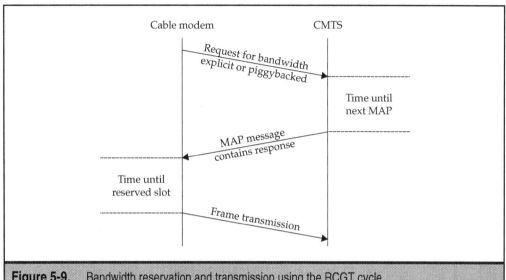

Figure 5-9.   Bandwidth reservation and transmission using the RCGT cycle

To summarize the process, upstream communication employs a bandwidth reservation mechanism that uses a request-grant scheme. The cable modem makes a bandwidth request to the CMTS for data that it needs to transmit. The request can either piggyback a data transmission from the cable modem or be made using a contention scheme. The CMTS grants the bandwidth through the use of a MAP message, and the cable modem transmits the packet during its grant period. The frame format of a DOCSIS frame is shown in Figure 5-10.

Security: DOCSIS Baseline Privacy   The specification employs two component protocols for security. The first is an encapsulation protocol for encrypting packet data across the cable network, which defines:

▼   The frame format for carrying encrypted data within DOCSIS MAC frames

■   The pairings of data encryption and authentication algorithms to form a set of cryptographic suites

■   A version of DES encryption

▲   Rules for applying the algorithms to a DOCSIS frame header

The second security protocol, called Baseline Privacy Key Management (BPKM), is used for providing secure distribution of security keys from the CMTS to the CM. The DVB/DAVIC set of specifications does not have security features.

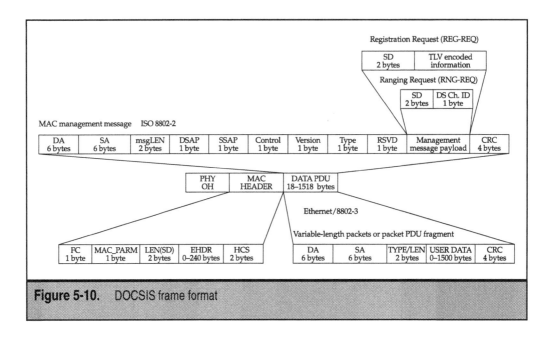

Figure 5-10. DOCSIS frame format

DVB/DAVIC Standard

In 1999, EuroCableLabs (ECL), the technical research branch of the European Cable Communications Association (ECCA), issued a final EuroModem specification as an alternative to DOCSIS. The specification is based on the DVB/DAVIC standard. A group of cable operators organized as the EuroModem Consortium and issued a tender for supplying EuroModem-compliant cable modems. After evaluating 24 responses, the EuroModem Consortium published a short list of 12 vendors from whom it expects to purchase up to 300,000 cable modems in 2001.

The DVB/DAVIC standard consists of a broadcast downstream channel, an interactive downstream channel, and an upstream channel. The downstream channel is defined in the ETSI ES 300 429 specification. It uses an 8 MHz channel and supports both 64-QAM and 256-QAM with MPEG-2 framing.

The upstream channel is defined in the ETSI document ES 300 800 and is also called the DVB Return Channel on Cable Networks (DVB-RCC) or DVB version 2 (DVBv2), which forms the basis for two-way communication. The standard has been ratified in ITU-T J112 Annex A. The upstream channel uses the ATM adaptation layer 5 (AAL5) framing for encapsulation of IP packets. The use of ATM means that the standard is able to use the features that are inherent to ATM, such as QoS. The upstream channel bandwidth is variable between 200 kHz and 4 MHz, yielding a maximum data rate of 6.176 Mbps.

In the DVB/DAVIC reference model, shown in Figure 5-11, the broadcast channel transmits broadband data, which is usually audio and video that are multiplexed into an MPEG-2 transport stream. The interaction channel provides the two-way transmission of

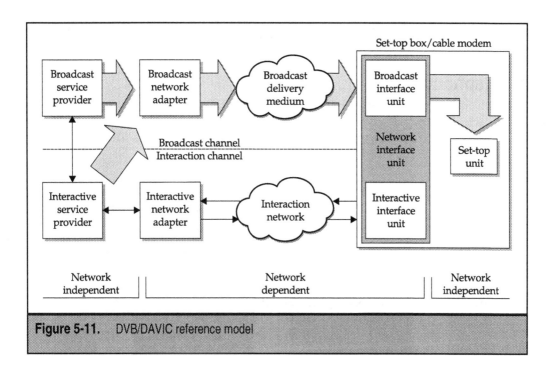

Figure 5-11.   DVB/DAVIC reference model

data. The downstream path of the interaction channel is used for control information, with the upstream being used for data. The specification also addresses two types of network interface units (NIUs): cable modems and set-top boxes (STBs).

DVB/DAVIC makes reference to *in-band* and *out-of-band*. In an in-band system, the downstream channel forms part of the broadcast channel. In an out-of-band system, the broadcast channel and downstream interaction channels are separate and FDM is used to separate them. Data from the subscriber uses the narrowband upstream channel.

The out-of-band dedicated interaction channel enables reliable communication between set-top boxes and the headend for services like VoD or pay-per-view (PPV). The STB is able to receive data at any time independently of the channel selected by the broadcast tuner. The typical use of this channel is to deliver system messages containing informational services on the condition of the system itself, program guides, emergency alerts, plus management-type commands to the STB. If an in-band approach were used (in other words, if the downstream broadcast channel were used), the information would have to be broadcast across all channels.

The out-of-band feature is one of the differences between the DOCSIS specification and DVB/DAVIC. Both standards can be used to communicate bidirectional IP traffic to either set-top boxes or cable modems. But a consensus in Europe is that for out-of-band downstream channels, DVB is the superior choice, since DOCSIS does not define a similar solution. For cable modems and for QAM data channels to set-top boxes, the two standards are essentially equivalent at a technical level. The differences between the two standards can be seen in Table 5-2.

Cable Telephony

The cable industry has invested billions in infrastructure upgrades since 1996. These upgrades, which allow cable companies the ability to offer high-speed Internet access and digital services, also position them to provide regular telephone services. The initial venture

		DVB/DAVIC	MCNS/DOCSIS	IEEE 802.14
Range	Downstream	70–130 MHz	50/54–	88–860 MHz
	Upstream	5–65 MHz	860 MHz 5–30 or 5–42 MHz	5–42 MHz
Modulation	Downstream	Differential	64- and	64- and
	Upstream	QPSK QPSK	256-QAM QPSK and 16-QAM	256-QAM QPSK and 16-QAM
Channel Width	Downstream	IB: DVB-C 7/8 MHz OOB: 1 MHz, 2 MHz	6 MHz	6 or 8 MHz
Data Rates	Downstream	1.544 Mbps, 3.088 Mbps,	30, 42	30, 42
	Upstream	n*8 Kbps 256 Kbps, 1.544 Mbps, 3.088 Mbps	160 Kbps– 10 Mbps	160 Kbps– 20 Mbps
Protocol	Downstream	IB: MPEG2-TS OOB: T1-like	MPEG2-TS	MPEG2-TS
	Upstream	Frame TDMA (minislots)	TDMA (minislots)	TDMA (minislots)
Security		None	RSA or DES	MKE or QKE

IB = In-Band, OOB = Out-of-Band

Table 5-2. Feature Summary of the Standards

into the residential telephone market has been through standard circuit-switched digital technology. Many companies, including AT&T and Cox, and a few others, have been successfully deploying cable telephone services in competition with ILECs since 1998.

The cable telephony market has been growing an average of 200 percent per year from 1998 to 2001 and is expected to experience an average annual growth rate of 65 percent from 2001 to 2005. There are many reasons for the growth of the number of cable telephony subscribers, including consolidated billing, the convenience of a single vendor for cable and telephone services, lower costs, additional features, and functionality.

Cable companies have been aggressive in their pricing of the service. In some parts of the country, cable telephony pricing was as much as 50 percent lower than that for the incumbent local exchange carrier, with deep discounts on additional features like call waiting. It is usually not necessary to purchase additional cable services to subscribe to cable telephone services, but additional savings can be had by bundling services. The aggressive competition of cable operators has resulted in their gaining market share at a very rapid rate.

To offer circuit-switched telephone service over a cable system, cable operators must install a telecommunications switch within the headend and negotiate interconnection terms with the local phone carrier so that cable telephony users can call other users that are on the regular PSTN network. The switch functions in the same way as a PSTN switch—a path is temporarily set up between the caller and the called party for the duration of the call. If the called party is also a cable telephone subscriber on the same cable system, the call never leaves the switch at the headend. If, however, the called party is on the PSTN or another cable provider's cable telephone service, the call is routed through the switch at the headend to the local phone service switch on the PSTN.

AT&T and Cox Communications account for the largest number of cable telephone subscribers. Although reliable telephony equipment for HFC networks is commercially available, many large cable companies have been cautious in deploying the service because of economic and operational barriers to doing so. Given the choice of installing dedicated equipment to offer circuit-switched cable telephony services, or equipment to provide two-way data services, most operators chose the path with the greatest demand—that of data services. It was a hedged bet, in light of uncertainty on where the market for cable telephony would be in a few years, so some operators chose a wait and see approach before making a commitment to cable telephony. If packet-based voice telephony, like Voice over IP (VoIP), becomes pervasive, then the decision to wait pays off handsomely because the investment made in the equipment to deliver data services can be leveraged to provide voice services.

Once the investment has been made in the infrastructure to support two-way data and digital services, it is a lot easier to layer incremental applications. If, however, it is determined that there is a need and a market for circuit-switched telephony, the decision to invest in the equipment to enable the service could be made at that time, on the basis of the supporting market data and projected return on investment. This was the reasoning that led many to take a more cautious approach in providing circuit-switched telephony across their HFC networks.

Deploying a separate voice telephony architecture from that of data is expensive. It requires different equipment and separate downstream and upstream channels, creating spectral inefficiencies. Cost of operations also increases because separate operational and management platforms are required with the appropriate staffing.

IP Telephony

IP telephony is viewed as a much more cost-effective way for cable operators to transport voice. It leverages the existing hardware and the service becomes just another data application. It will, of course, require some thought and investment to assure a level of Quality of Service exceeding that of other types of data applications. Other factors such as call management and billing must also be addressed, but that is viewed as a small price to pay. The investment made to provide QoS, a reliable service, and the other features required for call management and billing can be leveraged for other applications. IP telephony also has the potential to integrate with other data services, which makes it even more attractive to cable operators. Integration enables new value-added services such as integrated voicemail and e-mail without additional expensive equipment.

To provide IP-based telephone services, technical challenges need to be overcome and issues such as call management, signaling, billing, security, QoS, and provisioning must be addressed. The first version of the Data Over Cable Service Interface Specification (DOCSIS) standard was not designed for IP telephony and so did not adequately address a lot of these issues. To complicate matters further, if IP telephony is to span multiple networks, a standard for interconnection is also needed. To address these issues, CableLabs launched an initiative called the PacketCable Project to research and address these issues.

The technical issues of IP telephony are well understood as a result of the work done on VoIP. Circuit-switched telephony uses a connection-oriented approach that creates a point-to-point circuit between the caller and the called party. Creating a circuit between the two parties is important to provide the consistent level of latency that voice requires. One problem with this approach is the inefficient use of the bandwidth of the circuit. During a telephone call, the port that terminates the loop connecting the subscriber's phone to the switch cannot be used for anything other than the call that is in progress or to alert the subscriber that there is another call waiting. Three-way calling is possible, but the point is that no other application can utilize that port or the circuit. During periods of silence, the circuit must remain dedicated to the call in progress, a waste of bandwidth.

A packet-based approach is a lot more efficient in that it allows the link to be shared by multiple applications. The trade-off with the approach is that you sacrifice some control over latency. It is a lot more difficult to control when the link will be available for use, and so a critical packet may be delayed while it waits for the link to become free. This is one of the challenges of VoIP. Using packet-based technology, voice calls are subject to delays, which adversely affect the quality of the voice call. IP telephony requires strict QoS and prioritization to minimize delays.

Another technical hurdle that has to be addressed is the bandwidth requirement for VoIP. Using the Nyquist theorem presented in the preceding chapter, on DSL, we determined that the bandwidth requirement for voice is 64 Kbps per second. This is not a problem in a circuit-switched world, because the entire circuit is dedicated to the call. In a world where voice has to share the connection with data, it would not take much before the voice traffic consumes the resources of the connection.

To address this problem, compression techniques are used to increase the number of voice conversations that can be carried by a packet network at a given data rate. Standards have evolved to address the issues of compression. One such standard is the ITU G.729 compression method that is known as Conjugated Structured Algebraic Code Excited Linear Predictive (CS-ACELP)...now that's a mouthful and a half. This compression technique accurately represents voice pitch as well as amplitude using a digital voice signal that occupies only 8 Kbps of bandwidth. It should also be noted that bandwidth is also used for silence for uncompressed digital voice, and more advanced compression techniques attempt to diminish the silent intervals. Table 5-3 provides a summary of some other standards that are used for signal compression.

To compete with regular circuit-switched telephony, IP telephony must deliver toll-quality calls that can be achieved only if the packets are delivered within a predefined tolerance for delay. To achieve this, greater control over the size of the packets that

VoIP CODECs	Name	Compressed Rate (Kbps)	MOS
G.711 PCM	Pulse Code Modulation	64	4.4
G.726 ADPM	Adaptive Differential Pulse-Code Modulation	40/32/24	4.2
G.729 CS-ACELP	Conjugated Structured Algebraic Code Excited Linear Predictive	8	4.2
G.723 ACELP	Algebraic Code Excited Linear Predictive	5.3	3.5
G.723 MP-MLQ	Multipulse Maximum Likelihood Quantization	6.4/5.3	3.9
G.728 LD-CELP	Low-Delay Code Excited Linear Prediction	16	4.2
MOS—Mean Opinion Score, MSO above 4 is considered Toll Quality by ITU			

Table 5-3. Compression Standards

utilize the links must be enforced. If a packet is too long, it could blow the delay budget for a voice packet. For instance, if the speed of the upstream channel on a cable system is 768 Kbps, and a packet of 1,500 bytes is to be transmitted across the channel, it would take about 15 milliseconds to transfer the packet. That could blow the delay budget for a voice packet if the delay tolerance was set to 9 milliseconds. If the 1,500-byte packet were fragmented into three 500-byte packets, then the longest delay to transmit each packet would be 5 milliseconds, which is well within the tolerance of the defined delay budget for the voice packet. Fragmentation and reassembly of large data packets (not the voice packet) is a technique that is used to gain more control over delay.

An important distinction is needed here, the concept of using IP packets to transport voice across a data network is called Voice over IP (VoIP), which is not the same thing as IP telephony. VoIP addresses the mechanics of transporting voice across a data network instead of the Public Switched Telephone Networks. It does not address the requirement for Quality of Service, and it does not provide additional features like call waiting and Caller ID that would normally be available in normal switched telephone service. IP telephony, in contrast, attempts to improve the quality of transmission to the point that it matches a switched service. It employs techniques like QoS and other features to make the service as close to toll grade as possible, while also addressing issues of interconnection of VoIP to the PSTN network. This is the focus of the PacketCable initiative.

DOCSIS 1.1   In 1999, CableLabs issued two new specifications that allowed for advanced features and security in high-speed cable data modems. The first set of specifications was the second-generation version of the DOCSIS 1.0 specifications called DOCSIS 1.1 (see Figure 5-12). It implemented the features that were necessary for the support of IP telephony, including Quality of Service, fragmentation, enhanced baseline privacy, and some multicasting. The specification formed the basis for PacketCable, a CableLabs-managed project aimed at identifying, qualifying, and supporting Internet-based multimedia products over cable systems.

CableLabs certified the first set of DOCSIS 1.1 modems in October 2001, setting the stage for the delivery of a new suite of applications such as real-time voice, streaming media, and tiered data services. Cable modems from Texas Instruments and Toshiba were the first to be certified, and cable modem termination systems (CMTS) from Cadant and Arris were the first to be qualified. Several other vendors' products failed to make the grade.

The second announced set of specifications was the Baseline Privacy Interface Plus (BPI+). The new security specification incorporates digital certificate–based authentication in cable modems that further enhances the data privacy and service protection offered by the earlier DOCSIS 1.0 specification.

The PacketCable Initiative

The PacketCable Initiative develops interoperable interfaces for delivering real-time multimedia services over a two-way cable plant. The IP protocol and the DOCSIS 1.1 specification form the basis for PacketCable networks. The project began in 1997 when the need for an end-to-end architecture for such functions as signaling for services, media

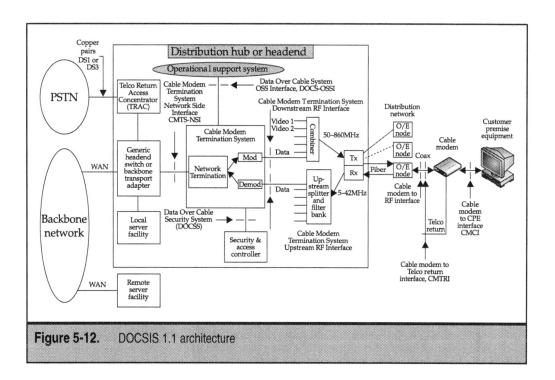

Figure 5-12. DOCSIS 1.1 architecture

transport at variable QoS levels, security, provisioning, billing, and other network management functions was identified.

The basic architecture, shown in Figure 5-13, defines what is known as a "SoftSwitch" architecture for Voice over IP with the core specifications describing how to provide the consolidated functions of a Class 5 central office voice switch on several general-purpose servers. Using a DOCSIS 1.1 IP foundation, the specifications are also designed to meet the needs of other real-time IP-based services such as real-time gaming, video conferencing, and other multimedia services.

The PacketCable specifications were submitted to the SCTE and ITU-T for adoption as standards. The initial set of specifications was approved in February 2001 by, and is available from, the SCTE. The specifications, known as "IPCablecom" within the ITU-T, were approved in March 2001.

The PacketCable Layers   PacketCable is a set of protocols that uses a three-layer model, as shown in Figure 5-14. Layer 1 is the transport layer, the layer where the DOCSIS 1.1 backbone network resides. On top of the DOCSIS data transport foundation are the core services, which PacketCable specifies as QoS signaling, capacity provisioning, security, network management, and billing interfaces. Additionally, on top of the core services platforms, stand the PacketCable applications layer specifications for call-control and custom feature signaling, media streaming codecs, and Public Switched Telephone Network (PSTN) gateways.

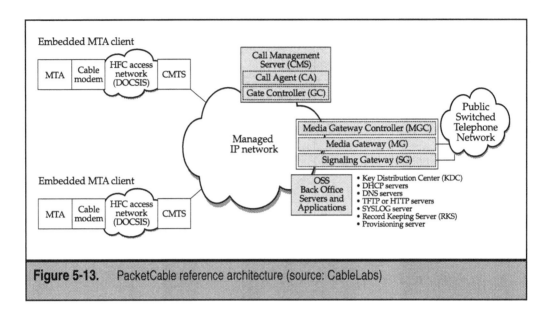

Figure 5-13. PacketCable reference architecture (source: CableLabs)

Finally, PacketCable QoS relies on DOCSIS 1.1, which provides for a dynamic Quality of Service. This means that resource reservations may be requested on a per-flow basis

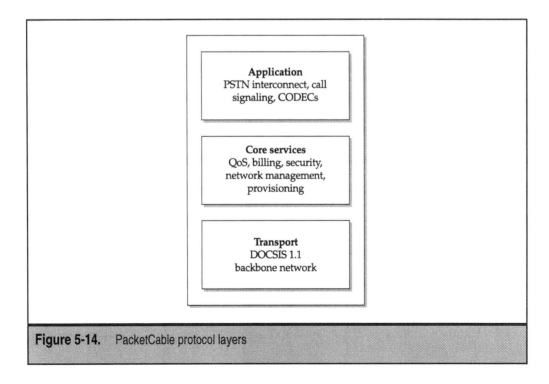

Figure 5-14. PacketCable protocol layers

and not on just a fixed basis. This enables different priority treatment to different traffic types, a key feature where voice is concerned.

Digital Audio-Visual Services over Cable

The U.S. Congress has set December 31, 2006, as a target date for broadcasters to switch from the analog TV system to digital delivery, a move that is supposed to offer consumers more programming choices and higher-quality sound and pictures. This move is being touted as the biggest technological advance in television since the days when color ousted the grainy black-and-white pictures. Between now and then, television will be aired in both analog and digital formats in some markets. After this date—or sometime thereafter, as the date will most likely slip—the spectrum used for analog television will be returned to the government. Cable operators will continue to distribute both analog and digital TV programs until analog television is no longer broadcast over the air.

Before the upgrade, the frequency band used for broadcast television was about 400 MHz. After the upgrade to a hybrid fiber-coax system, the available frequency range about doubled. The extra bandwidth above the original 400 MHz is being used for digital services. The math, however, does not appear to add up. If the original frequency band was about 400 MHz, which yielded about 55 channels, it follows that if the band is doubled, the number of channels should also double, to about 110 channels. But the last time I checked my digital program guide, it appeared there were well in excess of 300 channels. That is almost 500 percent more than what should have been available as a result of the upgrade. Where then did the excess capacity come from?

The answer lies in a family of protocols that we have made numerous references to: both in the DOCSIS specification (IEEE 802.14, DVB/DAVIC) and also in reference to DSL technologies. The protocol in MPEG (short for *Moving Picture Experts Group*) is so fundamental to the delivery of digital broadband services that it deserves a closer look.

MPEG-2 as a Broadband Enabler

MPEG refers to a suite of protocols developed by the Moving Picture Experts Group that were developed for audio and video encoding and compression for multimedia delivery. As is obvious from our previous references to various features of the MPEG protocols, its capabilities are not limited to audio and video but have found wider application as an enabler of broadband services. MPEG-1 was designed for coding progressive video at a transmission rate of about 1.5 Mbps and is the enabler of products and technologies such as Video-CD, CD-ROM, and MP3.

MPEG-2 is an international standard that is defined in ISO/IEC IS 13818-2. It is the protocol used for DVDs and is cable of handling SDTV, HDTV, and motion picture films. Claims of greater than 50:1 compression ratios have been made by various sources; the ratio, however, should be treated with caution, as the level of compression is dependent on the complexity and the type of data that is being compressed. Originally, an MPEG-3 standard was planned to specifically address HDTV requirements, but it was eventually abandoned and the focus shifted when it became obvious that the MPEG-2 standard was beefy enough for the job.

MPEG-2 Compression   High-quality compression is critical in the storage and delivery of audio and video content. Without it, the uncompressed size of the digitized content would overwhelm the network resources and blow storage budgets. A 5 MHz bandwidth analog TV picture after digitization is equal to a 270 Mbps digital stream, which requires a bandwidth of 140 MHz to transport. With numbers like this, the need for compression becomes evident. Video requires fast compression techniques to be able to capture and compress and decompress at 30 fps. As a result, most video compression techniques work better in hardware because software is a lot slower. The good news for cable operators is that it takes a lot more processing power to compress than to decompress, which translates into much more affordable decoders for equipment like set-top boxes.

Compression is the reduction of bit rates in a digital signal or the reduction of bandwidth in an analog signal while preserving as much data as possible. Compression also confronts another factor: it must meet the limitations of the application to which it is applied. For example, compression in text must allow the reconstruction of the original text in its entirety because the limitation of text is that it is of no use if some of the information is lost. This type of compression is referred to as *lossless* compression because the final version is an exact replica of the original, with no information loss. It achieves compression by removing redundant information that is regenerated during decompression.

In another application, such as audio or video, the output does not have to be an exact replica of the source for it to look like, sound like, and be recognized as the source, so some amount loss can be tolerated. Compression techniques that lose information during compression are called *lossy* compression. The output is not an exact replica but a lower-quality version of the original. Compression is achieved by removing noncritical information (the remaining information may also have lossless compression techniques applied to remove any redundant information). MPEG-2 compression is lossy and is based on two principles.

The first principle, called *spatial redundancy*, holds that human vision is less discerning of higher spatial frequencies; in other words, certain fine details cannot be discerned by the eye. Spatial compression compacts the frame by looking for patterns and the repetition of those patterns among pixels in the same frame. This technique is also called *intraframe coding*, a term that we will return to later, so take note. Also, remember that intraframe-coded frames have complete information about the original frame, which is used to reconstruct the frame for display at a later time. Spatial compression does not compress as much as other types of compression techniques. MPEG-2 uses *discrete cosine transform (DCT)*, quantization, and Huffman coding to derive the final compressed code for the frame.

The second principle, known as *temporal redundancy*, makes use of the fact that in successive frames of video, relatively few pixels change position from one frame to the next. Using this as a basis, if a frame has already been coded for transmission, only the information about the pixels that have changed in subsequent frames need be transmitted, so just the motion vectors for the changed pixels, not the complete later frames, are transmitted (see Figure 5-15). If a subsequent frame did not change at all, then even less data needs to be transmitted. The transmitted information in this case is a message stating that this frame is the same as the preceding one. The only time the intraframe process is repeated is when there is a new frame. This type of compression, called *temporal compression* or *interframe* coding, is a technique that achieves the greatest compression ratio.

A Word on Motion Prediction   Now that the stage is set, we will now explore the magic of MPEG-2. In order to exploit the fact that pictures often change little from one frame to the next, MPEG employs temporal prediction modes; that is, it attempts to construct frames using information from a previously constructed frame. The frame that is created using this forward predication technique is called a *P-frame* (predicted frame). P-frames are built from *I-frames* (intraframes) by using motion prediction. As was discussed earlier, intraframes are spatially compressed frames that contain complete information about the original frame: this makes them good reference points because they embody the compressed content of a source frame. To take one more step, if a P-frame is built from an I-frame—which we can have some level of confidence in—it should be safe to build a P-frame from a P-frame that was derived from an I-frame. The rule, then, is: a P-frame is built from either an I-frame or a P-frame using motion prediction.

Now the fun begins. We now have I-frames and P-frames. If we assume an I-frame and a P-frame are a set of beginning and end frames, we should be able to apply some magic and interpolate frames that fit between the I-frame and the P-frame by taking information from both. By doing this, we are smoothing the motion transition from I-frame to P-frame through the use of frames that are interpolated from the data of each. These interpolated frames are called B-frames, or bidirectional frames. They are bidirectional because they are generated from the information from the frames before and after them.

The order of compression levels obtained is this: I-frames are the least compressed, P-frames are about 66 percent of the size of the corresponding I-frame, and B-frames achieve about an 83 percent reduction of the I-frame (about 50 percent of the P-frame). The different picture types, I-, P-, and B-frames, typically occur in a repeating sequence called a *group of pictures*, or *GOP*. The GOP begins with an I-frame and then has P-frames

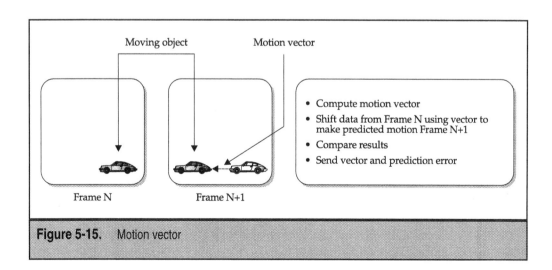

Figure 5-15.   Motion vector

spaced throughout. The remaining pictures are B-frames that were interpolated from the I- and P-frames. The end of a GOP is defined as the last frame before the next I-frame. Group lengths of 12 and 15 are common.

An obvious point is that for a B-frame to be created, both I- and P-frames must already exist. Consequently, bidirectional coding (B-frames) requires I- and P-frames to be stored, with the resulting GOP being assembled in an order that has nothing to do with the way the frames were received. The other consequence of this arrangement is that B-frames introduce delay because of buffering at the receiver.

Profiles and Levels   MPEG is applicable to a wide range of applications from simple to very complex. The many encoding tools available allow it to be adapted to the level of complexity needed to get the job done. The many tools, though, make it a bit unwieldy because of the thousands of options possible if all the different combinations of its features were used.

To make life simpler, the MPEG-2 standard is divided into profiles, and each profile divided into levels, as shown in Figure 5-16. A *profile* is basically a subset of the entire range of combined possibilities for a particular level of complexity. A *level* is a set of constraints such as maximum sampling density on parameters within the profile. There are 24 combinations but not all are defined. The basic rule is a decoder capable of handling a certain profile must be able to decode lower profiles.

▼ **Simple profile**   This profile does not support B-frames; hence, no picture reordering or buffering is necessary and this profile is suitable for low-delay applications. It has been defined only at the main level (simple profile at main level SP@ML).

■ **Main profile**   This adds the support of B-frames, improving picture quality at a given bit rate. Most MPEG-2 video decoders support the main profile,which is designed for a wide range of uses. Main profile at main level (MP@ML) is required by most broadcast applications, including SDTV.

■ **SNR profile**   This adds support for enhanced DCT coefficient refinement using signal-to-noise ratio scalability.

■ **Spatial profile**   This adds the support enhancement layers using the spatial scalability tool.

▲ **High profile**   This adds support for 4:2:2 sampled video.

MPEG-2 System Layer   In many ways, the magic of MPEG-2 is also its weakness. If an I-frame is corrupted, then the subsequent P- and B-frames are affected, an error that quickly propagates. The process also discards a lot of data that could be used in the event of an error. The decoder also needs management information such as the channel assignment of the stream and other system management messages. To facilitate this, a system layer is also defined.

MPEG-2 is a standard that consists of nine different parts. Part 1 defines the way in which one or more elementary streams (ESs) of video, audio, or data can be combined

	Profile and maximum total bit-rate (Mbps)				
	Maximum sampling density (Hor/Vert/Freq)	Simple Profile (SP)	Main Profile (MP)	SNR profile (scalable)	Spatial profile (scalable)	High Profile (HP)
High level (HL) (1920/1152/60)	—	MP@HL 80 Mbit/s	—	—	MP@HL 100 Mbit/s + lower layers
High–1440 (1440/1152/60)	—	MP@H–14 60 Mbit/s	—	MP@H–14 60 Mbit/s + lower layers	MP@H–14 80 Mbit/s + lower layers
Level	Main level (ML) (720/576/30)	SP@ML 15 Mbit/s	MP@ML 15 Mbit/s	SNR@ML 15 Mbit/s + lower layers	—	MP@ML 20 Mbit/s + lower layers
Low level (LL) (352/280/30)	—	MP@LL 4 Mbit/s	SNR@LL 4 Mbit/s	—	—
ISO 11172 (MPEG-1) 1.856 Mbps	—	—	—	—	—

Figure 5-16.   Profiles and levels

into single or multiple streams for transmission or storage. This is accomplished through the use of two data stream formats: the *program stream (PS)*, intended for storage and retrieval from storage media (the DVD standard uses this format); and the *transport stream (TS)*, the format most referenced in other broadband applications, combines multiple streams into a single stream.

We are more concerned with the transport stream, also termed *MPEG-2 TS*, as it is the mode used for data over cable standards, satellite, and ATM, among other transport systems. The TS uses fixed-length packets and is intended for non–error free environments. Each ES is input to an MPEG-2 processor, which accumulates the data into a stream of *packetized elementary stream (PES)* packets (see Figure 5-17). Each PES is then multiplexed into the constantly moving transport stream. If there is nothing to be sent, null packets are multiplexed into the stream.

The transport stream consists of a sequence of fixed-sized transport packets of 188 bytes each (shown in Figure 5-18). At the start of each packet is a Package Identifier (PID) that tells the receiver what to do with the packet. The receiver has to decide which packets are intended for the program that is currently being watched. It identifies these packets by the PID numbers. The receiver drops any packet in the stream that is destined for other programs. There are typically four types of PID: an audio PID, a video PID, a *Service Information (SI)* PID used for program information, and finally, a *Program Clock Reference (PCR)* PID that is used to synchronize video and audio packets. Clock references are usually embedded in the video stream so PCR PID are not as prevalent.

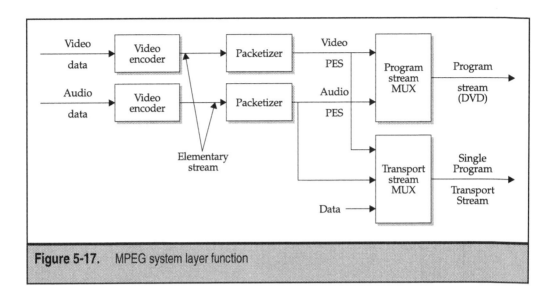

Figure 5-17.   MPEG system layer function

The Service Information packets—SI PIDs—tell the receiver about such things as the format of the transmission, program guide information, and multiple language selection. The

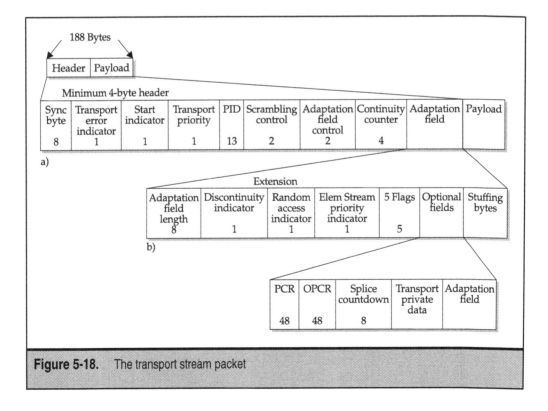

Figure 5-18.   The transport stream packet

receiver uses information from the SI to update its electronic program guide. The first part of the SI is the *Program Association Table (PAT)*, which is always transmitted with PID = 0. The PAT contains a list of *Program Map Tables (PMTs)*. The PMT has information about each program and the types of streams and their PIDs that are associated with each program.

We will try to summarize the flow. At a minimum, a single television program has one (though it may have many more) video and audio streams. Each of these streams is multiplexed into the transport stream (see Figure 5-19) and is identified with a unique identifier—the PID—before it is sent to the receiver. Each packet is associated with a specific program through a Program Map Table (PMT), which is simply a list of all the PID numbers of the different streams for a specific program and their corresponding program numbers. Finally, a Program Association Table (PAT) is used to keep track of each program number and the corresponding program.

The User's View Let's assume that a movie is being transmitted called *Broadband Tales*. This movie is being transmitted on channel 96 and has an audio stream and a video stream. The streams have been identified as audio = PID15 and video = PID23. The video and audio packets are multiplexed into the transport stream and become the payload of TS packets.

The subscriber decides to find out what's on television. She calls up her electronic digital program guide as provided by her new digital set-top box. The guide that was compiled from the information received from Service Information packet shows that *Broadband Tales* is showing on channel 96, so the subscriber switches to channel 96 (in digital TV, this is really a program number). At this point, an analog television set would have tuned in to a specific subband frequency for channel 96, but this is a digital system. Instead, the set-top box consults the PMT to see what PIDs are associated with program 96. It determines that PID23 is the video stream, and PID15, the audio stream. The set-top box picks up the streams identified by PID15 and PID23 and decodes them for presentation on the attached TV.

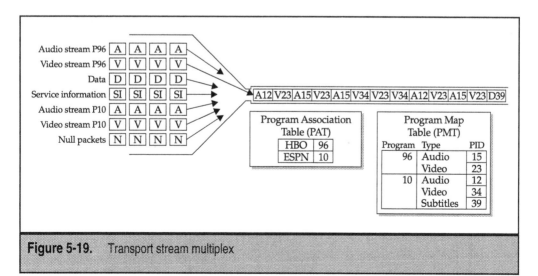

Figure 5-19.   Transport stream multiplex

At midnight, the PAT must be refreshed with a fresh list of movies. An SI packet with PID0 is sent with an updated Program Association Table.

Another user perspective on the MPEG-2 TS is that of the user who is accessing the Internet. In this application, the flows are exactly the same as for audio and video. The data packets from the Internet along with other video and audio packets are multiplexed into the transport stream and become the payload of the TS packet. A special PID value of 0x1FFE is used to identify MCNS/DOCSIS packets. The cable modem at the subscriber recognizes the packet and presents it to the user across an Ethernet connection.

MPEG-2 is a reliable set of protocols that has worldwide appeal. It is likely to remain a dominant standard for entertainment video because it has a large installed base of decoders. The standard places no restrictions on encoders, giving each vendor a degree of flexibility. The standard fully defines the functions of the decoder at different levels and profiles, so the encoder developer can be as creative as he wants as long as the encoder produces an MPEG-2-compliant bit stream.

In the data over cable market, MPEG-2 is also firmly established. Of the three standards available today for cable modems, MPEG-2 is the common choice for downstream transport. It is positioned to be a key enabler of broadband services.

SUMMARY

Cable companies are using their upgraded broadband networks to offer a wide array of new digital services in order to meet the competition posed by DBS, wireless cable, broadcasters, and telephone companies.

As of August 2001, an estimated 12.2 million homes have subscribed to digital cable, a service that offers extended channel offerings, CD-quality, and commercial-free music. The number of digital cable customers is expected to increase to 48.6 million homes by 2006.

The industry is now also expanding its competitive offerings to include business and residential telephone services delivered over its fiber optic infrastructure. At the close of 2001, at least nine of the nation's largest multiple system operators (MSOs) offer residential and/or commercial phone service in more than 45 markets, serving more than one million customers.

Once upgraded to a two-way HFC network, cable operators are able to offer converged services that include telephone service, digital audio and video, high speed Internet, and other advanced services to consumers. With the new fiber based network, the dream of a single connection delivering multiple services is a lot closer to reality than it was when in 1996 when the telecommunications act was passed. The new infrastructure has also enabled cable operators to explore new markets and as of November 2001, over 6.4 million subscribers have subscribed to high-speed data services.

History has shown that the cable industry has always been a tenacious group of operators. They have thrived in the face of competition and at times, regulatory challenges. In the broadband market, they lead DSL in the number of subscribers and are projected to maintain that lead into the foreseeable future.

CHAPTER 6

Wireless Access Networks

T he growth of the Internet to date is probably just a prelude to another growth phase that will be based on untethered access. The exact implications of this are yet to be determined, but it is a trend that, if it proves to be true, will impact telcos, cable operators, and other wired service providers.

The challenges associated with using wire as a transmission medium are well understood. Many of the characteristics of the medium that make it particularly challenging have been covered in previous chapters. However, there is another aspect of wired networks that we have not yet considered, one that directly impacts their reach and the bottom line of the service. A wired network's requirement of a physical connection to the customer's equipment is assumed, but this simple requirement carries with it a price tag that must be factored before any service can be offered or any decision to expand the network made.

Wireless networks have no such restriction, which makes them a natural choice for expanding the network into areas that are too hard to reach by wire. Many service providers have taken an interest in wireless technology because of the flexibility that it affords. It has even given birth to a new type of provider called a *wireless Internet service provider,* or *WISP.* What better way to enter into the service provider market than to offer a service that bypasses the traditional wired infrastructure? By taking this approach, WISPs minimize the capital outlay required for service start-up and positions providers to realize a faster return on investment with a smaller number of customers. Going the wired route requires a larger customer base and a longer lead time before a return on investment is realized.

This chapter examines the wireless technology, its appeal, and the different solutions that are available and emerging.

CHARACTERISTICS OF WIRELESS ACCESS NETWORKS

Broadband wireless is not the same as mobile wireless. Mobile wireless, like cell phones, allows users the service mobility while using the service. Broadband wireless is communications delivered from a ground antenna or satellite to buildings or fixed sites without a wired connection at high data rates.

The appeal of wireless networks lies in their reach without a dependence on a wired infrastructure. They enable the provider to deliver high-speed broadband solutions to places that lack a wired infrastructure, as well as those that have one. In our review of DSL and cable networks, we stated that a feature of these services that is heavily marketed is the fact that they are "always on." In reality, wireless is the best-positioned technology to stake a claim to that description. Unlike wired solutions, which require the subscriber to be physically connected to the network in order to be "always on," wireless networks are always on and the subscriber is always connected.

The ability to provide services over the air affords a competitive advantage in the speed to deliver, and the pricing of, services. This untethered connection has a broad appeal to both the consumer and the enterprise. It enables the delivery of business, personal, social, and entertainment services to areas that would not normally have been able to receive such services.

In business, the benefits of a ubiquitous network connection lie in the immediate access it provides to corporate networks outside of the office walls. The untethered feature of wireless extends the enterprise and aids productivity. Decisions can be made more quickly because information is more available and the lines of communication are always open.

Broadband wireless falls within two categories: fixed wireless and satellite. With a fixed wireless system, signals are broadcast from a fixed ground antenna to antennas mounted on homes and buildings. They make use of microwave signals with speeds up to 1.5 Gbps. Examples of these types of systems include the Multichannel Multipoint Distribution System (MMDS) and the Local Multipoint Distribution System (LMDS), both of which will be covered later in this chapter. Satellite solutions employ—as the name implies—satellites that broadcast their signals over wide areas to dishes mounted on buildings.

Spectrum Management

The expansion of the information economy to our airwaves is creating an unprecedented demand on the available radio spectrum, which is fast becoming a most valuable asset. The airwaves at any given moment are a superhighway of multiple wireless transmissions. The problem that this presents is that the radio spectrum is a limited resource and has to be treated as such.

If a company wishes to expand a wired infrastructure, it is just a matter of obtaining the appropriate rights of way and then buying and laying more wires. The bandwidth within the sealed environment of a fiber, copper, or coaxial cable is privately owned. The bandwidth available across the airwaves is not. This valuable real estate, which occupies between 300 kHz and 300 GHz, must be shared by commercial users, the military, emergency services, hospitals, civil aviation, and federal and local government, among many other entities. Spectrum management is important to ensure efficient and fair use of the airwaves.

In the U.S., the Federal Communications Commission (FCC) is responsible for spectrum management policies. The National Telecommunications and Information Administration (NTIA)—an agency within the U.S. Department of Commerce—provides the same function for federal government users. It is the responsibility of the FCC to regulate the spectrum in a manner that is in accordance with public interest, convenience, or necessity, and it is their responsibility to determine what wireless services go where.

Every country has different spectrum requirements and different processes to determine effective management of the spectrum. No single policy for effective spectrum management can be applied globally to all countries. Basic principles, however, have broad applicability for effective spectrum management:

▼   Maximize the efficient use of the radio spectrum.

■   Ensure that the spectrum is made available for new technologies and services, and that flexibility is preserved to adapt to new market needs.

- Develop a fair, efficient, and transparent process for awarding licenses.
- Make allocation and licensing assignments consistent with marketplace demands.
- Promote competition.
▲ Ensure that the spectrum is available for important public benefits such as safety and health.

The demands of new wireless technology has made the job of spectrum management a great deal more challenging than it has ever been. It is becoming increasingly difficult to forecast what new service will be launched in the future, much less to determine which frequency range will be efficient for a given service.

The pressures of demand have resulted in the FCC taking a more flexible, market-based approach to spectrum allocation instead of regulatory planning as it has always done. The new approach is more reliant on private-sector petitions to the FCC for specific spectrum requirements, and each request is considered in the context of its impact on existing users.

Frequency Allocation and Allotment

The management of the frequency spectrum is accomplished by dividing the entire radio spectrum into blocks (or bands) of frequencies. These blocks are then each allocated for a particular type of service. Each allocation can be further subdivided into bands, or allotments, that are designated for a particular service, and each allotment may be further divided into multiple channels.

As an example, the allocation for the land mobile service is divided into allotments for business users, public safety, and cellular users, each having a portion of the band in which to operate. Frequency assignment refers to the granting of the right to operate a radio transmitter on a certain channel, or set of channels, at a particular location under specific conditions.

Unlicensed Spectra   Not all spectra are licensed. A licensed spectrum requires an operator to be granted a license to operate within a certain band at a particular location under specific conditions and for a set duration of time. Anyone who wishes to transmit without a license can use the unlicensed band. This, however, is not a free-for-all, and operators are required to follow rules to limit interference—such as by using lower power levels of transmitters—while operating within the bands.

The 5 GHz unlicensed spectrum consists of four frequency bands, including two that overlap. The bands include:

▼ **The U-NII bands**   5150–5250 MHz, 5250–5350 MHz, 5725–5825 MHz
▲ **The ISM band**   5725–5850 MHz

The acronym U-NII refers to the *Unlicensed National Information Infrastructure*; the U-NII bands were set aside by the FCC in 1996; the ISM band, whose name is short for *industrial, scientific, and medical,* preceded the U-NII bands.

Lower emission limits were set within the first two U-NII bands to limit the amount of interference between users instead of requiring the use of spread-spectrum technology (see Table 6-1). By lowering emissions to limit interference, it was hoped that less expensive equipment would encourage manufactures to develop more cost-effective broadband communications equipment to operate within this space. The rules are different for the third U-NII band because it is defined within the existing ISM band, which makes it subject to more powerful ISM systems.

The three U-NII bands all have different power restrictions that are related to their applications:

▼ **Band 1—5.15 MHz–5.25 MHz**   Used for indoor applications such as wireless LANs

■ **Band 2—5.25–5.35**   Used for short outdoor links such as campus applications

▲ **Band 3—5.725–5.825**   Used for long outdoor links such as fixed broadband wireless access

The benefit of the unlicensed spectrum to a service provider lies in its low start-up costs. The bandwidth is free, and the cost of the equipment is typically lower than for operation in the bands that allow higher power levels and spread-spectrum technology.

Many vendors, among them Cisco Systems, are developing equipment for operation within this space, and new broadband services are being launched that leverage this equipment. The low start-up cost also means more competition, which creates competitive pricing pressures. Competitive pricing means lower prices for the customer, but the increased competition also creates the possibility of more traffic and congestion. This is the trade-off with using an unlicensed band.

Resolving Competitive Interests   One of the more challenging aspects of spectrum management is conflict of interest resolution. Because radio spectrum is a limited resource,

	5.15–5.25 GHz	5.25–5.35 GHz	5.470–5.725 GHz	5.725–5.825 GHz
U.S.	50 mW max 200 mW (EIRP)	250 mW max 1 W (EIRP)	—	1 W max 4 W or 200 W (EIRP)
Europe	200 mW(EIRP)	200 mW(EIRP)	1 W (EIRP)	25 mW (EIRP)
Japan	200 mW(EIRP)	—	—	—

Max: Peak power delivered to the antenna
EIRP: Equivalent isotopic power radiated; power transmitted by a directional antenna in its strongest direction

Table 6-1.   U-NII Bands Have Different Power Levels

there will almost certainly be times when there will be competition for a specific spectrum. It is the responsibility of the FCC to decide the winner and the loser. The underlying principle used in making this decision is to determine which use of the spectrum will best serve the public's interest. The method used to determine the winner has changed, but traditionally the factors listed in Table 6-2 have been used.

Criteria	Explanation
Public need and benefit	•Dependence of the service on radio rather than wirelines or fiber •Market demand for the service •Relative social and economic importance of the service, including safety-of-life and protection-of-property factors •Probability of establishment of the service and the degree of public support that is expected for the service •Impact of the new service on existing investment in the proposed frequency band
Technical considerations	•Necessity for the service to use particular portions of the spectrum, including propagation characteristics and compatibility with services within and outside the selected frequency band •Amount of spectrum required •Signal strength required for reliable service •Relative amount of radio and other electrical interference likely to be encountered •Viability of the technology
Equipment limitations	•Upper practical limits of the useful radio frequency spectrum and, in general, what higher limit can be expected in the future due to technological advance •Operating characteristics of transmitters, including practical limitations (that is, size, cost, and technical characteristics) •Receivers available and/or being developed, including their selectivity and practical usefulness for the intended service

Table 6-2.   Traditional Approach to Resolving Competing Interests

Market-Based Licensing

Today, in cases where more than one party applies for exclusive use of the same spectrum in the same geographic area, auctions are being used to decide who is granted the license. In the past, the FCC relied on lotteries and comparative hearings to make these decisions, but both these approaches have since been superseded by a more dynamic market-based policy that auctions available spectrum to the highest bidder. Comparative hearings, which are basically a process by which the qualifications of each competing applicant are examined, were found to be too time-consuming and resource intensive. Lotteries also had their drawback, as it was found that they encouraged speculative interest from parties whose only intention was to acquire licenses for resale.

Auctions are viewed as a superior approach because they assign licenses quickly to those that value them most, and they avoid the appearance of the government's making biased decisions. Furthermore, the public benefits a lot faster from the auctioned spectrum because the winning party has a vested interest in getting the most from their investment as quickly as possible. Finally, auctions generate revenue, whereas comparative hearings did not; the total revenue generated for the U.S. Treasury by 32 auctions as of February 2001 was $32 billion.

Four types of licenses are exempt from the auction process: public safety radio; digital television for incumbent broadcasters; noncommercial educational and public broadcast; and international satellite. The FCC is also directed to ensure that small businesses, minorities, women, and rural telephone companies have an opportunity to participate in the provision of spectrum-based services.

International Allocations

Obviously, the FCC and the NTIA cannot make frequency allocation decisions for the radio spectrum outside of the U.S. The spectrum, however, knows no geographical or political boundaries. The radiocommunications conferences of the ITU are the principal organization for international spectrum allocation. Their role becomes increasingly more important as new global services are launched. Their table of frequency allocations (outlined in Table 6-3) reflects a global consensus as to the requirements of their member states. The ITU also publishes the international radio regulations, which include the allotments and the rules for radio operation around the world.

The ITU further designates each allocation as either primary or secondary. A *primary* allocation receives priority in using the allocated spectrum. If there are multiple primary services within a band, they each have equal rights. An incumbent service has the right to be protected from any other services that may be launched at a later date. *Secondary* allocations—as the name implies—have a lesser priority than primary allocations. They are not allowed to create interference for, but may be subject to interference from, a primary service. Multiple secondary operators have equal rights.

Bilateral Agreements   Neighboring countries run the greatest risk of a clash in spectrum management policies. Bilateral agreements between neighbors are usually sought to coordinate use and resolve conflicts. The U.S. and Canada, for instance, have many bilateral agreements regarding spectrum use between them. These agreements help resolve issues like interfering television stations between the two countries especially in cities that are close to the border of the two countries.

In this regard, the ITU also serves as an intermediary for multilateral negotiations between member states. Countries typically submit spectrum proposals to the ITU for discussion at regional World Radio Conference preparatory meetings, where many administrations may be present. These proposals are also pursued outside of these meeting through direct negotiations between the countries. Table 6-3 displays the wireless spectrum allocation for current and planned major services.

Service	Bands	Description and Comments
Broadcast radio	535–1,710 kHz	AM band.
	88–108 MHz	FM band.
Broadcast TV	54–72 MHz	Channels 2–4.
	76–88 MHz	Channels 5–6.
	174–216 MHz	Channels 7–13.
	470–746 MHz	Channels 14–59.
	746–806 MHz	Channels 60–69.
Digital TV	54–88 MHz	Broadcasters have started transmitting digital signals in some markets. All broadcasters are expected to switch to digital TV by 2006, when the analog spectrum will be reauctioned.
	174–216 MHz	
	470–806 MHz	
Cellular phone service	806–902 MHz	
3G Wireless—proposed	1,710–1,770 MHz	
	2,110–2,170 MHz	
Personal Communications Service (PCS)	1,850–1,990 MHz	2G digital cellular service.
Wireless Communications Service (WCS)	2,305–2,320 MHz	Intended for wireless data services.
	2,345–2,360 MHz	
Satellite-delivered digital radio	2,320–2,325 MHz	XM Satellite and Sirius Satellite Radio are the only two licensed providers.

Table 6-3. Wireless Spectrum Allocation for Current and Planned Major Services

Service	Bands	Description and Comments
Multichannel Multipoint Distribution Service (MMDS)	2,150–2,680 MHz	Originally used for "wireless cable" service.
Unlicensed National Information Infrastructure (U-NII)	5.15–5.35 GHz 5.725–5.825 GHz	Reserved by the FCC for unlicensed use.
Direct Broadcast Satellite (DBS)	12.2–12.7 GHz	Major competitors to cable. DirecTV and EchoStar are dominant players.
Teledesic	18.8–19.3 GHz 28.6–29.1 GHz	Two-way digital satellite service scheduled for deployment by 2005. Marketed as "Internet-in-the-Sky."
Digital Electronic Message Service (DEMS)	24.25–24.45 GHz 25.05–25.25 GHz	High-capacity data service but limited distance. Teligent owns most of the licenses.
Local Multipoint Distribution System (LMDS)	27.5–29.5 GHz 31.0–31.3 GHz	Fixed wireless broadband access.
39 GHz fixed wireless service	38.6 GHz to 40 GHz	Last-mile communications; used by Winstar and Advanced Radio Telecom to extend their fiber networks.

Table 6-3.   Wireless Spectrum Allocation for Current and Planned Major Services *(continued)*

Fundamentals of Wireless Technology

When a radio frequency (RF) current is applied to an antenna, an electromagnetic field is generated and propagated through the air. These frequencies make up a part of the *electromagnetic radiation spectrum* (see Figure 6-1) called the *radio spectrum*. The U.S. government, for purposes of convenience, has given labels to different ranges within the spectrum, as shown in Table 6-4. The actual labels have no scientific or engineering significance other than a good way to classify different ranges within the spectrum.

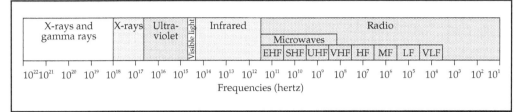

Figure 6-1. The electromagnetic spectrum

Description	Abr.	Frequency Range	Free Space Wavelength	Some Common Applications
Extremely Low Frequency	ELF	3 Hz to 30 Hz	100,000 km to 10,000 km	Detection of buried metal objects
Super Low Frequency	SLF	30 Hz to 300 Hz	10,000 km to 1,000 km	Electric power, submarine communication
Ultra-low Frequency	ULF	300 Hz to 3 kHz	1,000 km to 100 km	Telephone audio
Very Low Frequency	VLF	3 kHz to 30 kHz	100 km to 10 km	Navigation and position location
Low Frequency	LF	30 kHz to 300 kHz	10 km to 1 km	Radio beacons, weather broadcast stations for air navigation
Medium Frequency	MF	300 kHz to 3 MHz	1 km to 100 m	AM broadcasting
High Frequency	HF	3 MHz to 30 MHz	100 m to 10 m	Short-wave broadcasting
Very High Frequency	VHF	30 MHz to 300 MHz	10 m to 1 m	TV and FM broadcasting, air traffic control, mobile radio
Ultra-high Frequency	UHF	300 MHz to 3 GHz	1 m to 100 mm	TV broadcasting, radar, cell phones, mobile radio
Super High Frequency	SHF	3 GHz to 30 GHz	100 mm to 10 mm	Satellite, aircraft navigation, radar
Extremely High Frequency	EHF	30 GHz to 300 GHz	10 mm to 1 mm	Remote sensing, radio astronomy

Table 6-4. The Radio Spectrum and Some Common Applications

RF Propagation

RF transmissions are sometimes called *unguided transmissions* because the signal flows through the air instead of being guided, as is the case with wire transmissions. The propagation of these signals through the air falls within three different classifications:

▼   Ground wave

■   Ionospheric

▲   Line of sight (LOS)

Ground wave propagation follows the curvature of the earth, as shown in Figure 6-2, and utilizes carrier frequencies up to 2 MHz. Radio waves in the VLF (3 kHz to 30 kHz) band propagate in a ground or surface wave. These transmissions are good for very long ranges and are good for strategic communications. An example of ground wave propagation is AM radio broadcasts.

Ionospheric propagation is sometimes called sky waves or double-hop propagation because it utilizes the ionospheric layer of the atmosphere to refract the signal back to earth (see Figure 6-3). Radio waves in the HF (3 MHz to 30 MHz) band propagate in sky waves and are also good for long ranges. Higher frequencies suffer less from refraction and attenuation. The minimum frequency that will penetrate the ionosphere at a vertical incidence is called the *critical frequency,* so the objective with sky waves is to find the highest frequency that will be refracted (instead of penetrating) to determine the *maximum usable frequency (MUF).* The higher the refracting surface, the greater the range and the higher the critical frequency, which increases the MUF. During the day, the ionosphere contain more ions, which increase attenuation, so high-frequency communication is dependent on time of day and the weather patterns. Communications are usually better at night. An example of this type of propagation is Ham radios.

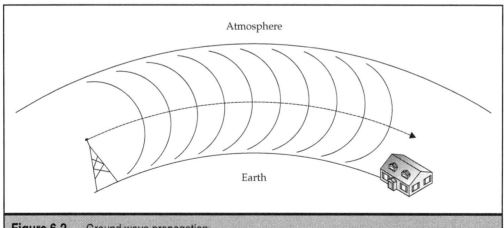

Figure 6-2.   Ground wave propagation

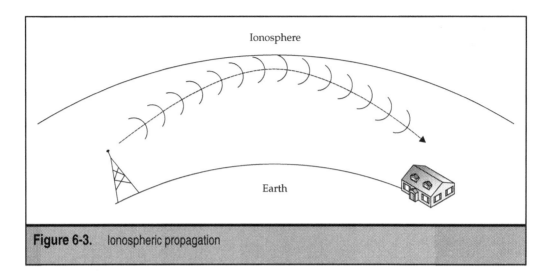

Figure 6-3. Ionospheric propagation

Line-of-sight (LOS) propagation, also called space waves, transmits the signal in a straight line directly to the receiver, as shown in Figure 6-4. Radio waves in the VHF/UHF (30 MHz to 30 GHz) bands propagate in space waves and travel shorter distances. The receiving station must therefore be in the line of sight of the transmitter. The curvature of the earth becomes a factor when designing these types of transmission systems. The height of the antenna and the distance of the receiver must be considered in order to compensate for the earth's curvature. Examples of LOS propagation are microwave transmissions.

Compensating for Fresnel Zone and the Earth's Curvature   Light and radio are forms of electromagnetic radiation operating at different frequencies, but visual and radio line of sight

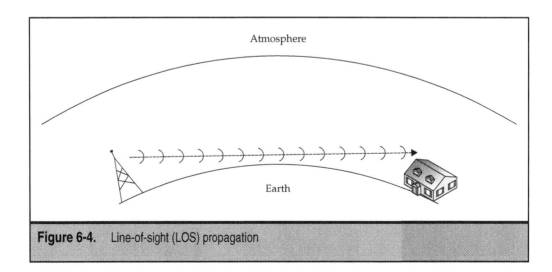

Figure 6-4. Line-of-sight (LOS) propagation

are two different things. Being able to see a distant building with a pair of binoculars does not necessarily mean that a wireless connection can be established.

Augustin Fresnel was a nineteenth-century French physicist who observed a curious thing. Light passing near a solid object is subject to bending, or *refraction*. This refraction of the light affects the signal strength of the original light beam depending on how close the object is to the beam. The effect is called the *Fresnel effect*, and both radio and light are subject to it. If a radio signal passes near an object like a mountain or a building, the object can affect the strength of the signal even though it does not directly obscure the visual line of sight. Refraction of the signal by nearby objects can create problems for a wireless connection. The area around the line of sight of the signal that must be kept clear of objects is called the Fresnel zone (see Figure 6-5). A tool for calculating the Fresnel zone and maximum range can be found at http://www.cisco.com/go/aironet/calculation. In most instances, raising the antenna above the obstruction is enough to compensate for the problems caused by the obstruction.

The curvature of the earth is also something that must be considered when designing a fixed wireless network. The curvature of the earth begins to come into play as the distance between the antennas begins to extend beyond seven miles. The antenna must be raised to compensate for the bulge of the earth. The additional height requirement to compensate for the bulge of the earth can be calculated from the following equation:

$H = D^2 / 8$; where H = Height and D = Distance between antennas in miles

The maximum effect of the earth's curvature and the Fresnel zone occurs at the midpoint of the link. To ensure the proper clearances, the earth's curvature, the Fresnel zone, and the height of any objects in the line of sight must be added together to get the appropriate height of the antenna.

The general practice within the industry is to keep 60 percent of the Fresnel zone clear of obstructions. Table 6-5 is based on a 2.4 GHz signal and is a quick reference chart for determining the height requirements of an antenna at different distances. It is based on midpoint worst-case scenarios. It assumes no obstruction in the line of sight; to include obstructions, the height requirement in the table must be added to the height of the obstruction.

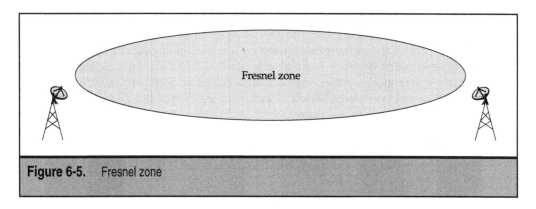

Figure 6-5.   Fresnel zone

Distance	Approximate Value "F" in ft F = 60% Fresnel Zone	Approximate Value "E" (Earth's Curvature)	Value "H" in ft (Mounting Height) Assuming No Obstruction
1 mile	10	3	13
3 miles	23	4	27
5 miles	30	5	35
8 miles	40	8	48
10 miles	44	13	57
15 miles	55	28	83
20 miles	65	50	115
25 miles	72	78	150

Table 6-5.   Fesnel Zone and Earth Curvature Matrix for 2.4 GHz Signal

Antennas

An antenna in a wireless system provides three main properties: gain, direction, and polarization.

Gain   *Gain* is a measure of how well an antenna will transmit a signal. The unit of measurement is either decibel-isotropic (dBi) or decibel-dipole (dBd). A *decibel* is a unit of comparison to a reference, so the difference between the two measurements is in the reference used.

An isotropic radiator—the reference used in the dBi measurement—is an antenna that sends a signal equally in all direction: up, down, and sideways. It is a theoretical model; no antenna functions this perfectly. The dBi measurement is therefore a measurement that compares the performance of an antenna with a theoretical isotropic antenna. An isotropic antenna is said to have a power rating of 0 dB—zero gain/loss when compared to itself—and is the measurement used by the FCC in its calculations.

A dipole antenna, in contrast, is a real antenna that has a different radiation pattern from an isotropic antenna. A dipole antenna radiation pattern is 360 degrees in the horizontal plane and 75 percent vertically. Dipole antennas have a gain of 2.14 dB over isotropic antennas because the beam is more concentrated. Some antennas are rated in reference to a dipole antenna. A dipole antenna has a gain of 0 dBd, which is equal to 2.14 dBi (in other words, 0 dB when compared to itself). Some documentation rounds up to 2.2 dBi. Care should be taken when reading antenna specifications, as sometimes reference is made to dB without specifying a reference. It is worthwhile to find out which reference model is being used.

Direction   *Direction* is closely related to gain. Antennas are usually unidirectional or omnidirectional. An antenna obtains more gain by concentrating its radiation in a particular direction (see Figure 6-6). Omnidirectional antennas concentrate energies equally in a horizontal direction. The tighter the concentration of the energy in the horizontal plane, the greater the gain. An omnidirectional antenna is different from an isotropic antenna in that it does not transmit vertically, whereas an isotropic antenna transmits in all three dimensions.

Polarization   All electromagnetic waves traveling in free space have electric [E] and magnetic [B] field components, which are perpendicular to each other and both of which are perpendicular to the direction of propagation, as shown in Figure 6-7. The orientation of the E vector is used to define the polarization of the wave; if the E field is orientated vertically, the wave is vertically polarized. Sometimes the E field rotates with time; when this happens, the wave is circularly polarized. Polarization of the wave radiating from an antenna is an important aspect when one is concerned with the coupling between two antennas or the propagation of a radio wave.

All antennas in a system must have the same polarization, which is specified by the governing authority in some countries. Most antennas are installed vertically or horizontally, but all antennas in a system must be installed in the same way, either horizontally or vertically, which is the same as saying they must all have the same polarization. If an antenna has its elements arrayed vertically—like a car antenna—it is vertically polarized;

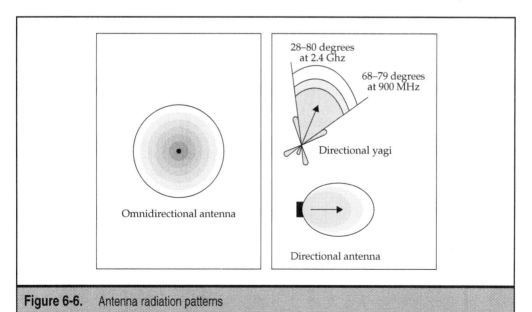

Figure 6-6.   Antenna radiation patterns

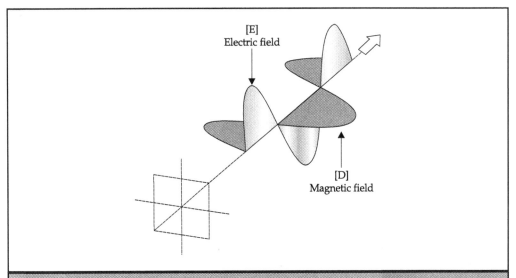

Figure 6-7. Polarization of electric and magnetic fields relative to propagation

similarly, if the elements are horizontal, it is horizontally polarized. Attempting to communicate between a station with a vertical polarization and another with a horizontal polarization results in weak signals.

Generally, if a station is to transmit in all directions, it will be vertically polarized. Most governments prefer that fixed-point systems use horizontal polarization in order to help isolate the signals from regular broadcast radio signals that are picked up by car radios using a vertical antenna.

Multipath Distortion *Multipath distortion* occurs when a receiving antenna receives reflected versions of a signal. Imagine a pool filled with water. If you drop a pebble in, it creates ripples or waves that travel in much the same way as a radio wave. Assume there is a receiver somewhere in the pool five feet from the point at which the pebble was dropped. As the waves travel outward, some will reach the side of the pool and will be reflected back toward the center of the pool. The receiver in turn receives the original wave and some of the reflected ones. This is the phenomenon that represents multipath distortion, as shown in Figure 6-8.

In areas where there are large amounts of RF-reflective surfaces, the receiver receives multiple copies of a signal, which combine to cause distortion in the receiver (see Figure 6-8). This type of interference can cause the RF energy to be too high. *Diversity antenna systems* are used to address multipath problems. Such a system uses two identical antennas that are spaced a short distance apart to cover a single area. When a signal is received on one antenna, the receiver switches to the other antenna and compares the signals. The receiver will then choose the antenna that has the best signal and will continue using that

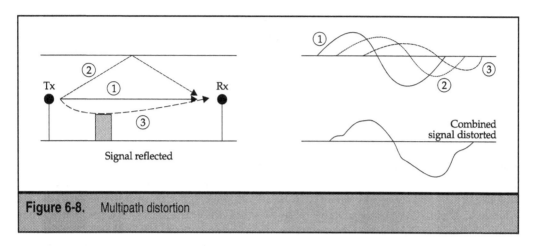

Figure 6-8. Multipath distortion

antenna for the remaining portion of the transmission. Note that the two antennas are never used at the same time.

Satellite Communication

Satellites are basically wireless systems that are interconnected to terrestrial infrastructure via terminals or antennas. The frequency band between 2.5 GHz and 22 GHz is where most satellite communication resides. At this frequency range, the radio waves are so short—10–100 millimeters—they are called microwaves, and they travel directly along the line of sight to their primary coverage area.

During the Second World War, letter designations given to different microwave frequency bands (see Table 6-6) were used to describe microwave radar systems. They have since become a fixture in the lexicon of microwave terminology and are now also used to describe satellite communication bands. The band-letter designation actually does make it a lot more convenient to explain the different frequency ranges that are used by various

Letter Designation	Frequency Band
L band	800 MHz to 2 GHz
S band	2 GHz to 3 GHz
C band	3 GHz to 6 GHz
X band	7 GHz to 9 GHz
Ku band	10 GHz to 17 GHz
Ka band	18 GHz to 22 GHz

Table 6-6. Microwave Letter Band Designations

applications. For instance, it is a lot easier to remember that the first commercial communications satellite used the C band instead of remembering that it used the band between 3 GHz and 6 GHz.

The earth stations of many telephone companies around the world used the C band for terrestrial microwave relay in the late 1960s. The power levels of C band satellites had to be limited so as not to interfere with these stations. Ku band satellites are allowed to use much higher power levels because few terrestrial communication networks were assigned to use this band.

Today, many different satellite systems are used for different purposes. Each satellite is optimized for the application it is designed to serve; the basic components of a satellite are shown in Figure 6-9. Some are designed to increase coverage of telephone service to areas that are not served by wired infrastructure; others have specific applications such as XM Satellite and Sirius that broadcast digital radio across the entire U.S. Our focus, however, is on those intended for broadband service. It is unlikely that satellites—or fixed wireless—will replace DSL, cable, and fiber; instead, they will serve to complement these technologies. The main problem that broadband satellite service addresses is getting high-bandwidth access to places without a high-bandwidth infrastructure.

Geostationary (GEO)   GEO satellite systems orbit the earth at the speed of the earth's rotation—a geosynchronous orbit—thereby keeping the satellite in a fixed position relative to the earth. These satellites operate at a fixed distance of 22,300 miles (35,786 km) above the earth's equator, and each covers about one third of the surface of the earth. Most of the

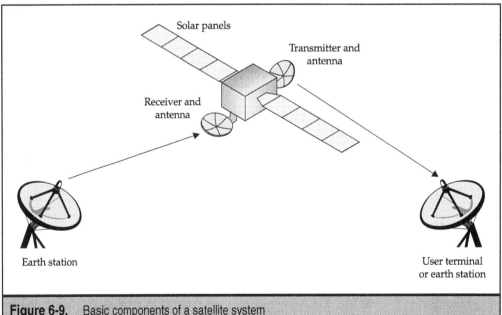

Figure 6-9.   Basic components of a satellite system

communication satellites in operation today are geostationary and are used for many different applications that include voice, video and data.

GEO systems are often used for meteorological purposes. A set of satellites called Geostationary Operational Environmental Satellites, or GOES, are operated to produce imagery used to watch weather patterns. DirecTV is also another example of a GEO system, which uses five (a sixth is scheduled for launch at the end of 2001) high-powered satellites at 101, 110, and 119 degrees west longitude to broadcast television service to the homes of consumers. Yet another example is XM Satellite, which uses two geostationary satellites to beam broadcast signals to two earth-station antennas for the Digital Audio Radio Service (DARS). Since the launch of the XM Satellite service, up to 100 new commercial-free radio stations are being broadcast from coast to coast of the U.S. See Figure 6-10 for examples of satellite positioning.

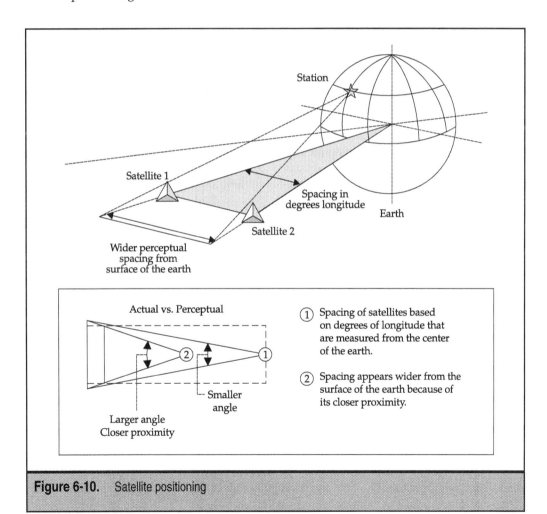

Figure 6-10. Satellite positioning

LEO and MEO   LEO systems orbit the earth 400–16,000 miles above the earth. These satellites were developed to deliver global near-real-time voice communication, which is made possible because of their low orbits that minimizes delay. These satellites actually move across the sky and take about two hours to completely circle the earth. The fact that a satellite is constantly moving means that it is in contact with a ground station for only a short time—5–10 minutes—after which the transmission must be relayed to another satellite or to another ground station.

LEO systems can also use intersatellite links that allow a transmission to be relayed from satellite to satellite until it gets to one that is in the correct location. Multiple satellites are usually used to ensure redundancy and to ensure that a transmission is not missed. LEO systems are expensive to operate, as they require many more satellites than a GEO system. Several operators of these systems either have gone out of business or have had very limited success—Iridium and GlobalStar, for instance—because of the expense of operating the service and competition from terrestrially based services.

Medium Earth Orbit (MEO) satellites are the hardest to classify even though they are given a classification of medium orbit. These systems operate at about 6,000–22,300 miles above the earth, so in fact they operate in ranges equivalent to both LEO and GEO systems (see Table 6-7). These systems are usually used for GPS-style surface navigation but are also proposed for communication systems.

VSAT Satellite Network   Very Small Aperture Terminal (VSAT) systems have been around for years and have been the way you or I could buy time on a satellite (assuming we could afford to). Using a VSAT, time on a specific satellite is purchased by an individual or organization and is usually used for linking remote offices.

A VSAT is a remote terminal in a satellite communications network. These antennas use a small dish that is normally 2.4 meters or smaller. In comparison, larger earth station antennas may be as large as 11 meters or more. A VSAT system allows communication between remote earth stations. A satellite acts as a reflector to redirect a transmission from a terminal to another terminal hundreds of miles away. The beam from the satellite is less concentrated than the beam from the terminal, and so transmissions from the satellite cover a much larger area that may have many terminals.

Various topologies exist for VSAT networks. A star topology is good for broadcasting, where a central station broadcasts a single transmission to the satellite for distribution to many smaller sites. This is the mode in which satellite TV works. Star topology, or point-to-multipoint, can also support two-way transmissions, with the remote terminal transmitting back to the central site. A terminal wishing to speak to another terminal in a star topology must send the communication to the central site, which in turn retransmits to the intended remote station. This requires two hops to the satellite, once to communicate back to the central site from the sending remote and a second time from the central site to the receiving remote.

A full-mesh topology facilitates point-to-point communication between any two remotes through a single hop to the satellite. The central site is not needed as an intermediary in this type of topology. A hybrid topology, which is a combination of the two topologies, is also possible. An example of an application of this topology includes instances

Feature	GEO	LEO
Application	Proposed for broadband services but latency is a major factor Good for high-speed information downloading and video distribution, broadcasting, and multicasting	Proposed for broadband services but has the advantage of less delay Good for high-speed networking, teleconferencing, and interactive applications
Delay	High (250 ms each way)	Negligible (6 ms each way)
Cost	Relatively lower Higher incremental cost to increase bandwidth (needs another very expensive satellite)	Relatively higher Lower incremental cost to increase bandwidth (needs the addition of cheaper satellites to the constellation)
Coverage	3–4 satellites would provide full global coverage	40–60 satellites for continuous service Teledesic originally proposed 800 satellites for their new service. This number dropped to 288, with each costing about $20 million
Tracking	Fixed large antenna aimed at the stationary serving satellite. Smaller dish antennas can also be used for fixed broadband services	From very small (pocket-sized) as is the case of GlobalStar phones. LEO systems use phased-array antennas, which are self-aiming boxes consisting of many smaller antennas. Several satellites are tracked using slightly different signals received by the array of antennas without physically moving the antenna
Lifespan	10–15 years	5–7 years

Table 6-7. GEO and LEO Feature Comparison

when remotes are allowed to transmit data only to the central site, but remote-to-remote voice communication is enabled through a mesh.

VSAT is viewed as an enabler of broadband communication using satellites. It is already used for corporate traffic between a head office and a remote site. When it is used as a satellite WAN connection, the LANs at both offices are linked through VSAT technology.

Satellite Antennas Since the minimum power required of a satellite transmitter is determined by its distance from the earth, it follows that a GEO system must be more powerful than a LEO system. In either case, the power of a satellite transmitter pales in comparison to that of a terrestrial radio station. When you consider that a GEO system can be placed in an orbit 55 times as distant as the lowest LEO orbit (a difference of 21,900 miles), you would imagine that the difference in power requirements would be of the same order of magnitude as the ratio of distances, 55:1. This, however, is not the case, as the design of the antenna system (or terminal) has a lot to do with the required power.

By concentrating the power of an antenna in a single direction using a concentrated beam, you can transmit a signal a greater distance for a given power level. Consider a garden hose with a constant water pressure: you gain more distance by restricting the water from spreading as it leaves the hose by decreasing the size of the hole. The smaller, concentrated beam of water gets a lot more distance than a dispersed beam even though the pressure for the water source has not changed.

The same principle holds true for satellite terminals. The curious difference between a satellite terminal and our garden hose is that the larger the reflector area of the terminal, the more concentrated the beam (in our water analogy this was reversed; a smaller hole does more to concentrate the flow of water than a larger hole). Doubling the diameter of a reflector antenna (or dish) reduces the area of the beam spot to one fourth of what it would be with a smaller reflector. This is called the *gain,* a principle that we have already covered. It tells us how much more power will fall on a particular area (square meter, square mile, or whatever) using the antenna compared to what that power would be if the transmission were evenly dispersed in all direction (isotropic dispersion). The larger antenna described would therefore have four times the gain of the smaller antenna. This is how the longer distance of GEO systems is compensated.

Another major difference between antennas for a GEO system and a LEO system consists of their respective capabilities to track the satellite. In the case of a GEO system, this is not a problem, as the satellite is always in the same position in the sky from the perspective of the earth station. The earth from the perspective of the satellite is also fixed in a given location. These antennas therefore are fixed and, once installed, remain trained in the optimum direction. LEO systems are a lot more challenging, as a satellite may remain in view of a ground station for only a few moments. The antenna must be able to track the movement of the satellite.

Finally, there are no power lines in the sky, so some other method of power generation must be available for satellites. Satellites usually rely on solar-powered panels to generate electricity. Batteries are also used to store this power for those times when the sun may be blocked (by an eclipse, for example).

Table 6-8 has been included to provide examples of different types of satellites systems, the features of each, the type of equipment as well as examples of different services that employ these systems.

System Type	Frequency Bands	Applications	Terminal Type/Size	Examples
Fixed Satellite Service	C and Ku	Video delivery, VSAT, telephony	1 meter and larger fixed earth station	Hughes Galaxy, GE American, Loral Skynet, Intelsat
Direct Broadcast Satellite	Ku	Audio and video to consumer homes	0.3–0.6 meter fixed earth station	DirecTV, Echostar, USSB, Astra
Mobile Satellite (GEO)	L and S	Voice and low-speed data to mobile terminals	Laptop computer with attached mobile antenna	Inmarsat, AMSC/TMI, ACES
Big LEO	L and S	Cellular telephony, data, paging	Cellular phone and pagers	Iridium, GlobalStar, ICO
Little LEO	P and below	Position location, tracking, messaging	Pocket sized and omnidirectional	OrbComm, E-SAT
Broadband GEO	Ka and Ku	Internet access, voice, video, data	Fixed 200 mm	Hughes Spaceway, Loral Cyberstar, Lockheed Astrolink
Broadband LEO	Ka and Ku	Internet access, voice, video, data, videoconferencing	Phase-array	Teledesic, Skybridge, Celestri, Cyberstar

Table 6-8. Satellite Service Comparison

BROADBAND WIRELESS SOLUTIONS

The tight economic climate of the beginning years of the 2000 decade has not been favorable for new ventures. Coming off the highs of the late 90s, when any project that appeared to be Internet related almost guaranteed a flock of private investments, the economic downturn in the new decade turned investors more cautious. New ventures had to have a solid business plan with a solid projection for profitability before they would be considered for funding by burnt venture capital investors. The demise of high-profile and high-promise ventures like Iridium and GlobalStar only added fuel to the fire.

In spite of this, Cahners In-Stat Group believes that the fixed wireless broadband access market will represent a substantial portion of the broadband access (wireline and wireless) market by 2005. The fixed wireless market, at the time of this writing, is segmented into three distinctly different markets based on the spectrum allocation of the service. These markets are the Local Multipoint Distribution System (LMDS), the Multichannel Multipoint Distribution System (MMDS), and license-free wireless services that comprise the Unlicensed National Information Infrastructure (U-NII) and the industrial, scientific, and medical (ISM) bands.

Figure 6-11 is an architectural reference model for fixed wireless systems. It depicts the components that are common to these systems. In addition to the physical equipment such as antennas, base station, indoor and outdoor units, the figure depicts the different interfaces required for fixed wireless systems. As shown, the RF specifications for MMDS, LMDS, 3G, and U-NII defines the interface between the base station and the receiving antenna.

Satellite services are also contenders for providing wireless broadband access. New ambitious services like Teledesic that hope to use 288 LEO satellites are being planned, but the failures of other satellite ventures have turned many cautious. Incumbents such as service providers of direct broadcast satellite (DBS) service already have a track record and history, so for them to expand into broadband services is less risky. Regardless, we are at the cusp of a significant turn in the way we view the world; we are about to become untethered from the network in a very pervasive way. Whether it takes one year or ten, this simple fact will be a positively disruptive influence on the way we interact, socialize, entertain, and conduct business.

For the remainder of this chapter, we will explore some of the more prominent technologies that are being developed or are already in operation today.

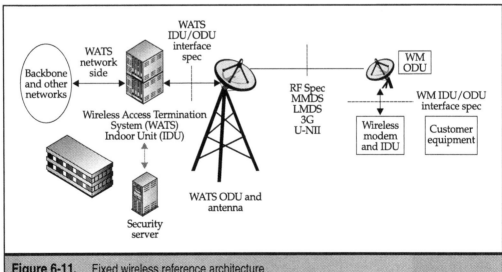

Figure 6-11.   Fixed wireless reference architecture

LMDS: Local Multipoint Distribution Service

LMDS, or *Local Multipoint Distribution System,* is a point-to-multipoint fixed broadband wireless service that can be used for two-way communication at data rates beyond 155 Mbps. Typical systems have data rates of 45 Mbps downstream and 10 Mbps upstream. The fact that a single transceiver can reach thousands of subscribers within a cell makes this service an attractive solution with an attractive price point for delivering broadband service. The incremental expense of the solution is realized at the time a subscriber subscribes and is limited to the cost of the subscriber equipment.

This is one of the main attractions of wireless from a service provider's perspective. A capital commitment is necessary only once the sale has been made. This is in contrast to wired solutions, where the infrastructure must already be in place, thereby requiring up-front capital investment before any revenue can be generated.

An LMDS system uses point-to-multipoint distribution of signals in cells that are roughly 2–3 miles in diameter and is a potential choice as a last-mile solution in areas where copper, cable, or fiber may not be convenient or economical. Some benefits of the technology as a fixed wireless broadband solution are:

▼ Lower deployment costs

■ Ease and speed to deployment

■ Return on investment realized a lot sooner

■ Capital outlay at the time the service is sold

■ Deployment feasible to hard-to-reach places and dispersed consumer base

■ Absence of regulation at local and state levels

▲ Cost 80 percent electronics, not labor and structural materials (no trenches to dig, no amplifiers to fix, etc.)

Band Plan and Auction

In February–March 1998, the FCC auctioned off 986 licenses to operators who planned to offer Local Multipoint Distribution System (LMDS) services in the 28 GHz and 31 GHz frequency bands; namely, the A and the B block, as shown in Figure 6-12. Each frequency block is allocated as follows:

▼ **Block A (1,150 MHz)**   The 28 GHz band: 27,500–28,350 GHz and 29,100–29,250 GHz; and the 31 GHz band: 31,075–31,225 GHz

▲ **Block B (150 MHz)**   The 31 GHz band: 31,000–31,075 GHz and 31,225–31,300 GHz

These licenses can be used for wireless local loop, high-speed data transfer, video broadcasting, and two-way communications.

ILECs and cable operators were restricted from owning Block A licenses in their authorized service areas for a total of three years. They were, however, allowed to apply for

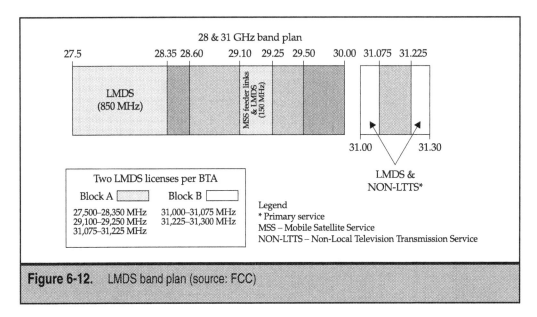

Figure 6-12.   LMDS band plan (source: FCC)

ownership of Block B licenses. This was done to ensure fair competition and avoid the risks of an ILEC blocking a potential lower-cost telephone service by holding the Block A license.

The auctioning of the LMDS bands was significant because it was the largest block of bandwidth ever to be auctioned. In a government review, it was stated that one system developer hoped to deliver 76 digital broadcast video channels while setting aside 1.5 Gbps for interactive data channels using 850 MHz of downstream bandwidth. Across the border, Canada also awarded licenses for 1 GHz of bandwidth in the 28 GHz range for a service called LMCS, *Local Multipoint Communications System,* and in Europe the 40 GHz range is used.

Architecture and Standards

An LMDS system can be built on an *Asynchronous Transfer Mode (ATM)* infrastructure. The typical system consists of a base station or hub and a base radio unit that has upstream and downstream channels to the microwave equipment (see Figure 6-13). It includes a *network operation center (NOC)* where all the network management systems and all the different feeds from other networks like PSTN, video, data, and other networks converge. The NOC can be colocated with the base station or be connected via fiber cables. The customer equipment consists of a radio unit and an antenna system.

For the remainder of the chapter, we will refer to base station as the *hub,* and the subscriber CPE as the *sub.* The two components of the sub are also referred to as the indoor unit (IDU) and the outdoor unit (ODU). The IDU consists of the radio unit, or wireless modem as it is sometimes called, which hosts all the interfaces necessary for attachment to the customer-owned network or equipment, a network management agent for communicating back to the provider's network management platform, and a power supply. The ODU refers to the antenna (a dish that is usually about a foot in diameter) and the transceiver.

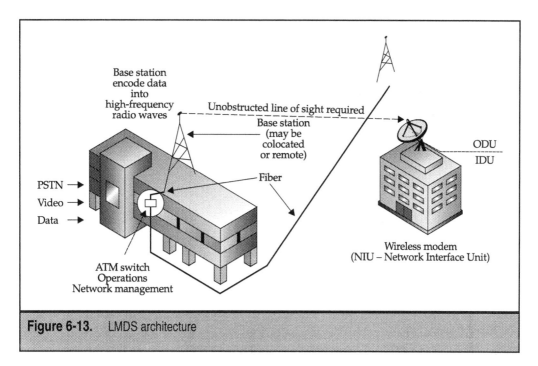

Figure 6-13. LMDS architecture

Connections to the PSTN and other network providers of data, video, and other services are achieved through the use of fiber cables, though other methods such as wireless, satellite, and private and public trunk cable are possible. The exact architecture of the components varies, as different providers have developed solutions to address specific business needs. Different manufacturers have different features and combinations of components, which makes it somewhat difficult to do anything but generalize about the different elements. We will therefore limit our discussion to the broader technology, features, and interfaces that are common to all implementations.

Standards Standards are becoming increasingly important as LMDS systems gain in popularity. Several standards initiatives are underway by different working groups that include the DVB, ATM Forum, ETSI, ITU, and IEEE 802.16 committee. As systems become more prevalent, interoperability issues become more pronounced. Balancing the need for standards in the early phases of development of a new technology can be tricky. Being too early in publishing standards may discourage vendor innovation, which has traditionally been the driver for technology achievement and optimization. Premature standards could also lead to costly solutions and suboptimal solutions.

Being too late also has its drawbacks. Standards that are published late in the lifecycle of a technology create issues of backward compatibility when a vendor goes to upgrade the network to leverage new features that may be included in the new standards. This fine line between being too early and being too late has always been a challenge, and the decision on when to introduce a standard is usually predicated on judging when the technology is mature enough to benefit from a standard and how quickly the technology is being adopted.

The approach of the working groups has been to leverage preexisting technology and standards where possible and to rely on equipment vendors' creativity in reducing costs, developing new features, and optimizing performance. The following areas for standardization are being considered:

In the area of the technology itself:

▼ Frequency and band plan for each specific geography

■ Frame format

■ Transmission interleaving type

■ Channel width

■ Modulation scheme

■ Multiple access protocol

■ Signaling and control

■ Performance (BER, delay, etc.)

■ Operational functionality

■ Security

■ Redundancy

▲ Power control, EMI, and other physical attributes

In addition to focusing on the technology itself, the standards committees are also focused on the interfaces between IDU and ODU, which include:

▼ Intermediate frequency

■ Frame formats

■ Customer type

■ Structure type (single family residence versus MDU versus MTU)

■ Powering

▲ Physical specifications

The customer interface must also:

▼ Be adequately defined

▲ Include physical interfaces to other transmission media—RH11, RJ45, T1, E1, ATM, and so on

Technical and Design Considerations

The frequency range of an LMDS system allows for the design of much smaller antennas with excellent directivity. These focused antennas limit, or effectively eliminate, multipath, because fewer signals are dispersed—so there is less to reflect. This allows more energy-efficient modulation techniques such as QPSK to be used. The directivity of the antenna also translates to lower power levels that promote frequency reuse.

Being able to use a modulation technique like QPSK impacts the affordability of the equipment. QPSK is supported by mass-produced ASIC-based demodulators and effects the most productive frequency reuse plan for cellular deployment. There is, however, some trade-off in the cost owing to the fact that LMDS systems usually use an ATM or Frame Relay switching technology that provides high performance but makes it more expensive on a subscriber basis.

In order to maintain a cap on the subscriber equipment costs, lower power levels are used, limiting the effective range of the transmission to about 2.5 miles. By limiting the distance between the hub and the subscriber transceiver, any potential for multipath is further reduced, hence the earlier reference to its effective elimination.

LMDS uses many small microcells that are similar to those used for cellular phone communication. The antennas—transmitter and receiver equipment—within each cell are fixed within the cell and do not roam from cell to cell in the way a cellular phone does. The use of cells allows frequency reuse, which is the process whereby a frequency can be reused in another cell within the coverage area. This is an important aspect of LMDS, as it allows a more efficient use of the available bandwidth. The spectrum is a limited resource that comes at a high price for most commercial networks, so any method that squeezes more from the investment is a good thing.

The use of microcells also promotes flexibility in the use of bandwidth, which can be effectively increased by decreasing the size of the cell and multiplied by increasing the number of cells. Before we delve into the issue of cell design, we will digress briefly to discuss a debate over different methods for dividing the available bandwidth for upstream and downstream communication.

FDD Versus TDD   With a fixed amount of bandwidth, it is necessary to employ a method for dividing the available bandwidth between upstream and downstream communication. As is usually the case in the early stages of the developmental cycle of a technology, different camps favor different techniques. This is the case with LMDS and other newer wireless technologies: a debate on the more efficient way of dividing the bandwidth between the two channels is ongoing, and it probably will be for a very long time even after it has been settled. We will not attempt to decide which technique is better; many smarter people have tried and cannot agree. We will make reference to the debate only to enlighten you on the options that are being considered.

The established model for wireless communication is *frequency-division duplexing (FDD)*. It is the technique that has been used for years in cellular telephones. A cellular telephone uses the frequency range of 824 MHz to 849 Mhz for the signals that originate from your phone; those that are intended for the phone—those that you hear—use the range 869 MHz to 894 Mhz. The method used for dividing the bandwidth between the send and receive channels is FDD. A newer approach to bandwidth division is *time-division duplexing (TDD)*, which allows a single channel to be used for both downstream and upstream communication through the use of different discrete time slots.

The essence of the debate is one of efficiency. TDD proponents argue that FDD is inefficient in the use of bandwidth because data and video applications are predominantly asymmetric in nature (more data flowing in one direction over the other). The upstream channel, which is predefined, is therefore underutilized, resulting in precious and expensive bandwidth being wasted. If TDD is used, the bandwidth can be given to the direction of flow that needs it most. This is achieved by dynamically allocating more time slots to the direction of flow that has the most traffic. So, during a large download, the available bandwidth in the downstream channel would increase and the bandwidth for upstream traffic would decrease by the same proportion.

The argument also extends to issues of applications like FTP. If FDD is being used and the file transfer is from the subscriber to a remote server, the upstream bandwidth is the channel that requires the lion's share of the capacity. If the channels were divided for an asymmetric flow, the file transfer would flow across the lower of the two channels. If both channels have the same bandwidth, then we are back to the original concern of wasting bandwidth, but this time in the downstream channel. TDD has no such limitation, because either channel can dynamically adapt to changing requirements.

FDD proponents argue that there is a price and performance trade-off when using TDD. TDD requires precise timing and sophisticated software to dynamically manage the bandwidth. As a result, it takes time to turn the line around. Additionally, all subs in the coverage area need to use a common time reference for defining the time slots.

FDD has been around for many years and is a well-understood, mature, and cost-effective solution that is used in many different wired and wireless applications. It has a very strong following because of its entrenchment and its wide installed base. To compete, and to promote the benefits of TDD, a coalition has been formed called the TDD Coalition. Table 6-9 provides a list of the current members of the coalition for further reference and research.

Member Name	Web Address
Adaptive Broadband	http://www.adaptivebroadband.com/
Aperto Networks	http://www.apertonetworks.com/
ArrayComm	http://www.arraycomm.com/
BeamReach Networks	http://www.beamreachnetworks.com/
CALY Networks	http://www.calynet.com/
Clearwire Technologies	http://www.clearwire.com/
Harris Corporation	http://www.harris.com/
IPWireless	http://www.ipwireless.com/
LinkAir	http://www.linkair.com/
Malibu Networks	http://www.malibunetworks.com/
Radiant Networks	http://www.radiantnetworks.com/
Raze Technologies	http://www.razetechnologies.com/

Table 6-9. TDD Coalition Members

Cell Design Considerations   Smaller cell sizes improve coverage, increase the available bandwidth, and improve the reliability of the service. The downside to smaller cell sizes is the infrastructure costs. An antenna, a transceiver, and potentially a tower must serve every cell (antennas may be mounted on roofs to save the cost of the tower). When planning coverage, the following considerations must be given to each cell:

▼ **Cell size and the number of cells**   A smaller cell size is directly related to improved performance and increased bandwidth. The lower power levels needed in smaller cells limit issues of multipath and improve performance and reliability of the service. When deciding on the cell size, the total coverage area must be considered, which relates directly to the number of cells that will be required to serve the coverage area. Too many cells can become too costly, as a base station is needed to serve each cell. Another consideration is population density: the greater the number of subscribers, the more traffic generated. Congested cells are better served by creating another cell. Line-of-sight issues also come into play when deciding on the size and number of cells. The topography, the vegetation, how developed the area is, and the types of buildings are all factors that may limit the coverage of a single cell.

■ **Cell overlap**   Another way to improve performance is to overlap cells—a 15 percent cell overlap is typical—to increase the potential number of users with a clear line of sight. Subs in areas of overlap may select the signal that is strongest. The risk with an overlapping multicell approach is the potential for cochannel interference arising because LMDS systems employ aggressive frequency reuse in each cell. The trade-off between cochannel interference and hub diversity must be considered when designing the coverage area.

■ **Weather effects**   Coverage simulation models show that the coverage of a cell decreases with adverse weather conditions such as rain or snow. As rain attenuation increases with distance, subs further away from the hub are more subject to the attenuation caused by rain. The system design should provide adequate availability during times of rain. The proposed design margin for outages caused by rain is 0.1 percent, or in other words, the system should be available 99.9 percent of the time with a margin of 0.1 percent for rain outages. The coverage area is therefore limited by weather patterns, a factor that must be considered when deciding the maximum size of the cell. To serve larger areas, especially in places that have a lot of rainfall, multiple cells are required with some overlap to ensure adequate coverage through hub diversity.

▲ **Transmission power and antenna height**   System performance can be enhanced by raising the height of the antenna. Intermediate objects such as trees can create problems for the signal. Raising the antenna helps to clear the obstruction. The power level of the signal is another way to improve performance. Weather conditions such as rain may create problems that could be overcome by increasing the power of the signal. Another factor that benefits from increased power is the distance of the sub from the hub. Boosting the power helps to extend the range of the cell. Too much power, however, may cause problems in another cell, especially when frequencies are being reused. Care needs to be taken to ensure that the power levels are optimized for the best performance.

Frequency Reuse   A goal of a good system design is to reuse as many frequencies as possible in each cell in order to maximize capacity without creating such other problems as cochannel interference. Using highly directive antennas and cross-polarization aids in optimizing the system for frequency reuse.

The objective of a cell site is usually to provide 360-degree coverage of the area immediately around it, a goal that appears to be in conflict with the need to be as directive as possible. The more directive the antenna gets, the less coverage provided. To accomplish the objective of achieving full coverage of the entire 360-degree area around the antenna while maintaining a narrow antenna beam to limit multipath, the area around the cell site is divided into sectors, which are each covered by highly directive antennas. The cell is commonly divided into 4, 8, 12, 16, or 24 sectors, with each antenna providing coverage for 90-, 45-, 30-, 22.5-, or 15-degree angles.

To gain the capability to reuse frequencies in a cell, the antennas within each sector use different polarizations. The picture in Figure 6-14 shows an example where the cell has been divided into 12 different sectors. The horizontal and vertical lines represent horizontal or vertical polarization. What is evident from the picture is a frequency reuse pattern of 6; with this scheme, six times the bandwidth is available in the cell than would

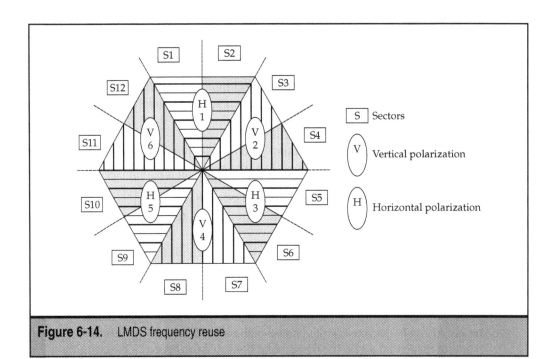

Figure 6-14.   LMDS frequency reuse

have been possible had the reuse pattern not been used. Different combinations are possible that create different reuse patterns.

LMDS Challenges

Despite the advantages of LMDS compared to wired solutions, the roll out services has been slow. Other than the challenging business cycle and tight financial markets, LMDS is also faced with some technical challenges, a few of which are listed below.

▼ Leaves, branches, trees, buildings, topography can block or attenuate LMDS signals. A clear line of sight is needed between the hub and the sub, a fact that severely limits the coverage area of a given cell in any urban area. Using smaller cells and using overlapping cells are both techniques that can be used to overcome this limitation.

■ LMDS is very susceptible to rain fade, as raindrops absorb the millimeter length waves of the signal. This makes it less suitable in areas with heavy rainfall or snow. The effect of rain fade can be mitigated through the use of FEC and adaptive power control.

▲ Standards are needed to ensure interoperability.

MMDS

The Multichannel Multipoint Distribution System, MMDS, is a point-to-multipoint fixed wireless technology that operates in the 2.5 GHz range of the spectrum in the U.S. and Canada and in the 3.5 GHz in many international markets. Sometimes called Wireless DSL, the technology is viewed as a viable solution for providing broadband services to a widely dispersed area. MMDS systems are able to achieve speeds of up to 10 Mbps.

Unlike LMDS, which is limited in reach, MMDS is able to serve a 35-mile area with hubs that are typically located on mountaintops and other high places. A single tower can provide coverage to a huge and heavily populated area at a much lower cost than LMDS. A single MMDS hub can cover an area that would require 50 to well over 100 LMDS hubs.

The relatively longer wavelength also makes MMDS less susceptible to interference from the weather and vegetation. Its wide coverage area also makes it a cost-effective solution for reaching dispersed populations in rural areas—a market that would be much more expensive to cover using LMDS because of the impact of terrain on its millimeter waves.

A New Application for an Old Idea

The technology was originally intended to be a wireless cable solution, an oxymoron that borders on amusing. In 1970, the FCC created the Multipoint Distribution System, MDS, which used the spectrum of 2.1 GHz to 2.7 GHz. Licensed operators were authorized to use the band for data or television programs within 30 miles of a community.

The service never really had a chance from the get-go because of high equipment costs, unreliable technology, and very strong competition and lobbying from cable operators. For more than a decade, the service remained dormant, only to be revived in the early 1980s as technology for transmitting pay-per-view programs to cable headends and the hospitality industry. In addition to the MDS spectrum, another service, called ITFS, *Instructional Television Fixed Service,* was used—and still is used today—by schools and universities to deliver instructional courses throughout different regions (see Figure 6-15).

The service as we know it today, MMDS, is a result of the Telecommunications Act of 1996, which provided for conversion of eight channels of the ITFS service to the new MMDS in order to stimulate competition with terrestrial cable services. MMDS service providers were authorized to lease the remaining ITFS channels during periods when they were not being used.

Since the act, the service has found a new lease on life, and instead of being used in competition with cable operators, it is viewed more as an Internet access methodology and an enabler of broadband communications. Instead of targeting an entrenched service group, the technology is being used to reap the low-hanging opportunities by tapping into the revenue stream that Internet access provides. Its wide coverage area allows it to serve large office complexes, hotels, motels, condominiums, and residences within a 15-mile radius.

The challenge is in the return path; being designed for one-way communication—as is cable service, its original intended competitor—the system was never designed to provide a return path. To facilitate upstream communication, a telephone return path or *wireless local loop (WLL)* technology is used.

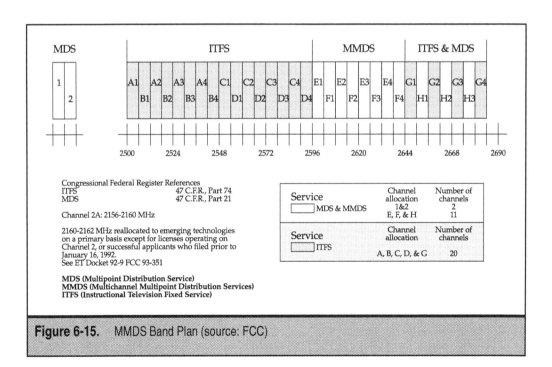

Figure 6-15. MMDS Band Plan (source: FCC)

In 1998, the FCC approved two-way rules that allow flexibility within the band plan for upstream communication. In August 2000, they began accepting applications for two-way services. The fact that the spectrum is shared between MDS, MMDS, and ITFS makes it more challenging to implement because of the potential for interference for current ITFS licensees. An additional challenge is to guide the service through the transition from one-way to two-way transmission without creating disruption to existing services. The bandwidth is already so limited that carving out an upstream channel will require some creativity and planning.

NOTE: Wireless Local Loop (WLL) is a technology that uses wireless to replace the last-mile PSTN copper links to the customer premises. There are many drivers for the technology, some of which include freedom from copper wires, competitive local phone service, and telephony build-out for rural areas. Several technologies are being explored for WLL, from CDMA to VSAT technology. The market is basic telephone service (narrowband) and so does not fit within the scope of our discussion.

Architecture

An MMDS system consists of a headend where all the digital and analog services converge, a base station for broadcasting the signal, the customer equipment, and the return channel (see Figure 6-16). The customer equipment consists of a wireless modem that has a LAN interface (usually Ethernet), a radio, and an antenna. The radio and the antenna may be combined into a single compact unit. The antenna (or dish), which is usually about a foot in diameter, is mounted on the outside of the building. The hub (base station and antenna tower) is usually located in a place that provides the widest unobstructed coverage of the coverage area. Signals are fed to the tower from the headend through a direct wire connection.

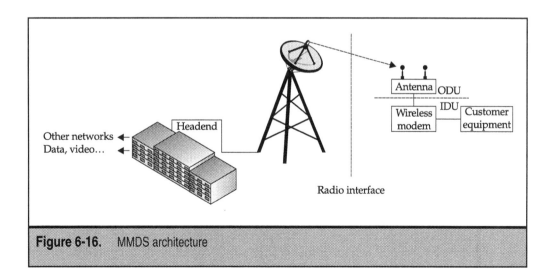

Figure 6-16.   MMDS architecture

In July 2000, six companies joined to form the Wireless DSL Consortium. The founding members were ADC Telecommunications, Conexant Systems, Gigabit Wireless, Intel, Nortel, and Vyyo. The consortium was formed in recognition that the industry was in desperate need of standards for wireless broadband solutions. Their intent was to define, develop, and implement a set of open wireless interfaces for broadband wireless products operating in the MMDS service segment.

In order to jump-start the process, the consortium adopted the Data Over Cable Service Interface Specification (DOCSIS) with enhancements compensating for MAC and physical layer requirements for wireless operations. It is interesting that DOCSIS was chosen—an endorsement of its capabilities and its emerging role as a standard for multiple media. The features of DOCSIS were covered in Chapter 5. LMDS and MMDS are compared in Table 6-10.

MMDS Challenges

▼ Service providers must upgrade their infrastructure or deploy new infrastructure to facilitate a return channel. Industry experts generally accept that the upgrade will mean smaller cell sizes, so instead of a single cell serving distances up to 35 miles, the distance may be reduced to 8–12 miles. The smaller cell size means a greater cost burden for delivering the service.

■ The service is less susceptible than LMDS to interference from environmental factors, but it is limited to a direct line of sight. The greater distance served by the antenna does allow it to be placed in an area that overcomes a lot of these obstacles, but nevertheless, the requirement for a direct line of sight is always a factor influencing coverage and deployment.

■ The number of TV channels has been significantly increased from 33 to 198 through the use of digital technology and compression techniques. The fact that the channels are shared creates the potential for capacity constraints as the number of subscribers grow. Unlike cable service that has a wide spectrum to play with within the confines of the cable, all wireless services have a limited bandwidth in which to operate.

▲ The spectrum allocation is not clear-cut. MMDS uses channels assigned for MDS, MMDS, and ITFS. This arrangement limits flexibility and dictates the need to negotiate market rights with the license holders of ITFS spectrum.

U-NII Solutions

In January 1997, the FCC set aside 300 MHz of spectrum in the 5 GHz band for an unlicensed service called Unlicensed National Information Infrastructure. Three bands were defined: two were defined in the 5.15–5.25 GHz and 5.25–5.35 GHz ranges; the third was defined in the 5.725–5.825 range.

The unlicensed bands were set aside to speed up the deployment of wireless broadband access for businesses, schools, and hospitals. In recognition of the potential for interference, the FCC defined rules that limit the power levels of antennas. These power

Feature	MMDS	LMDS
Frequency range	2.5–2.7 GHz Shared arrangement with ITFS	28.5–29.5 GHz and 31–31.3 GHz Dedicated
Speeds	Up to 10 Mbps	> 155 Mbps
Propagation characteristics	Medium to long ranges (~35 miles)	Short ranges (~2–5 miles)
	Not affected by environmental conditions like rain	Affected by environmental conditions like rain, fog, snow
Line of sight	Requires direct line of sight. The line-of-sight requirement is removed when VOFDM modulation is used (planned)	Requires direct line of sight
Antenna	~11 times the size of an equivalent LMDS antenna	~1/11th the size of an equivalent MMDS antenna
	Less directivity (broader beams) More multipath interference because of broader beam	More directivity (narrower beams) Less multipath (almost none) because of highly focused directivity
Common cell architecture	Single cell No frequency reuse, limited bandwidth available	Multiple microcells Larger bandwidth that varies by the size of the cell and frequency reuse
Two-way communication	Limited, as it has been approved only since 1998	Good two-way solution aided by high bandwidth availability
Modulation	QAM, OFDM Future MMDS technology will be based on VOFDM	QPSK
Cost	~$1,500 per mile	~15,000 per mile
Main service providers	XO Communications, Winstar	MCI Worldcom, Sprint

Table 6-10.   MMDS and LMDS Comparison

levels are strong enough to transmit data up to 25 Mbps or beyond, but they limit the range of the transmitted signals to about three miles.

Using this band, services providers can deploy new services at a faster rate and at lower start-up costs. For holders of wireless broadband licenses, the unlicensed spectrum can be used to supplement the limitations of the existing licensed service, or it may be used as an interim solution before the main service is launched. In essence, the major benefits of U-NII are flexibility, lower costs, and speed of service deployment.

The flip side of the coin is the fact that the foregoing benefits have the potential to attract many operators in this space. While this is good from the perspective of competitive service pricing generated by competition among many competitors, the added traffic creates the potential for interference between different service providers' service offers in spite of the power limits imposed by the FCC. Interference means performance degradation, and if the service degrades to the point of annoyance or if it makes the service unusable, what then is the point of cheaper prices? These are valid concerns, but as is the case with most technological challenges, modulation techniques like VOFDM have emerged to address these problems.

VOFDM: Vector OFDM

Vector Orthogonal Frequency Division Multiplexing (VOFDM) is built on OFDM modulation techniques. It is a radio frequency (RF) technology that delivers two-way data, voice, and video—wireless communications with speed and quality comparable to cable networks. VOFDM uses spatial processing for fixed broadband wireless solutions. As we discussed in Chapter 2, OFDM is a modulation technique where multiple carriers (or tones) are used to divide the data into subchannels across the available spectrum. Each tone is considered to be orthogonal (independent or unrelated) to the adjacent tones and, therefore, does not require a guard band.

By coding across different frequencies and varying time within each subchannel, OFDM is able to overcome random and impulse noise as well as cochannel interference. Vector OFDM (VOFDM) adds spatial processing to the mix to make the modulation technique more nearly impervious to interference created by multipath-reflected signals.

Under normal conditions, multipath is undesirable. The reflected signals once received by the receiving antenna have the propensity to create interference. VOFDM actually employs a technique called *spatial diversity* where two or more antennas are used to take advantage of multiple signals created by multipath. A dual-feed antenna receiver is installed at the customer premises to capture incoming signals from diverse paths and combine them to achieve the highest possible signal-to-noise level. By combining relatively weak reflected signals as shown in Figure 6-17, the system delivers an output to the user that is on par with the quality achieved with a direct line-of-sight transmission.

In the Broadcom implementation of VOFDM, the receiving station has two receivers that are connected to two different antennas. A spatial combining circuit inside a chip combines the received signals to get the sum of the two signals. The placement of the antennas is important to get the full benefit of spatial diversity. In the Broadcom solution, they are placed at least three feet apart. The range of the system is about 6–10 miles.

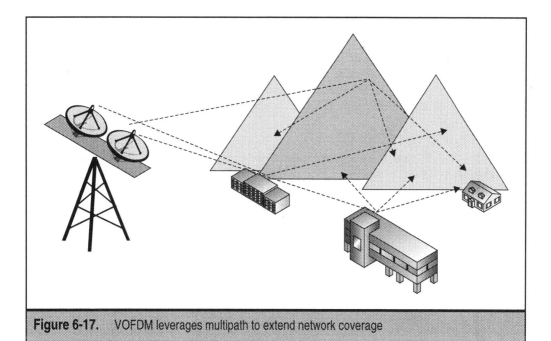

Figure 6-17. VOFDM leverages multipath to extend network coverage

NOTE: Lucent Technology has a competing technology, called BLAST (Bell Labs Layered Space-Time), which allows multiple signals to be transmitted at the same time within the same frequency. Normally, each signal requires its own frequency. BLAST requires several different transmit and receive antennas to separate the signals. In the announcement of the technology, it was stated that a BLAST prototype was constructed using 8 transmitting antennas and 12 receiving antennas. The prototype increased the speed tenfold.

In September 1998, Cisco Systems acquired VOFDM technology when it bought Clarity Wireless. VOFDM, which has since been adopted by the Broadband Wireless Internet Forum (BWIF), is designed to work well in rural, suburban, and highly urban areas in both line-of-sight and non-line-of-sight environments for both downstream and upstream channels. It takes advantage of multipath signals in areas where the line of sight to the receiver station is limited.

By overcoming the line-of-sight hurdle, the technology opens up many possibilities for the deployment of wireless networks—the provisioning costs are lowered, the coverage area is increased, interference and multipath fading issues are minimized, and the reliability of the service is improved. It provides a level of performance, scalability, provisioning, and security that is comparable to wireline DSL or cable networks.

The technology is frequency independent and offers improved spectral efficiencies by increasing throughput within a minimum RF spectrum. Speeds of up to 22 Mbps per downstream 6 MHz channel and up to 18 Mbps upstream are achievable using the technology. It is being promoted as the basis of a new broadband wireless access standard by the BWIF—a 38-member coalition that includes companies such as Cisco Systems, Broadcom, Texas Instruments, National Semiconductor, and Toshiba.

The proposed new standard combines the VOFDM physical layer (PHY) technology with DOCSIS 1.1 cable modem media access control (MAC) technology. A goal of the coalition is to deliver fixed wireless solutions that compete in price and performance with DOCSIS cable modem solutions.

Equipping the U-NII Market

The Cisco WT-2700 line of equipment is designed for the broadband wireless access markets. The family of equipment uses VOFDM to combine multipath signals to re-create the original signal, minimizes the line-of-sight requirement, and increases the coverage of a single area.

Security is addressed through the use of a 56-bit hardware-assisted implementation of the data encryption standard (DES) and DOCSIS media access control. The security feature set allows wireless access to be integrated into a VPN solution. As a contributor to the DOCSIS standard, Cisco has designed the line of equipment to integrate into a cable or wireless broadband headend solution. Specifically, the WT-2710 system is designed for operation within the U-NII band only, and the WT-2750, for the MMDS bands. U-NII and MMDS are compared in Table 6-11.

Feature		MMDS	U-NII
Frequency band	—	2.5 GHz	5.7 GHz
Channel size	Multipoint	1.5 MHz, 3 MHz, or 6 MHz	1.5 MHz, 3 MHz, or 6 MHz
	Point-to-point	—	6 MHz or 12 MHz
Coverage	Multipoint	Up to 35 miles	Up to 7 miles
	Point-to-point	—	Up to 20 miles
Non-LOS deployment (when VOFDM is used)	Multipoint	Yes	No*
	Point-to-point	—	Yes
License requirement	—	Yes	No
Speed of deployment	—	Lower	Higher
Cost of deployment	—	Higher	Lower

* While the technology is capable of non-line-of-sight operation in U-NII multipoint solutions, it is not recommended because of FCC restrictions.

Table 6-11. MMDS and U-NII Solution Comparison

3G as a Broadband Access Solution

To the average person, the mention of 3G (third-generation) wireless technology evokes mobile cellular communication. Less publicized is its potential as a fixed wireless solution—the technology is capable of speeds up to 2 Mbps in fixed mode.

First-generation wireless refers to the analog cellular transmissions that predate the digital types of service that are common today. Second-generation (2G) wireless is the current digital cellular and personal communications that we use today for voice communication and short text messages.

Third-generation wireless technology is capable of supporting circuit and packet data at high bit rates: 144 Kbps for high mobility (vehicular) traffic; 384 Kbps for pedestrian traffic; and 2 Mbps for fixed indoor traffic. It is a worldwide standard that is also known by the ITU designation of IMT-2000, the *International Mobile Telecommunications 2000 initiative.*

Of the different wireless broadband access technologies, 3G is better positioned than most to make the wireless broadband promise a reality. It has the benefit of global awareness, it is backed by standards, and it shares a commonality of design with existing wireless services that paves an easier path for making the transition into new services. In contrast, most other options, including MMDS and LMDS, are virtually unknown and require standards. With 3G, there is already a point of reference in the mind of the consumer, who understands wireless communication in terms of mobile cellular communications and therefore finds it easier to appreciate 3G as a logical next step for a technology he or she uses daily—the cognitive association is already there.

The business hurdles, too, are a lot less challenging with 3G. The technology requires a lot less marketing if it is positioned as the next generation of wireless service. Other wireless access options—with maybe the exception of DBS—requires a more extensive marketing plan because of the need to educate on new terminology and positioning of the product. For existing wireless operators, it is easier to upgrade than to build a new infrastructure.

So what's the delay? In the U.S., wireless companies must be able to access a significant amount of spectrum that is currently dominated by government and other commercial interests.

3G Spectrum

In October 2000, President Bill Clinton executed a memorandum that directed the Secretary of Commerce to work with the FCC to develop a plan to select a spectrum for 3G wireless. The memorandum also dictated some other key dates, but the main point was to enable the commission to identify spectrum for 3G systems by July 2001 and begin auctioning licenses by September 2002.

In July 2001, an agreement between the FCC and the Commerce Secretary was reached to postpone the deadline, and in September of the same year, the Commission added mobile allocation to the 2500–2690 MHz band. This allocation made the band available for advanced mobile and fixed terrestrial wireless services, including 3G. Incumbent users of the spectrum—ITFS and MMDS service licensees—were not relocated, nor were their licenses modified.

In that same month—September 24—after very aggressive lobbying from incumbent license holders, the FCC announced that they would no longer be considering the 2500–2690 MHz band for immediate 3G use. The story goes on, but the brief history recounted so far provides a good indication of why 3G services have been delayed. In short, we cannot decide on what portions of the spectrum to allocate for the service.

In an effort to offer some indication of when some of these issues may be resolved, we will take a brief look at the current state of affairs.

On October 5, 2001, the FCC, the NTIA, the Department of Defense (DOD), and other U.S. federal agencies released the "New Plan to Identify Spectrum for Advanced Wireless Mobile Services." The new plan addresses the potential use of the 1,710 MHz–1,770 MHz band and the 2,110 MHz–2,170 MHz band for advanced wireless services.

The plan also specifies that the federal government incumbents in the 1,710 MHz–1,770 MHz band "will be assessing their future spectrum needs in light of new national security demands" following the September 11, 2001, terrorist attacks. The government plans to complete their assessment by spring 2002, and the NTIA has proposed to postpone the September 2002 deadline for the auctioning of the spectrum until September 30, 2004.

NOTE: Wireless planners for 3G services have a strong interest in the 1,700 MHz band because many other countries have slated it for 3G. Its use would help reduce deployment costs, as it would facilitate mass production of chipsets regardless of where they are used. It would also aid in developing global roaming. The buzz term for this interest in a common band is *globally harmonized spectrum.*

Unfortunately, we will be waiting for some time before 3G services are launched in the U.S. In the meantime, wireless operators are announcing interim data strategies. *GPRS,* the *General Packet Radio Standard,* and *EDGE, Enhanced Data Rate for Global Evolution,* are examples of these interim solutions generically called 2.5G (2.5-generation) networks. GPRS, which evolved from GSM, transmits up to 115 Kbps, and EDGE, a TDMA solution, transmits up to 384 Kbps. The CDMA version of the 2.5G offering is HDR, *High Data Rate,* and is capable of data rates up to 1.4 Mbps in fixed mode.

DBS: Direct Broadcast Satellite

The *Direct Broadcast Satellite (DBS)* service is arguably in the best position to grab market share in the broadband wireless access market. Companies in this group, which includes DirecTV, USSB, EchoStar, and BSkyB, are now old hands at delivering direct-to-home digital broadcasting, a mass-market service that has experienced a rapid take-up. Multichannel broadcasting of subscription television is now a multibillion-dollar business.

The development of DBS services from the perspective of new service launch has been in lockstep with cable operators. As the largest competitors to cable, both industry groups maintain a competitive posture to gain market dominance. It should, therefore, not be surprising that DBS companies have been forging ahead in the rollout of two-way high-speed services. Initially, as with cable, the focus was on Internet access with a telephone return path. While the target market—Internet access—has not changed, a satellite return path is now available in addition to telephone returns. Progress, however, has been very slow—data services are still in their infancy, with access rates of about 500 Kbps and about 180,000 U.S. subscribers at the end of 2001.

EchoStar was one of the first DBS companies to offer two-way communication using a satellite return path, through its stake in StarBand (formerly known as Gilat-2-Home)— a partnership between EchoStar, Gilat Satellite Networks, and Microsoft. They launched a high-speed Internet access service in 2000, using a single antenna capable of receiving EchoStar's video signal as well as two-way high-speed Internet access. The service offers asymmetric speeds of 500 Kbps downstream and 150 Kbps upstream. Another service, DirecWay, formerly DirecPC, the satellite Internet access division of Hughes Network System, launched their service late in 2001.

DBS companies have been wary of the consolidation that has been happening in the cable industry. DBS operators have gained market share in cable markets at a rate that has surprised many that thought the service would succeed only in areas that did not have cable. Any change within the cable industry that has the potential to increase the cable operators' dominance is viewed with caution.

In 2001, EchoStar made an unsolicited bid for Hughes that was worth $26 billion. On October 28, 2001, General Motors and its subsidiary Hughes Electronics, together with EchoStar, announced the signing of an agreement that would allow the spin-off of Hughes from GM and the merger of Hughes (owner and operator of a substantial satellite operation) with EchoStar. The combined company would use the EchoStar name with a combined 16.7 million subscribers, which is about 17 percent of the subscriber television market, with cable operators controlling more than 80 percent. The Hughes DirecTV brand would continue to be used for the consumer satellite pay TV service—DirecWay is the brand name for the broadband service.

DBS Architecture

DBS uses geostationary satellites operating in the Ku band with a 12 GHz downlink and a 14 GHz uplink. The architecture of the service for broadband wireless is very straightforward. It uses a "bent-pipe" approach, a term used to describe the signal path when satellites are used: signals go up and are reflected to the target earth station.

The basic architecture of the service is very similar to cable: a headend receives feeds from the different services and networks and encodes them into MPEG for digital transmission (see Figure 6-18). The composite signal is transmitted to a geostationary satellite, which in turn broadcasts it to the subscriber antennas. The subscriber antenna is connected to a receiver that connects to the television set.

For Internet access, requests for a web page are sent from the subscriber to the provider's headend (DBS companies sometimes use the term *network operations center (NOC)* when the service is being used in the context of data). The NOC is connected to the Internet via a wireline facility. The response from the Internet is sent to the NOC, which combines the data along with other television feeds for transmission to the satellite and back to the subscriber. (A detailed discussion of MPEG and its transport service was offered in Chapter 5.) The operation of DBS from a broadcast perspective uses the same techniques discussed.

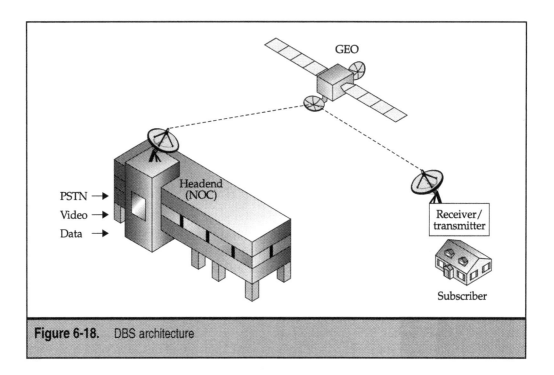

Figure 6-18. DBS architecture

A feature that distinguishes DBS from other satellite C band services is the size of the antenna. The large dish that is used by C band services has a lot to do with the power limitations imposed on satellite transmissions in that band. As mentioned previously, many telephone companies around the world use the C band for terrestrial microwave relays, so power levels of satellites operating in this band had to be limited to prevent interference. The Ku band, where DBS operates, does not have this limitation, so higher power levels are permitted. The higher power level allows smaller dishes.

Satellite Topologies

Irrespective of the orbit that is used, GEO and LEO satellites use one or a mixture of three different topologies depending on the treatment of the payload. Satellites can either transparently retransmit received signals or use on-board processing.

In transparent mode, either a bent-pipe star topology (point-to-multipoint) or a bent-pipe point-to-point system is used.

The *star topology* is characterized by a large gateway earth station, which transmits one or more high data-rate forward links to a number of smaller user terminals (see Figure 6-19). Each broadcast contains MAC-level addressing that identifies the intended receiver.

In the return path, the remote user terminals transmit bursts at low to medium data rates.

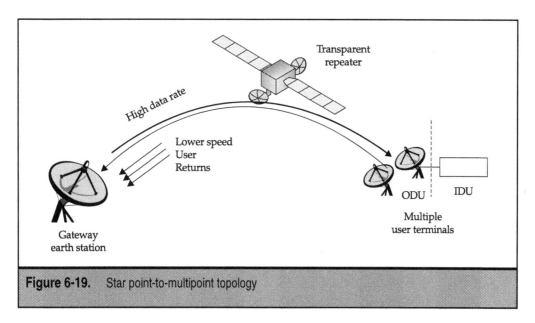

Figure 6-19. Star point-to-multipoint topology

The *two-way point-to-point* topology uses a dedicated two-way connection between a user terminal and the large gateway earth station (see Figure 6-20). It is based on the VSAT concept and supports both asymmetric and symmetric data speeds.

For satellites that use *on-board processing,* the satellite plays a more significant role that aids in the switching function (see Figure 6-21). A wireless trunk is established between

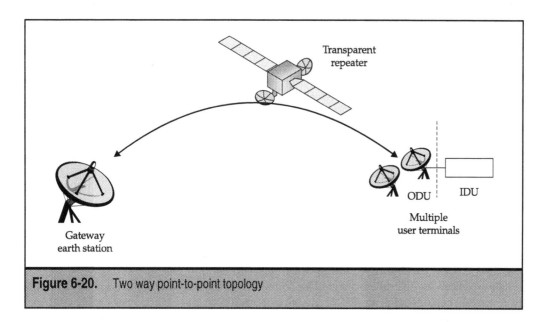

Figure 6-20. Two way point-to-point topology

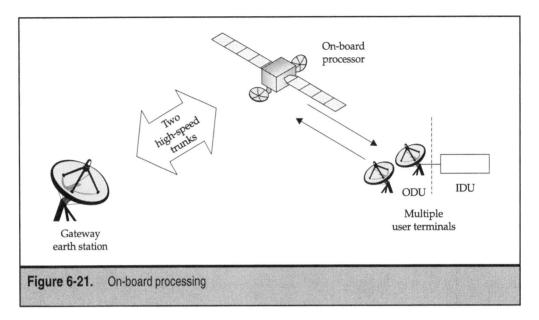

Figure 6-21. On-board processing

the satellite and the large gateway earth station. The gateway station transmits a multi-plexed stream up to the satellite that de-multiplexes it into individual streams that are forward-transmitted to different geographical regions or different regional pockets of user terminals.

The return path signals from the multiple geographies or regional pockets of users are multiplexed into a single downlink trunk between the satellites and the gateway.

GEO and LEO Planned Ventures

There are several bold projects for high-speed broadband service using LEOs. Bold be-cause most people are aware of LEO ventures because of the names that have come and gone like Iridium. Regardless, projects like Teledesic and Skybridge hope to be able to im-prove on current satellite data rates by launching LEO satellites and related ground sta-tions in direct competition to DSL and cable.

These ventures are not cheap—they run in the billions of dollars—and the launch dates have slipped further into the future. Many of these ventures are planning to target the enterprise market initially, rather than the consumer, in the hope that businesses will be more inclined to pay a premium for the service than will consumers for their lower-margin Internet access.

If these projects ever get off the ground, they do have the potential to offer speeds competitive with wireline solutions. Till then, LEO services' "crash and burn" reputation will prove to be a stumbling block. Strong marketing and a strong business plan will take precedence over the technological possibilities.

In an effort to succeed, projects like Teledesic have scaled back their original plans from an 800 constellation to 288. If the focus remains on the business case and ventures continue to fine-tune their plans, the likelihood of success greatly improves.

Teledesic Network

Teledesic, self-described as the Broadband Internet-in-the-Sky Network, is a high-capacity broadband network that uses LEO satellites (see Figure 6-22). The planned reach is 100 percent of the earth's population and 95 percent of the landmass with the capacity to support millions of people. The list of investors in the venture include Craig McCaw, Bill Gates, Motorola, His Royal Highness Prince Alwaleed Bin Talal Bin Abdul Aziz Alsaud, Abu Dhabi Investment Company, and Boeing. The targeted launch date is 2005.

Using fiber optics as the guideline for service quality, the Teledesic network is designed for high bandwidth, low latency, low error rates, high availability, and "fiber-like" QoS. The network components consist of:

▼   A *ground segment* that provides a gateway to other service providers and corporate networks, network operations, and network control

■   A *space segment* that provides the satellite-based switching functions and communication links to the user terminals

▲   *User terminals* that provide the interface between the satellite and the user network and equipment

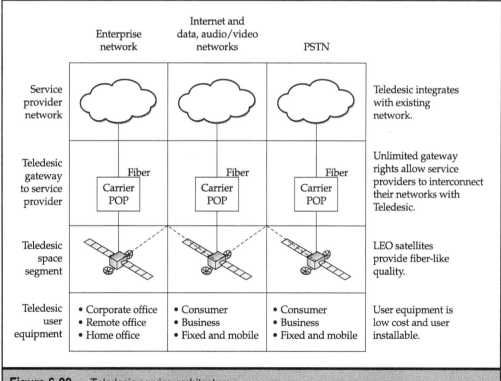

Figure 6-22.   Teledesic service architecture

The network will support speeds of up to 64 Mbps from the network to the user (downlink) and up to 2 Mbps for the uplink. Terminals designated as broadband terminals will have 64 Mbps symmetrical data speeds.

Technology   The network uses on-board satellite processing to aid in switching. Each satellite in the constellation is a switching node and has intersatellite communication links with other satellites in the same and adjacent orbital planes. The meshing of satellites—a "geodesic" network—makes for fault tolerance, faster routing, and lower latency.

Using a fast packet-switching scheme, the network employs a protocol that is similar to the Open Shortest Path First (OSPF) protocol. Each packet has a destination address of the target terminal, and each node (satellite) independently selects the path with the least delay. With this packet-based approach, multiple packets for a given address may take different routes through the network. The destination terminal buffers and reorders the packets (see Figure 6-23).

The packets are of a fixed length and contains the destination address, sequence information (for reordering), error control, and the payload.

The service operates in the Ka band, downlinks operating between 18.8 GHz and 19.3 GHz, and uplinks between 28.6 GHz and 29.1 GHz. As with LMDS, frequencies in this spectrum are susceptible to rain and other environmental factors. To compensate, the satellites operate at a high angle of elevation above the horizon. The significance of this approach is that instead of transmitting against the fall of the raindrops, the satellite transmits in the direction of the fall, where there is less surface area to the drop. By employing this approach, the Teledesic network hopes to achieve an availability of 99.9 percent or better.

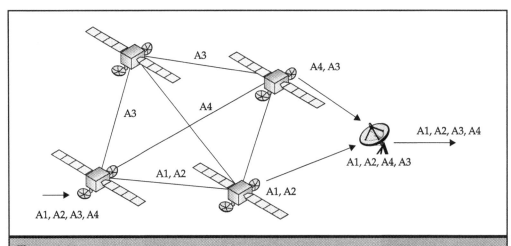

Figure 6-23.   Teledesic distributed adaptive routing

Skybridge Network

Skybridge plans on using a combination of GEO and LEO satellites to deliver a full range of IP-based services. The GEO service, planned for 2001, is designed to provide IP-multicasting services for applications like video-conferencing and distance learning. The LEO service, planned for 2004, will use 80 satellites to deliver high-speed integrated digital services. Features of this network are shown in Table 6-12.

General Features	
Partnership	Alcatel, Toshiba, Boeing, Thomson. CNES, Starsem, COM DEV, EMS Technologies, SRIW, Level 3, Loral, Mitsubishi, Northrop Grumman, Qualcomm, Sharp, and SNECMA
Satellites	GEOs and 80 LEOs
Planned deployment	2004 for two-way broadband services using LEOs 2001 for IP-based multicast services using GEOs
Hub and Satellite	
Operation	Multiple transponders with a capacity of over 100,000 per hub C, Ku, and/or Ka band
Switching	At the hub site
Available bandwidth	Outbound: DVB up to 45 Mbps and over 60 Mbps with 8 PSK Inbound: Multiple TDMA carriers with an aggregate rate up to 30 Mbps per transponder
Data transmission	Multicast: 64 Kbps to 8 Mbps per multicast session Point to point: Burst rate dynamically allocated. Downloads can be shaped between 50 Kbps and 2 Mbps depending on grade of service Uploads: Between 64 Kbps and 1 Mbps depending on terminal equipment
User Terminal	
Outdoor unit (ODU)	Antenna: 0.9 m to 1.8 m Transmit power: Up to 2 W and 5 W in C band
Indoor Unit (IDU)	L band interface to ODU 10/100Base-T LAN interface

Table 6-12. Features of the Skybridge Network

SUMMARY

Wireless communication offers the greatest potential for broadband access because it offers the greatest freedom. However, it also poses the greatest challenge because the medium that it uses is a finite resource, is publicly owned, and must be shared by many groups for many purposes. It is used by technologies that save our lives, protect, entertain, communicate, and support many other applications. Who, then, should, and by what measure do we, decide which application is more important, which should go and which should stay? The answer is, we cannot; we just have to find an equitable way of sharing. Other access media do not have these challenges—they exist within their own sealed domains.

For these reasons, new technologies must be efficient in the way they utilize the spectra in which they operate. Technologies tend to be more expensive when they need to get more from a lot less, and because there is less, there is a constant need to find new ways to be more efficient, and the quest for efficiency leads to further change. The problem that this introduces is knowing when the right set of feature and function has been reached and when the focus should be shifted to developing standards to ensure interoperability. All these reasons contribute to a more deliberate approach to defining and publishing new standards. And standards are what wireless broadband access would benefit from most.

Many wireless technologies are emerging for broadband access, but they are not as far along as wireline solutions. We will, however, eventually see many of the technologies discussed become more commonplace because as market projections show, a market exists for these technologies and where there is a market, entrepreneurs have always found a way.

Network Professional Library
Broadband Networking

Table of Contents

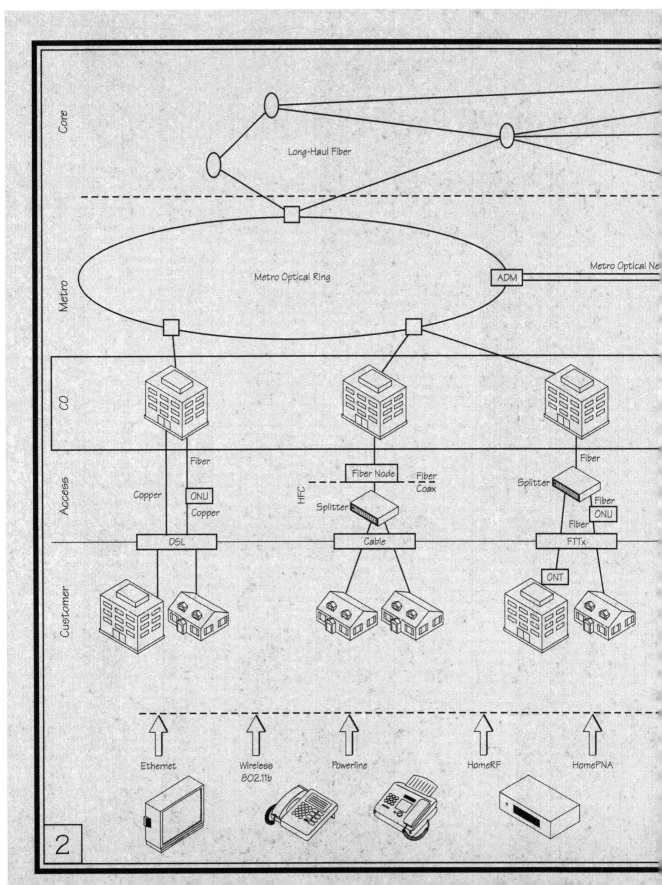

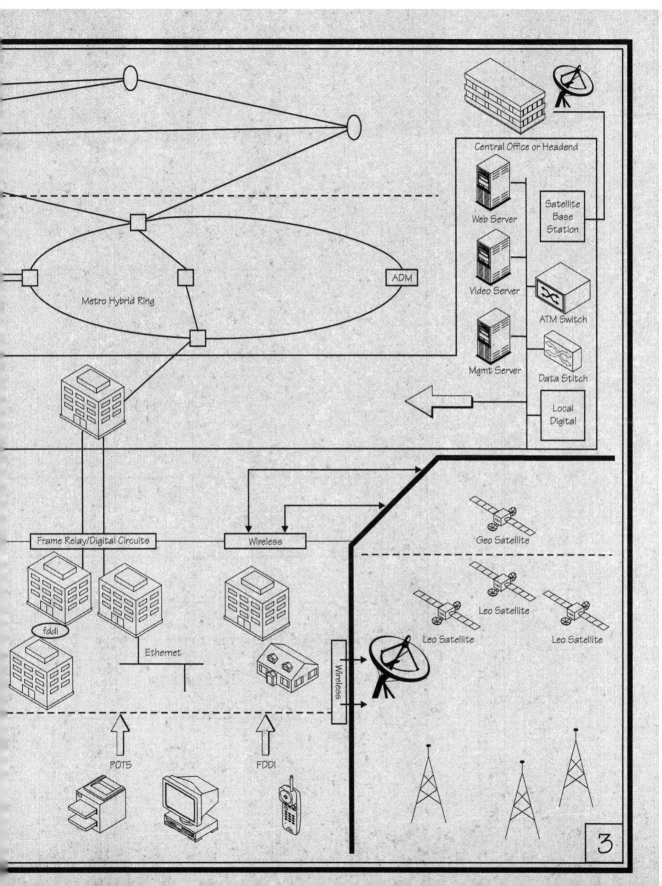

Central Office or Headend

Web Server

Satellite Base Station

Video Server

ATM Switch

Mgmt Server

Data Stitch

Local Digital

ADM

Metro Hybrid Ring

Frame Relay/Digital Circuits

Wireless

Geo Satellite

Leo Satellite

Leo Satellite

Leo Satellite

fddi

Ethernet

Wireless

POTS

FDDI

3

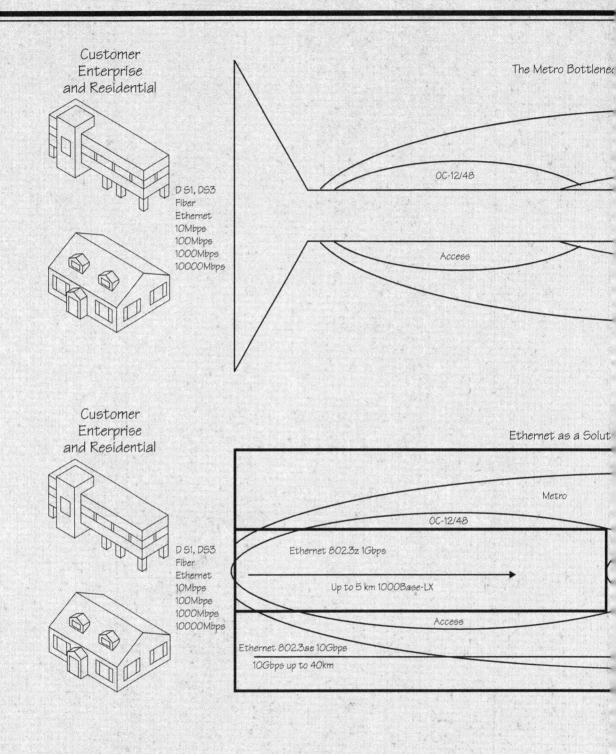

Customer
Enterprise
and Residential

D S1, DS3
Fiber
Ethernet
10Mbps
100Mbps
1000Mbps
10000Mbps

The Metro Bottlenec

OC-12/48

Access

Customer
Enterprise
and Residential

D S1, DS3
Fiber
Ethernet
10Mbps
100Mbps
1000Mbps
10000Mbps

Ethernet as a Solut

Metro

OC-12/48

Ethernet 802.3z 1Gbps

Up to 5 km 1000Base-LX

Access

Ethernet 802.3ae 10Gbps

10Gbps up to 40km

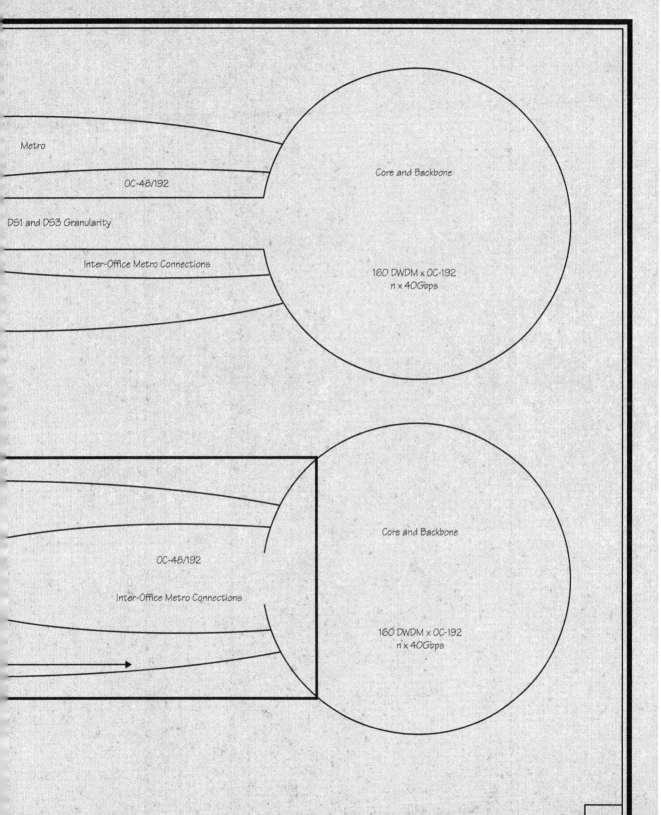

Metro

OC-48/192

DS1 and DS3 Granularity

Inter-Office Metro Connections

Core and Backbone

160 DWDM x OC-192
n x 40Gbps

OC-48/192

Inter-Office Metro Connections

Core and Backbone

160 DWDM x OC-192
n x 40Gbps

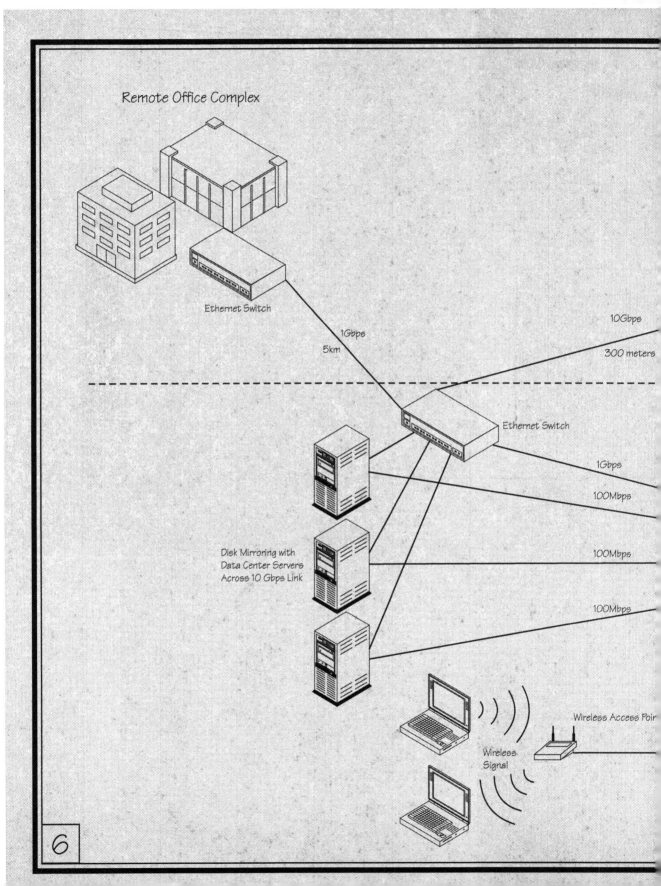

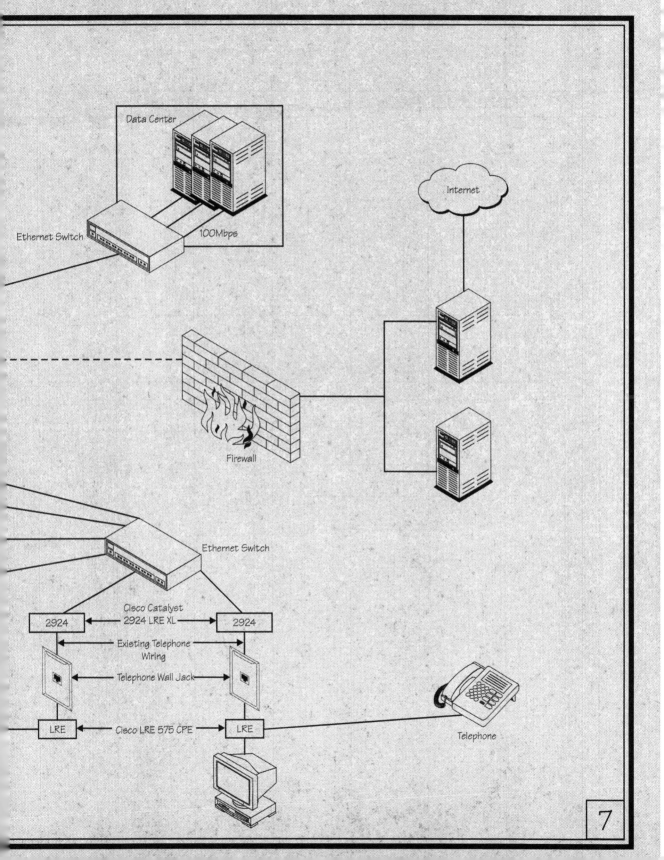

Data Center

Ethernet Switch

100Mbps

Internet

Firewall

Ethernet Switch

Cisco Catalyet
2924 LRE XL

2924

2924

Existing Telephone
Wiring

Telephone Wall Jack

LRE

Cisco LRE 575 CPE

LRE

Telephone

Electromagnetic Spectrum for Telecommunications

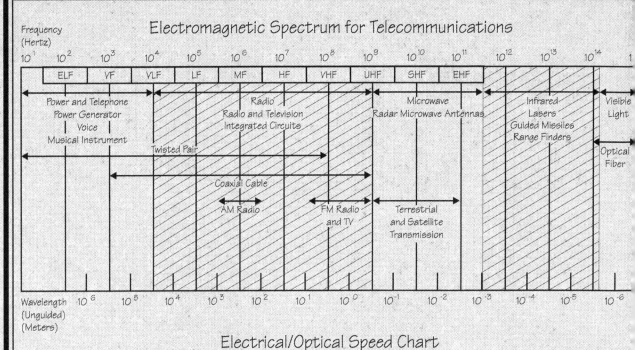

Frequency (Hertz)

10^1 10^2 10^3 10^4 10^5 10^6 10^7 10^8 10^9 10^{10} 10^{11} 10^{12} 10^{13} 10^{14}

| ELF | VF | VLF | LF | MF | HF | VHF | UHF | SHF | EHF |

Power and Telephone
Power Generator
Voice
Musical Instrument

Radio
Radio and Television
Integrated Circuits

Microwave
Radar Microwave Antennas

Infrared
Lasers
Guided Missiles
Range Finders

Visible Light

Optical Fiber

Twisted Pair

Coaxial Cable

AM Radio

FM Radio and TV

Terrestrial and Satellite Transmission

Wavelength (Unguided) (Meters)

10^6 10^5 10^4 10^3 10^2 10^1 10^0 10^{-1} 10^{-2} 10^{-3} 10^{-4} 10^{-5} 10^{-6}

Electrical/Optical Speed Chart

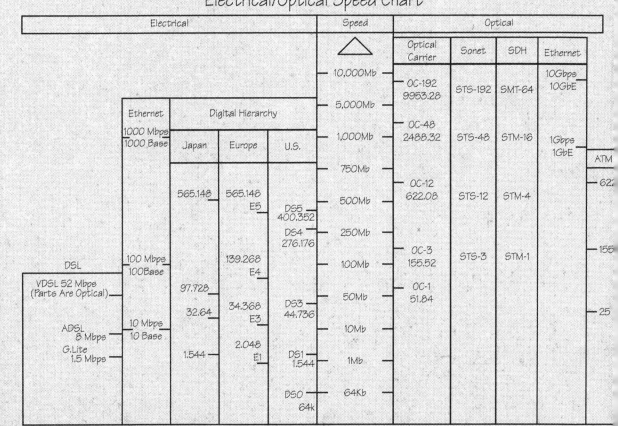

Electrical					Speed	Optical			
Ethernet	Digital Hierarchy					Optical Carrier	Sonet	SDH	Ethernet
1000 Mbps 1000 Base	Japan	Europe	U.S.		10,000Mb	OC-192 9953.28	STS-192	SMT-64	10Gbps 10GbE
					5,000Mb				
	565.148	565.148 E5			1,000Mb	OC-48 2488.32	STS-48	STM-16	1Gbps 1GbE
			DS5 400.352		750Mb				
					500Mb	OC-12 622.08	STS-12	STM-4	
			DS4 276.176		250Mb				
DSL	100 Mbps 100Base	139.268 E4			100Mb	OC-3 155.52	STS-3	STM-1	
VDSL 52 Mbps (Parts Are Optical)						OC-1 51.84			
	97.728	34.368 E3	DS3 44.736		50Mb				
	32.64				10Mb				
ADSL 8 Mbps	10 Mbps 10 Base								
G.Lite 1.5 Mbps		2.048 E1	DS1 1.544		1Mb				
	1.544								
			DS0 64k		64Kb				

ATM
622
155
25

CHAPTER 7

FTTx Access Networks

During the heyday of the late 1990s, when the growth of the Internet was the subject of many conversations, venture capital funding flowed freely and everyone wanted a piece of the pie of the great promise. Billions of dollars were invested in laying fiber across the U.S. and the oceans. At the end of the decade, there was an over-supply of fiber in the core of the network.

The excess capacity, when combined with the large amount of content and the growing demand for affordable bandwidth for homes and businesses, created what is now known as the "metro service bottleneck." An abundant supply of bandwidth exists in the core of the network, but the infrastructure that connects the users is still not adequate to make use of its availability. The metropolitan area is now the focus of upgrades, and fiber rings are being installed in metro areas. Eventually the bottleneck will move to the edge of the network, the portion that is used to connect our homes and our offices. This is the business rationale for the technologies and services we have discussed so far. Cable, DSL, and wireless broadband access are all attempts to alleviate the bottleneck at the edges.

A side benefit of the optical build-out is the R&D that was focused on developing smaller and more affordable optical components. The advances in technology have resulted in more affordable components that are now making their way into the distribution and access portions of the network. It is very possible that, by the year 2010, we will look back at 2001 and 2002 as the pivotal years when fiber became more commonplace as an access technology.

In this chapter, we will take at a look at another approach to addressing the bottleneck, FTTx.

FIBER TO THE X ACCESS NETWORKS

FTTx, fiber to the x, refers to the installation of optical fiber cable directly, or much closer, to the customer. The *x* is a wildcard that is used to generically describe different applications of the technology. The more specific uses of the term describe where the fiber terminates; common variations include to the curb (FTTC); to the home (FTTH); to the business or building (FTTB); to the neighborhood (FTTN); and to the cabinet (FTTCab). Another, more recent description is fiber to the office (FTTO).

▼ **Fiber to the curb (FTTC)**   It brings the fiber close to the customer premises but does not actually extend as far as the customer. Copper is generally used for the connection to the customer. *Fiber to the neighborhood (FTTN) and Fiber to the cabinet (FTTCab)* use the concept of bringing the fiber closer to the customer premises by running the fiber to a street cabinet or another location within a few hundred feet of the fiber/wire handoff.

■ **Fiber to the building (FTTB)**   Sometimes referred to as *Fiber to the basement* or *Fiber to the business*, it brings the fiber even closer to the customer than FTTC solutions. The fiber extends to a location within the building where it is handed off to the local building wiring, which is normally copper, though it may also be fiber. MDUs and MTUs are ideal for FTTB solutions.

▲ **Fiber to the home (FTTH)**   The ultimate form of fiber access, it extends the fiber from the provider network directly into the home.

Many of the features of FTTC were covered when we reviewed VDSL. We will therefore focus our attention on FTTH and FTTB solutions. Remember that the main difference in the three approaches is the point at which the fiber terminates. With FTTH and FTTB, the fiber extends all the way to the customer premises, so in theory, it has everything that FTTC has plus a bit more—the network interface to the customer.

The biggest inhibitor to unimpeded high-speed transmission to customers is the bandwidth of the last mile. Whether the medium is copper or coaxial cable, higher-speed signals attenuate much more rapidly with distance, and each transmission medium has its own set of characteristics that make it less appealing than fiber optics. Incumbent service providers recognize this fact and the obvious benefits of fiber.

Cable operators have upgraded, or are in the process of upgrading, their coaxial trunks to fiber trunks, resulting in hybrid fiber-coax (HFC) networks. Traditional telephone service providers are also leveraging the benefits of fiber through the introduction of VDSL, where fiber is run to a central location in the neighborhood, FTTCab, and connects to copper loops that extend to the subscribers. In both instances, the solution stops short of running the fiber to the customer premises; the existing wiring infrastructure is used to do that. In these applications, the fiber is a means of consolidating different feeds and backhauling it to the operator's headend or central office.

Background

In 1986, BellSouth began deploying *fiber in the loop (FITL)* with the intent of initiating trials for fiber to the home. The FTTH trial deployments began in Orlando, Florida, with the delivery of video and telephone service. After several years, the trials, which were conducted in several trial locations, came to an end and BellSouth concluded that the solution was too immature and too expensive for general deployment.

The focus was shifted to FTTC, as this was deemed more cost-effective because multiple customers could share the fiber and the ONU. What's more, the technology had a lot in common with FITL. As a result of these trials and the emergence of standards (Bellcore TR-909), FTTC became a viable solution that was generally deployed by BellSouth for telephone service in 1995.

The capability of the FTTC infrastructure was expanded in 1998 to include high-speed data and video services. The evolution of the FTTC solution into a full-service system prompted BellSouth to increase the deployment of fiber in its distribution network over the next two years.

NOTE: The Bellcore TR-909 is a set of specifications for fiber in the loop (FITL).

The work on FTTH was not dead. Other companies in the U.S., and especially in other parts of the world, continued to evaluate its feasibility through trials using different technologies. BellSouth continued its work on FTTH through cooperation with other international telecommunications companies that started an initiative called the Full Service Access Network (FSAN). As a result of this effort, a new specification was developed for an optical network that used passive components, which was called a PON, a passive optical network.

FTTH Council

Today, fiber has the lowest penetration of all the access media, but few disagree that it represents the best medium for forging broadband connections. It does not take much imagination to see the benefits of an all-fiber network, which would be like connecting every home, every business, and every building to an endless supply of bandwidth. In recognition of the potential, a number of companies—including Corning, Alcatel, and Optical Solutions—from different industry sectors formed the FTTH Council in 2001.

The council held its first meeting in August 2001, when 43 companies attended; within two weeks, the number of member companies increased to 56. At the time of this writing, there are 61 members from the telecommunications, computing, networking, application, and content and service provider industries. The mission of the council is to educate, promote, and accelerate FTTH and the resulting quality-of-life enhancements.

The formation of the council is very timely. Independent trials of FTTH solutions have been ongoing for the past few years, and the technology is now being deployed in some U.S. markets. The lessons learned from independent trials are usually not shared among different companies for the benefit of the industry. This is where the council should prove helpful. By bringing together experts from different related fields, the industry ends up being the winner, and if the industry wins, so do the companies that operate within it.

It is too early since the council's formation to see an immediate benefit. But by using the success of other industry coalitions as a benchmark, this group should help bring the required focus to this emerging market. Their focus should also help reduce the cost of the equipment required for FTTH deployments and the development of standards.

FTTx Facts and Myths

There are common misconceptions about FTTx, the first being the belief that by installing fiber, you immediately gain access to an almost unlimited supply of bandwidth. In practice, the amount of bandwidth that can be obtained from a fiber optic cable is similar to the amount of oil that can be mined from an oil well that has an almost unlimited supply of oil.

The amount of oil that can be obtained is purely a matter of economics. There is a point at which the cost to drill very deep oil beds exceeds the economic benefit of drilling. As time goes by and the cost of drilling decreases—through technology improvements, for instance—it makes sense to go back to those beds that were not economically feasible to exploit.

Likewise in the case of bandwidth, the deciding factor on the amount of bandwidth that will be available to the user is the cost of the equipment to extract that bandwidth. The cost to tap the bandwidth must be balanced against the price the user is willing to pay. So in summary, if fiber extends to the home or business, business factors will prevail over technological possibilities. The customer will get only the amount of bandwidth that makes economic sense in terms of the price being charged for the service.

Misconception number two is that fiber means all digital and digital pictures are of a better quality than analog. This is true only when the analog signal is distorted as a result of interference or some other factor. A good analog signal presents a picture that is hardly distinguishable from one created digitally. In fact, digital signals can be too sharp and look unnatural because of the compression that is usually required. Also, the analog image is always the point of reference; the intent is to make the digital image as good as the analog. An uncompressed digital signal needs about 15 times more bandwidth than does an analog signal. Finally, a fiber medium does not necessarily imply all digital content.

Another common misconception is that fiber is too expensive to extend to the home or business. This is true if separate strands of fiber were being run between the provider and the customer, as is done with copper loops. The use of analog transmitters and passive optical networks (PONs), which allows the fiber to be shared among users, significantly reduces the cost to the point that it is no longer an issue on a per-subscriber basis.

The Passive Optical Network

From the beginning of development of optical networks, developers have envisioned the coexistence of optical and electrical signals and so have always striven to make the transition from one to the other as smooth as possible. Optical fiber has been used for years, to haul traffic over long distances between local distribution networks that were predominantly based on electrical signals and wire. These local distribution networks are now being targeted for upgrade, and fiber is becoming the natural choice because optical splitters offer low maintenance, do not need power, and are small enough to fit into a manhole.

At the heart of FTTx is the *passive optical network (PON),* a system that uses fiber optics to connect customers by extending the fiber infrastructure closer, or all the way, to the customer premises. The term passive describes the fact that optical transmission has no power requirements once the signal is traversing the network. With PONs, the signal is transported by lasers and requires no electric power, thereby eliminating the need for expensive powered equipment between the provider and the customer.

PONs utilize light of different colors (wavelengths) over strands of glass (optical fibers) to transmit large amounts of information between customers and their intended destination, and between the service provider and the customer.

Another common name for PONs is BPON, or *broadband passive optical network.* A PON consists of fiber optic cable, an *optical line termination (OLT),* passive optical splitters, and an *optical network unit (ONU),* which is placed at or close to the customer.

The OLT is located in the service provider's main facility—the local exchange, for instance—and interfaces with the networks that provide the users with services such as data and video. The fiber is connected to the OLT and cascades outward toward the customer and runs deep into different neighborhoods, as shown in Figure 7-1.

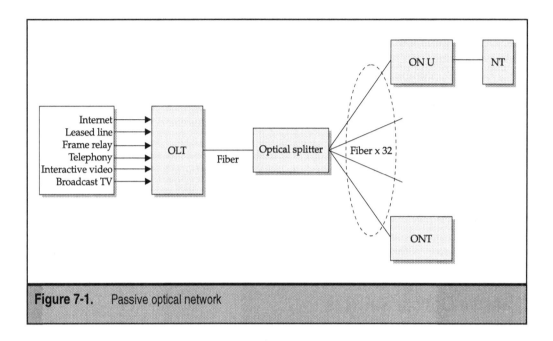

Figure 7-1.   Passive optical network

Optical splitters are used to split the main fiber feed into many different feeds that travel to different parts of the coverage area. At the end of the fiber run is an ONU, which may be combined with a network termination unit (NT) to produce an ONT, *optical network termination*. Up to 32 ONUs may be connected to an OLT.

The actual placement of the ONU/ONT is what differentiates the different FTTx options. In FTTC applications, it is normally placed in a cabinet on a street close to the locations being served, and in FTTH and FTTB, units are placed at the customer location. Additionally, the use of an ONU or an ONT—an integrated ONU and NT—is governed by where the PON terminates, as depicted in Figure 7-2.

An ONU is used in those cases where the network termination must be separate from the ONU, as is the case when using VDSL. In these cases, an ONU is used to terminate the fiber in a cabinet on the street close to the subscriber; VDSL is then used across the copper link to the customer premises, where it terminates on an NT. This is also sometimes the case in FTTB applications.

For FTTH, and in some instances for FTTB, an ONT is used because the fiber terminates at the point where the connection is handed off to the customer's internal wiring.

The OLT can generate its own light signals, or it can take a SONET signal (such as OC-12) from a colocated SONET cross-connect. The signal is then broadcast through one or more subscriber ports to either an ONU or an ONT. The ONU/ONT then converts the optical signal into an electrical signal for use in the customer premises.

A PON can deliver up to 622 Mbps downstream and up to 155 Mbps (OC3 speed) upstream. This bandwidth can be shared among several subscribers, which makes it a lot more cost effective than earlier approaches where discrete strands were used per customer. The network uses a point-to-multipoint topology, which makes it ideal for appli-

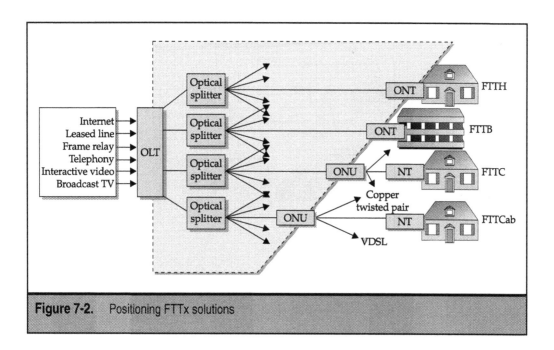

Figure 7-2. Positioning FTTx solutions

cations where the same information is transmitted to multiple users, as is the case for digital television services. A bus or ring topology was ruled out because of the disruptive potential of a single user using these topologies. In a PON, problems at a single-user site do not affect other users of the system.

Full Service Access Network Initiative

The G-7 (Group-7) initiative was a group of seven telecommunications companies that was formed to address the different access network requirements of different countries. The objective of the group was to develop a common cost-effective system for deployment. In 1995, the group was renamed the *Full Service Access Network (FSAN),* and it has been studying passive optical networks as a common platform for all telecommunications companies.

The group, which now consists of 21 members, is not a standards body, but rather submits specifications to standards organizations like the ITU for consideration and incorporation within standards. It is divided into different workgroups, each having specific areas of focus. One such group, the *Optical Access Network (OAN)* workgroup, created an ATM passive optical network specification, which was presented to various standards bodies including the ITU, ETSI, and the ATM Forum. The ATM PON was determined to be a cost-effective solution for deploying FTTH, FTTB, and FTTC solutions.

In 1998, the ITU adopted the ATM PON specifications, which were published in G.983.1, a broadband optical access system based on PON (see Figure 7-3). The PON as defined in ITU G.983.1 has two interfaces options: a 155 Mbps symmetrical PON interface and an asymmetrical interface operating at 622 Mbps downstream and 155 Mbps upstream.

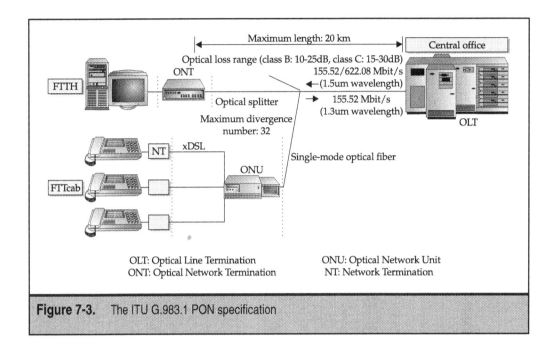

Figure 7-3.   The ITU G.983.1 PON specification

NOTE: The first standard that actually described PON in detail was G.982. This early description focused on early ISDN-type equipment. Today, there are five different PON standards, with the last three additions published in 2001, as shown in Table 7-1.

Modes of Operation   The symmetrical mode of operation allows a limited number of MPEG-2 video feeds to be delivered to the customer. This mode is typically used in a point-to-point delivery scheme where a separate video channel is transported to each ONU. Obviously, this quickly erodes the available bandwidth and limits the number of customers that can be served in this fashion.

Using the asymmetrical interface, the 622 Mbps bandwidth can be used to deliver over one hundred 6 Mbps video channels. If 32 customers are attached to an OLT, this equates to about three video channels per customer in an FTTH application, a much more efficient use of the system. The downside is the limited amount of bandwidth that would be left over for data services. To address this, the FSAN OAN group defined a solution that overlays wavelength-division multiplexing (WDM) on top of the G.983.1 standard to produce an additional downstream band.

The specification was adopted by the ITU and published in February 2001, in G.983.3. In this new standard, the additional band is called an *enhancement band (EB),* and the original downstream band has been renamed to the *basic band (BB).* Figure 7-4 illustrates the different bands that were defined in the new revision of the standard.

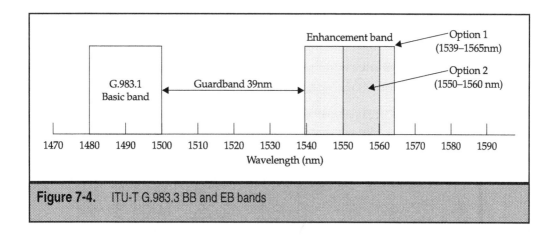

Figure 7-4.   ITU-T G.983.3 BB and EB bands

VDSL is dependent on FTTCab, and as a result, a subcommittee of FSAN was formed called FS-VDSL, *Full Service VDSL.* The objective of this committee is to promote standard-ization and the deployment of cost-effective, full-service VDSL networks (see Table 7-1). More information on VDSL technology can be found in Chapter 4.

Recommendation	Year/Description
G.983.1	November 1998—Defines broadband optical access systems based on passive optical networks.
	--Describes the physical layer and ATM as a carrier service. This is the specification created by the FSAN Optical Access Workgroup. Some documentation sometimes refers to this standard as aPON, short for ATM PON.
	--Wavelengths for single fiber: Downstream Wavelength: 1,480 to 1,580 nanometer (nm) Upstream Wavelength: 1,260 nm to 1,360 nm
	--Wavelengths for dual fiber: Upstream and downstream: 1,260 nm to 1,360 nm
G.983.2	June 2000—Defines ONT management and control interface specification for ATM PON.
	--Describes the management functions and operations channel protocol and message set (OMCI) between the OLT and ONT.

Table 7-1.   ITU G.983.1 Standards

Recommendation	Year/Description
G.983.3	February 2001—Defines an additional wavelength band to the downstream direction of a broadband passive optical network (B-PON). --This new standard adds an additional wavelength band to the downstream direction of a PON. Before this addition, only the two wavelengths specified in the G.983.1 standard were available; one for downstream and the other for upstream. The third wavelength band, the *enhancement band (EB)*, allows further discrimination between downstream traffic types. --The upper limit of the original downstream band (now called the *basic band: BB)* was reduced from 1,580 nm to 1,500 nm. --The additional downstream wavelength (EB) is: 1,539 nm to 1,565 nm. --The are two EB options, 1,539 nm to 1,565 nm and 1,550 nm to 1,560 nm, which can be used for different applications.
G.983.4	November 2001—Defines a broadband optical access system with increased service capability using dynamic bandwidth assignment (DBA). --This standard improves the efficiency of the PON by dynamically adjusting the bandwidth among the optical network units (ONUs) that are near end users or in homes. --DBA offers two major benefits. First, the more efficient utilization enables network operators to add more customers to the PON. Second, customers can enjoy enhanced services, such as those requiring bandwidth peaks beyond the traditional fixed allocation.
G.983.5	November 2001—Defines a broadband optical access system with enhanced survivability. --These protection options provide for fault tolerance features that improve reliability and availability of the network.

Table 7-1.   ITU G.983.1 Standards *(continued)*

Operation

A passive optical network uses several different access protocols, which include ATM, Gigabit Ethernet, wavelength-division multiplexing (WDM), and time-division multiplexing

(TDM). Our focus in the chapter is on the ITU G.983 standards that address the workings of ATM. We will be taking a look at some of the other protocols in a later chapter.

In ATM PONs (aPON), virtual circuits (VCs) are established between two end points across the PON. A virtual circuit means that the connection between the two endpoints is logical and not physical.

These VCs are bundled into virtual paths (VPs) for fast switching through the carrier network. The unique pairing of a virtual path indicator and a virtual channel indicator (VPI/VCI) identifies each unique virtual connection through the network.

PONs can be viewed as using a dual star topology, the first star being established on the side of the OLT that interfaces with the service network (the side that provides the service feeds) and the second star at the point of the passive optical splitter.

The job of the splitter is to take the downstream optical wavelength and replicate it optically so that each ONU/ONT receives a copy of the original signal. Using this approach, a single laser at the OLT serves up to 32 different ONU/ONTs. The maximum distance between the OLT and the ONU/ONT is 20 km.

The OLT acts as an ATM edge switch that switches data between the different interfaces on the service side and the passive side—the side that interfaces with the customers. The fiber that feeds from the OLT to the neighborhood is split into different fiber feeds through the use of an optical fiber splitter. The typical split is 32 fibers per OLT. All users that are served by a splitter share the bandwidth of the single fiber that connects to the OLT.

Upstream and Downstream Channels   Upstream and downstream communications occur over different channels, which can be different wavelengths on a single fiber cable or the same wavelength over two different cables. When a single cable is used, the wavelengths are 1,480 nm to 1,580 nm downstream and 1,260 nm to 1,360 nm upstream. When two cables are used, the wavelengths for downstream and upstream are the same—1,260 nm to 1,360 nm.

With G.983.3, WDM is overlaid on top of the original specification and the original downstream channel is reduced to 1,480 nm to 1,500 nm (BB). The new enhanced band (EB) is defined as 1,539 nm to 1,565 nm.

A TDM scheme is used to allocate bandwidth in the downstream channel. The OLT communicates with all the connected downstream ONU/ONTs by broadcasting ATM cells to all connected ONU/ONTs. The receiving ONU/ONTs compare a 28-bit destination address, which is the unique paring of VPI and VCI, within the cell with its own address to recognize the cells that are intended for it. If there's a match, the ONU/ONT copies the cell, removes it from the network, and sends it to the customer premises.

The contention for bandwidth in the upstream channel is handled via a TDMA protocol (a grant protocol). The OLT has a burst-mode receiver that is used to receive cells from remote ONU/ONTs. The varying distances of ONU/ONT are managed through a distance ranging protocol, and cells from different ONU/ONTs are encrypted with separate encryption keys.

Device and Data Management   The aPON specification uses two types of fixed-length cells that are 53 bytes long (the size of an ATM cell). Cells of the first type—data cells—carry the user data, signaling information, and OAM information (operation and management). Cells of the second type—PLOAM, for *physical layer operation and management*—carry grants that are used for orderly contention access and information about the physical infrastructure.

The specification defines seven types of grants:

▼ Data grants

■ PLOAM grants

■ Divided slot grants

■ Reserved grants

■ Ranging grants

■ Unassigned grants

▲ Idle grants

During symmetrical operation, the OLT divides the upstream and downstream channels into frames and time slots. Time slots are 53 bytes long, and frames are a combination of time slots. In the downstream channel, a frame consists of 56 time slots—54 for data and the remaining 2 for PLOAM cells. The PLOAM cells are equally spaced within the frame, as shown in Figure 7-5.

In asymmetrical operation, aPONs use frame time slot configurations that are four times greater on the downstream channel to accommodate the 622 Mbps speeds.

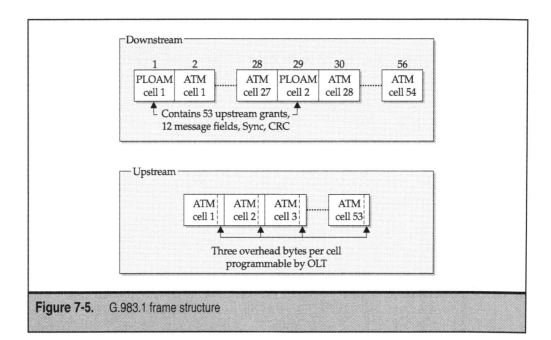

Figure 7-5. G.983.1 frame structure

Upstream frames are 53 time slots long, but each time slot is 56 bytes each (in contrast to 53 bytes for the downstream channel). The extra 3 bytes are overhead bytes that the OLT can use for different functions such as gathering data from ONU/ONTs on traffic queue depth for the enforcement of QoS.

The proper operation of the PON is dependent on time allocation. In the downstream path, the use of time-division multiplexing makes it simpler to control the time allotment and use of the channel, especially since the only contender for the time is the OLT. Even though the outgoing data is destined to multiple ONU/ONTs, the OLT is the final and only arbiter of when to send the data.

In the upstream mode, things get a bit more complex. Time-division multiple access (TDMA) suggests that multiple users will be vying for the same bandwidth, so some method of control is needed to ensure that they do not collide. What makes this especially tricky is the fact that the ONU/ONTs are all different distances from the OLT, which means that they will be different time intervals away from the OLT. So, in order to coordinate transmission, each ONU/ONT needs to have a sense of a common time or be synchronized in time.

To achieve this, a process called *ranging* is used. Basically, the OLT determines the various distances of the ONU/ONTs and assigns an optimal time slot to each ONU/ONT based on its distance from the OLT. The OLT uses the PLOAM cells to transmit grant messages to inform ONU/ONTs of the TDMA time slots to which it is allocated.

When an ONU/ONT needs to transmit data, it waits for the OLT to send a PLOAM cell, which can contain up to 27 grants that can be read by any ONU/ONT. The ONU/ONT checks the data grant number in the PLOAM cell against its own; if they match, it uses the appropriate time slot to send its data. On receipt of the cell, the OLT switches it to its appropriate destination, which may be on the service WAN side or on the passive side if the destination is a user on the PON.

Boarding New Devices   As new devices come active, the aPON is able to detect them through an automatic routine. Every once in a while, the OLT creates a gap in the upstream time slot and sends a ranging grant, which provides special permission for anyone to join. Any device can answer these grants.

When a new ONT sees this message, it responds with its unique serial number. The OLT uses this information to transmit the information needed for the ONT to join the network. This information consists of the network ID of the aPON, a PLOAM grant, and a data grant, which are used by the ONT to access the network.

Security   As with cable access networks or any other network that uses a point-to-multipoint topology, security is always of particular concern because any device can read packets on the network. The fact that a broadcast mode of transmission is used means that all downstream users get to see the content of the packet. The same is true in a PON: every ONU/ONT is able to see the content of every packet. Without security, the PON becomes an easy domain for someone up to mischief to hijack a packet.

The specification for PONs addresses this through the use of a method called *scrambling*. The payload of each cell is encrypted using a 24-bit key, which by today's standards is considered weak. These keys are periodically changed, which improves security.

QoS   ATM as a protocol has the ability to discriminate between different connections and apply different levels of Quality of Service. Even though the G.983.1 standard assumes an ATM mode of operation, the standards G.983.1 through G.983.3 treated all connections as equal. As a result, no QoS could be enforced across the PON.

The new standard announced in November 2001, G.983.4, addresses this by introducing the concept of *dynamic bandwidth assignment (DBA)*, a feature that improves the efficiency of the PON by dynamically adjusting the bandwidth among the ONU/ONTs. The introduction of DBA enables the network to accommodate more customers on the PON and allows the users to benefit from enhanced services like QoS or bursting beyond their fixed bandwidth allocation.

The mechanics of the operation involves the OLT requesting information from the ONU/ONT regarding their queue depth. With this information, the OLT is able to assess the type of service that was purchased and weigh it against existing network conditions. Based on the results of this comparison, the OLT can issue a grant.

Availability and Reliability   Any network that is meant for public use must be highly reliable with high availability. A function of PLOAM cells is to provide physical management functions for the network. ONTs are able to give status reports about themselves to the OLT. The OLT can also fine-tune lasers and measure power levels by setting bits in the PLOAM cells. These are all good features, as they serve to optimize the performance and reliability of the network, but they do little to ensure availability.

The new G.983.5 specification defines a broadband optical access system with enhanced survivability. Redundant features are added that allow an ONU/ONT to connect to two different OLTs, which greatly enhances fault tolerance.

Business Drivers for FTTx

The metro service bottleneck is an issue that needs to be addressed before the abundance of bandwidth that is available in the core can be utilized. Once this is resolved, the issue becomes the access speeds of the local connections. Various solutions such as cable and wireless are being rolled out to address the access portion of the problem; the metropolitan portion is being addressed through the upgrade of the infrastructure to fiber optics.

Utilizing fiber all the way to the home or office, through FTTx, is seen as the mother of all solutions, as it provides the infrastructure on which a full suite of services can be delivered with the appropriate Quality of Service. Additionally, it provides the basis for delivering applications without concern for bandwidth limitations. In short, it is networking nirvana, the end-to-end infrastructure that is able support all the services and applications we will ever need. With an all-fiber infrastructure, the network drives the application; the application does not drive the network, as has always been the case.

Traditionally, most telecommunication fiber rings used Synchronous Optical Network/Synchronous Digital Hierarchy (SONET/SDH) technology. The networks are optimized for long-haul and metropolitan interexchange applications that use optical-to-electrical-to-optical conversion at each node. These solutions are too expensive and do not present the best cost-effective platform for local network access.

Optical networks that require active components require the nodes that act as a regenerator to be built before the fiber can be used—an approach that requires an up-front investment. The ideal solution for an access network is one that does not require a big up-front investment, but one that can be deployed incrementally as the service is sold. Passive optical networks are seen as a cost-effective solution because of the affordable components that are used and the fact that ONU/ONTs can be added incrementally as the demand grows. Figure 7-6 shows CIBC World Markets estimates of the growth rate of the PON equipment market by residential and business applications.

Other benefits of PONs that make for a good business plan include the following:

▼ Fiber is less expensive than copper to maintain. Profit margins are improved because operational costs are less than for a wire infrastructure.

■ A single strand can be used from the OLT to support multiple users.

■ The point-to-multipoint topology means that a single OLT interface is shared among users instead of dedicated to a single user. For G.983.1, the split is 32:1 (32 users per OLT interface); other PON systems can get as much as 64:1.

■ Bandwidth upgrades can be done remotely instead of by dispatching a technician (eliminating the need for expensive truck rolls).

■ Less network design is needed up front to anticipate future bandwidth needs, as the design can be easily changed at a later date.

■ Both the OLT and the ONU are able to interface to different network types, IP, Ethernet, ATM, etc., thereby allowing for easier integration with a variety of common architectures and protocols.

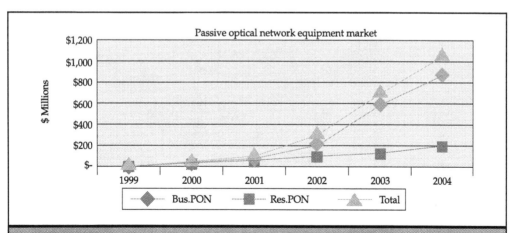

Figure 7-6. Projected PON equipment market (source: CIBC World Markets estimates)

- Longer distances can be spanned with less distortion than over wire solutions. Fiber also does not have to contend with the problems of electromagnetic interference.

▲ A reel of fiber weighing 4.5 lbs has the capacity of 200 reels of copper weighing 1,600 lbs.

As an ATM-based system, a PON is able to transport different types of traffic and can adapt to any type of service or application. Traditional legacy telephone services can ride transparently across the infrastructure, as can any other service, from a T1 connection to an Ethernet-to-Ethernet LAN application, as shown in Figure 7-7. This flexibility and transparent transport capability make a PON a good platform to deliver multiple services. The infrastructure handles a digital television feed as well as a data download.

The business case for FTTx depends on its proximity to existing fiber rings. CIBC World Markets statistics assess that 76 percent of businesses are within one mile of existing fiber rings and 50 percent of residences are within two miles. New housing developments and office parks will most likely be the target of FTTH and FTTB service. For older communities, the gap will probably be bridged with the use of FTTCab and a technology like VDSL.

U.S. FTTx Activities

FTTx is still not a popular technology in the U.S. It has not yet made the headlines in a significant way, and it has yet to capture the imagination of Wall Street. On the surface, it would appear that nothing significant is happening in the area of FTTx, but in reality it may just be a sleeping giant.

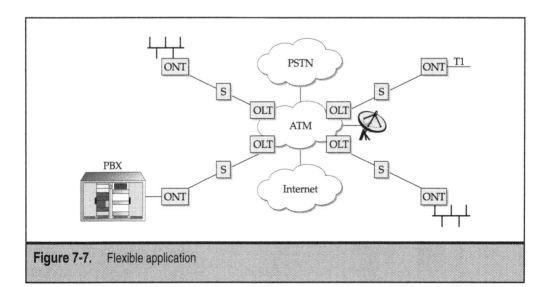

Figure 7-7. Flexible application

The major telephone companies are expanding fiber deeper into neighborhoods. New housing developments and office parks are being equipped with fiber, and rural communities are also benefiting. In all probability, the technology will appear to be dormant or nonexistent until something of significance causes it to leapfrog into the consciousness of the public. When that happens, it will appear to have emerged from nowhere, but the infrastructure will be in place in many areas and the service will be generally available.

Throughout the U.S., several local utilities, competitive local exchange carriers, and local community administrations are implementing FTTH solutions. As we wrap up this chapter, we will turn our attention to some of the various projects and ventures that are emerging throughout the states. Table 7-2 is a list of a few of the projects that are either in progress, in planning, or currently operational. To gain a better understanding of the motivation and business reasons for deploying FTTH service, we will briefly review a project that is currently being rolled out in a small rural county in the Northwest.

Operator/Supplier	Status/Description
Marconi	Operational/City of Palo Alto City of Palo Alto Utilities introduced a fiber to the home (FTTH) network utilizing Marconi's FTH-1000 system.
Verizon/Marconi	Announced January 2002/Brambleton, Virginia A planned community will have a fiber optic network linking homes in the first phase of the development, giving Brambleton residents access to the latest in voice, video, and data services the day they move in.
WINfirst	Operational/Sacramento, California
WINfirst	Planned WINfirst has received franchises for Sacramento, San Diego, Los Angeles, Dallas, Houston, Austin, San Antonio, and Seattle.
BellSouth	Operational/Dunwoody, Georgia Announced Offering: Internet access through a 100 Mbps interface (actual speeds available to customers during the trial will be limited to tariffed BellSouth consumer high-speed data offerings) --120 channels of digital video entertainment --70 channels of analog video entertainment --31 channels of CD-quality digital audio service

Table 7-2. U.S. FTTx Activity

Operator/Supplier	Status/Description
Competisys (utility company) and World Wide Packets partnership	Announced 2001/Poppy Meadows development, American Canyon, California The company plans on offering an integrated bundle of voice, video, and data services over a Gig-E metro network, which Competisys says gives it more bandwidth than most current broadband alternatives, at a retail price of about $115 per month.
Home Town Solutions	Announced 2000/Morris, Minnesota CLEC partnership Federated Telephone Cooperative and Agralite Electric Cooperative. Town of about 6,000 people served by incumbent US WEST, now Qwest. Offering FTTH to residential customers to deliver dial tone, long distance, high-speed data, and video services. Projection: need about 65 percent of town to subscribe to be profitable.
Nex-Tech (subsidiary of Rural Telephone)	Announced 2000/Norton, Kansas FTTH network, about 80 percent of population signing up before the network was even built. Services include a bundle of local and long-distance telephony, 56 Kbps dial-up Internet and cable TV for about $56 a month. High-speed Internet access will cost an additional $49.95 a month.

Table 7-2.   U.S. FTTx Activity *(continued)*

FTTx: a Solution of Choice for Grant County   The Grant County Public Utility District (PUD) has been providing electricity to Grant County for over 50 years. Located in the rural eastern section of Washington State, PUD—a consumer-owned utility since 1938—produces hydroelectric power from dams located on the Columbia River. With a population of about 75,000 people, only a small percentage of the power produced is consumed locally. Most of the power produced is exported to other parts of the state and sold to other power companies in the surrounding states.

This arrangement generates a lot of cash, which PUD decided to invest back into the local community. Operating in a rural county, with an inadequate telecommunications infrastructure, PUD began looking at fiber. An initial motivation for building the infrastructure was to automate the reading of electric meters around the county and cut down on the number of miles traveled and the expense of truck rolls. The county is about 50 miles east to west and 100 miles north to south, with approximately 36,000 homes. The largest town has a population of about 15,000 people.

This initial need to automate meter reading eventually led to the idea of leveraging the infrastructure to provide meaningful services that would benefit their customers. So PUD embarked on building the Zipp network, an all-fiber network that will connect each home. Each house is connected to the network free of any charge to the homeowner, and the fact that the homes are connected to a high-speed infrastructure attracts service providers.

Before the infrastructure was built, some people did not even have basic telephone service. At the time of this writing, there are six Internet providers, one video provider, and one telephone provider. A customer wishing to purchase a new service contacts the local service provider directly. The service provider enters into an agreement with the customer and approaches PUD to provide the service through a wholesale agreement with the service provider.

PUD projects six years to complete the build-out of the network. At the end of 2001, 6,000 homes were passed with fiber—about one-sixth of the homes in the county—at a cost of about $22 million. The projected cost for 2002, when another 6,000 homes should be passed, is another $20 million.

Before the infrastructure existed, there were very limited options for basic telecommunication services that are widely available in most cities across the U.S. Today, Grant County is equipped with a fiber infrastructure that extends to the home. This high-speed infrastructure has attracted service providers, and customers are benefiting from more choice and a lower cost per service. PUD, too, has benefited, and the utility now has the benefit of automated meter reading and a new wholesale market for fiber.

PUD's vision of the network is one where farmers are able to monitor fields remotely from their homes, a student is able to find a scholarship online, and a health care system is able to benefit from improved communication with other hospitals. They also see the network as an investment in future business for the county, as new businesses move into the area to benefit from the infrastructure. New business means new jobs, and new jobs mean a boost to the local economy.

SUMMARY

One aspect of the FTTx solution that positions it for success is the fact that the major telecommunications companies around the world do not need to be sold on the concept. The concept originated with them. The cost benefits of the solution are extremely attractive, as fiber presents the lowest operational costs of all transmission media.

The year 2001 was an active year for the technology, one in which enhancing standards were announced. It is highly probable that 2002 will be the year when we will begin seeing more products and services announced, bringing with them an awareness of the technology. The potential of the technology should not be discounted, especially when one considers its backers.

From a networking perspective, FTTx presents the best of the best features, no electro-magnetic interference, less attenuation over longer distances, and all the bandwidth we would ever need. Using fiber as a solution, the network is no longer the bottleneck, assuming the cost of the components keeps up with the requirements of the applications.

The uptake so far has been slow; the technology is not widely publicized. Many professionals working within the networking industry have never even heard of it. However, chances are that one day it will surprise us, and we will ask, from whence did this option suddenly appear?

CHAPTER 8

Digital Transmission
and Frame Relay

So far, we have explored technologies that have a wide applicability across different user bases. While the need for lots of bandwidth in support of converged services is now becoming more evident in homes, large corporations, college campuses, and industrial factories have always had to address these issues.

These entities have always been required to link remote offices and campuses for the sharing of information, workgroup collaboration, voice, video, data, and security services. Over the years, the need became even more critical as the speeds of local area networks (LANs) improved.

The dilemma faced by many enterprises was the fact that local bandwidth that was available for applications was significantly greater, and a lot more affordable, than bandwidth that was available to link remote sites. This meant that the bandwidth between two remote sites, the wide area network (WAN), carried a much higher premium per kilobyte than what was available locally.

This disparity meant that the WAN link, being a premium resource, had to be utilized as efficiently as possible. As a result, any improvements—regardless of where in the network—that helped to get more from the link were viewed as positive. Applications were designed with this in mind, and methods that bundled or multiplexed different services to share a common WAN link were explored.

Common access technologies that were used for WAN links—and are still used today—include "build-your-own solutions," and solutions that share a common carrier infrastructure such as Frame Relay access to the carrier network. There are, of course, many other technologies and solutions, but these were the most common. Building your own typically entails using leased lines with time-division multiplexing or statistical multiplexing.

CORPORATE ISSUES

Before deciding on an appropriate WAN strategy, corporations must assess and address the costs associated with the different WAN access methodologies. Deciding whether to build using leased lines or to buy a solution like Frame Relay is a matter of weighing the cost of ownership and operation against the benefits of paying a service provider to offload some of the burdens of operation. In either case, the cost—whether it results from a capital outlay for the purchase equipment or a charge from a service provider for a bundled service that includes the equipment—will include the hardware, software, and operational and management support.

The decision to build is often predicated on the need to retain control. A key contributor to this decision is the application that will be using the link. If the application has specific performance requirements such as low latency, a bundled service may not be able to meet these objectives, whereas a point-to-point leased line (private circuit) connection can. Building instead of buying carries with it the fully burdened cost of developing, deploying, operating, and managing the solution.

In contrast, buying a service allows the corporation to share the cost of operation and management of the infrastructure with other users of the service. The cost to the corporation, therefore, is a lot less than that of building and operating it. The trade-off with this approach is that the customer gives up control and must compromise on service and performance levels that the service provider is willing to meet.

Common to both approaches is the concept of a point of demarcation. Whether a private circuit solution is employed or a solution like Frame Relay is purchased, there is always a point of demarcation where the provider hands off responsibility to the customer. It is a boundary between the two networks. Even when a build-your-own private circuit solution is being used, there are components in the engineering of the circuit that must be maintained by the carrier and so fall on their side of the demarcation.

Point of Demarcation

The point of demarcation has traditionally been used as a barrier to protect the public network from damage as a result of something that happens on the customer side. For voice service, the point of demarcation is very clear because it is legally defined. In data communication, it is less defined.

The relevance of demarcation to the business customer is its direct correlation to where their scope of responsibility ends (or begins). The scope of the customer's responsibility directly impacts cost. Conversely, the greater the scope of responsibility, the less are the benefits that may be derived when compared to the costs. These two elements—costs versus benefits—are why the point of demarcation may be an important consideration to a business customer.

The typical point of demarcation in a data network is at the CSU/DSU, *channel service unit/digital service unit*, a digital interface between the end-user equipment and the telephone network. What makes the demarcation for data networks a lot less loose than that for voice networks is the fact that the CSU/DSU can either be provided as part of a service or provided by the end user.

In instances where customers are building their own solution, they will typically supply the equipment and the point of demarcation becomes the point at which the CSU/DSU plugs into the telephone network. If the service is one where the solution is being bought, Frame Relay for instance, the point of demarcation varies widely from solution to solution.

In a fully managed solution, the service provider usually provides the CSU/DSU and a router. In this instance, the point of demarcation between the customer and the provider is typically the interface between the provider's router—typically the Ethernet interface—and the customer's network. The customer supplies the cable that connects the interface to the LAN. The benefit of this arrangement is that the service provider assumes full responsibility for configuration, support, operation, and management of the solution up to the router.

If the solution is halfway between being a fully managed service and just the basic provision of a leased line—an arrangement called a *bearer service*—the point of demarcation is usually at the CSU/DSU, which is provided by the provider. The customer in this instance provides the router and the cable that connects to the CSU/DSU, which is provided by the service provider. There are other variations on a bearer service, varying from the service provider's supplying a Frame Relay circuit and the customer supplying the DSU/CSU to the service provider providing a DSU/CSU and a *frame relay access device (FRAD)* with limits on the scope of configuration and management of the solution.

To complicate things further, the service provider of a service like Frame Relay may not necessarily be the provider of the circuit. So there is a further point of demarcation between a fully managed service and a bearer service, the point where the service provider equipment touches the telephone network. Figure 8-1 shows the typical points of demarcation.

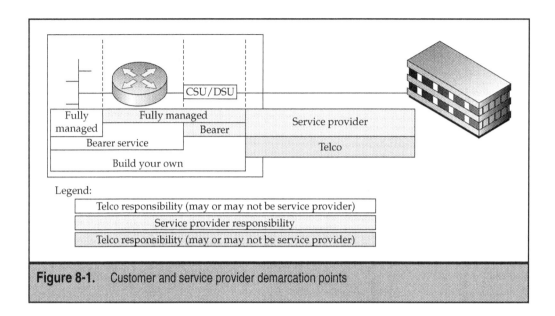

Figure 8-1. Customer and service provider demarcation points

The CSU/DSU does not have be a stand-alone piece of equipment. The functionality of the CSU/DSU is frequently integrated into another device such as a router or a multiplexer. As a side note, throughout this chapter generic reference will be made to a router as the intelligent device that sits in front of the CSU/DSU. It should be noted that this reference is being made in order to simplify the explanation; other devices such as switches and multiplexers can occupy this position.

PRIVATE CIRCUIT SOLUTIONS
AND DIGITAL TRANSMISSION

The need to communicate between two sites may extend beyond a single application. For instance, an interoffice link may be needed to tie two LANs together, to share a single mainframe application, and to transport voice traffic. The simplest solution would be to utilize different links for each application. But considering the cost of each link, it is a lot more cost effective to use a single connection for all applications.

Other than the cost of installing and operating three different links for three different applications, a solution that employs multiple links could result in the inefficient use of a link if the application that uses it does not tap its full potential. An underutilized link is a waste of a valuable resource, and as an expensive resource the potential of the WAN link must be fully exploited across all applications.

Multiplexing

A common and simple way of sharing a single high-speed link between multiple applications is to multiplex the different application traffic into a single traffic stream across the link. A multiplexer takes the traffic from different applications and multiplexes into a single traffic stream; the combined stream is transmitted across the high-speed link to another multiplexer on the other side of the link. The remote multiplexer demultiplexes the traffic into the different traffic types and sends each to the appropriate destination device.

The use of a multiplexer is an efficient way of sharing a high-speed link between many different applications. Muxes use three different techniques to allocate bandwidth: frequency division, time division, and statistical time division.

Frequency-Division Multiplexing

Frequency-division multiplexing (FDM) is an analog technique that uses different carrier frequencies for each channel. A guard band is left between adjacent channels to ensure that the channels do not interfere with each other. As an example, if three applications were to share the link, the multiplexer would create three different channels, A, B, and C, each having its own carrier frequency. All channels may be equally wide, or one channel may have a greater bandwidth if an application requires it.

Assume the channels are equally allocated, and each application needs a bandwidth of 6 kHz, and a guard band of 300 Hz is used, as is depicted in Figure 8-2. If the starting frequency is 12 kHz, channel A would have a frequency range of 12–18 kHz, channel B would be 18.3–24.3 kHz, and channel C would be 24.6–30.6 kHz, with a 300 Hz buffer between any two channels. The bandwidth required to transmit all three applications would be 18.6 kHz.

Frequency-division multiplexing does not require all channels to terminate at the same location. Multidrops of the link may be created to connect another location. This technique has pretty much been superseded by time-division techniques that are less prone to electromagnetic interference and are more suitable to digital communication.

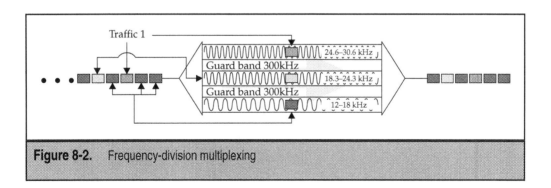

Figure 8-2.   Frequency-division multiplexing

Time-Division Multiplexing

Time-division multiplexing (TDM) is a technique that divides the available channels into time slots. Each input port of the multiplexer (with its own form of application traffic) is sampled at a very fast rate. The sample rate must comply with the Nyquist theorem, which says the sample rate must be twice the highest signal frequency in order to reconstruct the original signal.

To better understand the concept, imagine a machine that is constantly checking different mailboxes to see if there is anything to send. If there is something to send, it takes it and places it into predefined slots on a conveyor belt that is moving at a constant rate. If a slot passes by every second and it takes three seconds to place the contents of three mailboxes in separate slots on the conveyor, the content of each mailbox will be placed in every third slot. If a mailbox has nothing in it, there will be nothing to put in the passing slot; the slot will therefore go by empty.

Applying this image to TDM multiplexing, each interface on the mux, which has a different form of application traffic, is sampled, and if there is traffic, it occupies a time slice and is transmitted. If there is no traffic, the time slice is transmitted empty. Therefore, the link is being allocated at a set ratio between all applications. If each interface is being given equal sample time, then the bandwidth is being allocated equally to all applications (see Figure 8-3).

NOTE: Applying the Nyquist theorem to a 4,000 Hz passband, it would take 8,000 samples to capture a 4 kHz signal. Assuming an eight-bit sample, it would take 64 Kbps (8 bits×8,000) to communicate the digitally encoded signal in real time over a circuit. This value formed the basis on which T1 carrier circuits were designed, since they were designed to carry voice that was digitized. The 4 kHz signal (for a voice channel) is sampled and multiplexed using time-division multiplexing into 24 channels.

A benefit of this approach is that everything is predictable, so any application that likes a predictable flow or a predictable latency—like voice traffic—benefits from this technique. The downside is the fact that the link is not used to its fullest potential. If an application is not active, then the percentage of the bandwidth allocated to the traffic is not

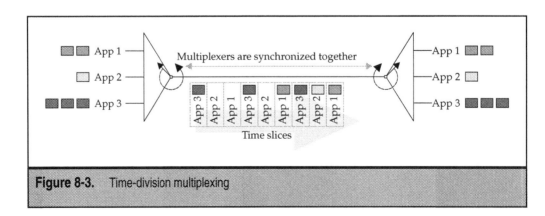

Figure 8-3. Time-division multiplexing

being used even if the other applications could benefit from it. Bandwidth being an expensive resource, other means of bandwidth allocation evolved, which attempted to fully utilize the bandwidth at all times regardless of how many applications were active at any given moment. This type of multiplexing is called statistical time-division multiplexing.

Statistical Time-Division Multiplexing

Statistical time-division multiplexing (STDM) uses more intelligence in the multiplexers. Using a statistical approach, the multiplexer is able to identify when an interface is idle and allocate the time to the other active interfaces (see Figure 8-4). Besides benefiting bandwidth efficiency, this approach also enables more applications to share the link, because the mux statistically compensates for the idle time. The ability to support more applications across a link is true when the traffic pattern is bursty in nature, as is LAN traffic. If the traffic pattern is a constant bit stream, this benefit is less applicable. With the added features of compression, priority schemes, and automatic speed detection, among others, STDMs offer a greater level of functionality than other muxes.

Not everything benefits from a statistical approach. Voice and video, for instance, perform better with a constant and predictable flow that is better served by TDM muxes.

The Digital Data Rate Hierarchy

Throughout the U.S. and the world, digital time-division multiplexing is used to offer a set of different discrete data rates for leased-line circuits. In the early 1960s, Bell labs developed a set of interfaces and data rates to transport digitized voice over twisted pair copper wires. The different data rates had a hierarchical relationship, with the higher data rates being a multiplexed version of the lower-rate data streams.

The lowest data stream was a DS0 signal, a 64 Kbps stream, which formed the foundation of the hierarchy. The DS0 was designed as a voice channel that was optimized for the transport of digitized voice. The 64 Kbps figure was calculated using a sample of two times a 4 kHz voice passband in accordance with the Nyquist theorem.

The hierarchy was based on different multiples of the DS0 signal with extra overhead bits added as frame separators. The next level in the hierarchy was a multiplex of 24 DS0

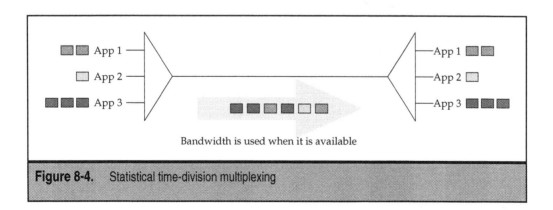

Figure 8-4.   Statistical time-division multiplexing

data streams to form a DS1 signal, also called a T1 carrier. Eight extra bits are added to the multiplexed stream to give a total data rate of 1.544 Mbps (24×64,000 = 1.536Mb + 8 bits = 1.544Mb).

The DS*x* designation describes a signal and speed format, and the *T*, a carrier class. The *North American Digital Signal Hierarchy* is based on T carriers: T1, T2, T3, T4, and T5, with data rates of 1.544 Mbps, 6.312 Mbps, 44.736 Mbps, 274.176 Mbps, and 400.352 Mbps respectively (see Table 8-1).

The European Digital Signal Hierarchy is based on an E carrier. Table 8-2 is a summary of the different carrier classes and data rates throughout Europe and Asia. The European digital hierarchy was defined after the American version, and as a result, it did not need overhead bits because of improved electronics. (Europe's hierarchy was developed much later than the U.S. and uses a more efficient multiplexing scheme.)

The optical domain also has a similar hierarchy. The North American Synchronous Optical Network (SONET) and the international Synchronous Digital Hierarchy (SDH) are also based on a TDM approach for deriving a digital hierarchy of speeds. Table 8-3 provides a summary of the available speeds in the optical domain. The designation OC-*x* indicates the optical carrier, with the number representing *x* being a multiple of the OC-1 rate of 51.84 Mbps.

The digital hierarchy defines the different levels of multiplexing. For example, the first level of the hierarchy multiplexes a number of DS0s into a single DS1 digital signal, which is then placed on a carrier (T1). Subsequent levels are a multiplex of the previous signal as shown in Figure 8-5. The end result is a predefined increment of data rates that is available from one level to another.

The impact of this approach to the business customer is that it creates different fixed data rates from which to choose that are multiples of the DS0 signal. To move from one speed to another often means moving to the next higher carrier class, which may have much more bandwidth than is needed. The other option is to install another circuit of the same class as the one that needs the upgrade.

Hierarchy	Levels of Multiplex	Signal Name	Structure	Carrier Class	# DS0s	Data Rate
	0	DS0			1	0.064
Level 1	1	DS1	24×DS0	T1	24	1.544
(Intermediate)		DS1C	2×DS1	–	48	3.152
Level 2	2	DS2	4×DS1	T2	96	6.312
Level 3	3	DS3	7×DS2	T3	672	44.736
Level 4	4	DS4	168×DS1	T4	4032	274.176
Level 5	5	DS5	336×DS1	T5	5760	400.352

Table 8-1. North American Digital Data Rates Hierarchy

Hierarchy	Levels of Multiplex	Carrier Class	# DSOs	Data Rate
Europe				
Level 1	1	E1	32	2.048
Level 2	2	E2	128	8.448
Level 3	3	E3	512	34.368
Level 4	4	E4	2,048	139.268
Level 5	5	E5	8,192	565.148
Japan				
Level 1	1		24	1.544
Level 2	2		96	6.312
Level 3	3		480	32.064
Level 4	4		1,440	97.728
Level 5	5		8,192	565.148

Table 8-2.   Europe and Japan Digital Data Rates Hierarchy

In many cases, installing a second circuit was not convenient, as it did not fit into the strategy of leveraging a single connection for all applications. A second line meant splitting the traffic between the two lines, and that usually led to inefficiencies. The alternative meant installing the next higher class and using only a portion of the bandwidth, but this approach carried a higher price per megabyte if the full data rate was not to be used. The answer to this dilemma was found in inverse multiplexing.

Optical Carrier	SONET	SDH	Data Rate
OC-1			51.84 Mbps
OC-3	STS-3	STM-1	155.52 Mbps
OC-12	STS-12	STM-4	622.08 Mbps
OC-48	STS-48	STM-16	2488.32 Mbps
OC-192	STS-192	STM-64	9953.28 Mbps

Table 8-3.   SONET/SDH Data Rates

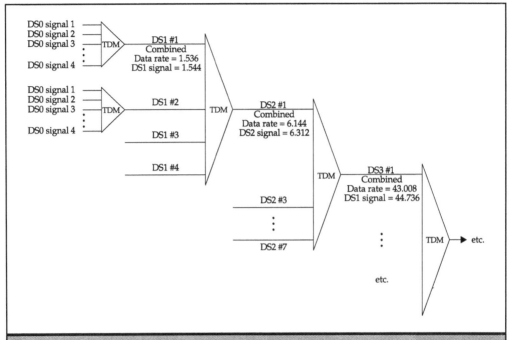

Figure 8-5. Different levels of multiplexing used to provide higher data rates

Inverse Multiplexing

Inverse multiplexing (imux) is a term that is used to describe a scalable wide area solution that employs multiple discrete circuits. The concept is opposite to traditional multiplexing, where multiple data streams are combined into a single stream that is transmitted across a common link, with the remote multiplexer demultiplexing the combined stream into its original parts. Inverse multiplexing reverses this process; a single data stream is transported over multiple high-speed circuits that act as a single logical circuit.

Imuxing provides a flexible and incremental way for increasing bandwidth. As an example, if the bandwidth has to be increased using a single T1, the next available circuit size from a carrier may be a T3. This does not mean that the entire capacity of the T3 circuit would have to be used, only the amount of bandwidth that is required. With an imux solution, the bandwidth can be upgraded incrementally by installing one or more single T1s to fit the specific requirement. Figure 8-6 shows an example of the operation of an inverse mux.

In an imux solution, a single circuit failure does not take down the entire link, as is the case with a single large circuit. With an inverse multiplexer, the multiple wide area links appear as a single high-speed link to the router, and the multiplexer is able to delete a failed circuit dynamically without disruption to the router. The terms "NxT1" or "NxE1," where *n* is the number of circuits in the imux link, is used to describe the number of dis-

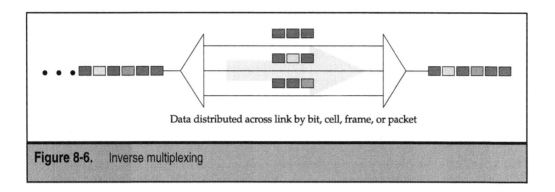

Data distributed across link by bit, cell, frame, or packet

Figure 8-6. Inverse multiplexing

crete links that are being used. The total bandwidth available to the connection is the aggregate of the number of links in use.

Some benefits of imux solutions include:

▼ The ability to grow or decrease the bandwidth at a rate that better matches the actual need. If bandwidth requirements are expected to grow at a steady pace, it may be more cost effective to upgrade by adding another T1 as required versus installing a DS3 from day one.

■ That it is more cost effective than buying a bigger circuit. Though the cost of a T1 or a DS3 varies with different carriers, a rule of thumb that generally works for determining when it makes sense to upgrade to a DS3 versus an imux solution is to upgrade when the total number of T1s needed exceeds 7. Using more than seven T1s begins to be less cost effective when compared to the price of upgrading to a DS3.

■ Their usefulness where higher-speed circuits may not be available. In these instances, being able to aggregate lower-speed circuits circumvents the problem.

▲ That multiple links inherently provide more fault tolerance than a single link. An outage on a single link would take the entire connection down. With multiple links, a single link failure will degrade service but not stop it. Additionally, multiple links could be installed with diverse paths, or from different providers, thereby adding another level of redundancy.

Types of Imuxes   There are many types of multiplexing techniques, but only three are industry standards: Multilink Point to Point Protocol (MLPPP), Inverse Multiplexing over ATM (IMA), and Multilink Frame Relay (MFR). All three share the following features:

▼ Use of sequence numbers to guarantee the order of data delivery.

■ Provision of some level of transparency that allows the group of links to appear as a single composite high-speed link to higher protocols.

▲ Achievement of a level of resiliency by providing features that delete failed links, restore links, or add new links.

Multilink PPP, or *MLPPP*, is the first inverse multiplexing technique to be standardized. The Internet Engineering Task Force adopted it in 1990. The deployment of this scheme is typically in a router, though it can be deployed as a stand-alone system. It uses sequence numbers to ensure the exact reordering of packets across the links. As a solution within a router, it has the propensity to use a lot of CPU cycles; using it as a stand-alone solution (behind the router) is a lot more efficient.

Besides MLPPP, other load-sharing techniques are used in routers to distribute the traffic across multiple links. Whole packets are distributed over multiple links, a packet at a time. Load sharing is similar to inverse multiplexing in that it allows multiple links to be shared by a common traffic stream. The main difference between MLPPP and other forms of load sharing is its dependence on some other higher layer to do the reassembly of the packet. Many routing protocols, such as OSPF and EIGRP, allow for load sharing on a per-destination or per-packet basis but are dependent on the TCP layer to reassemble the packets at the final destination. In contrast, MLPPP uses sequence numbers to ensure proper reassembly on the end of the MLPPP link. Figure 8-7 shows the points within the connection where the packets are reassembled using the two techniques.

Inverse Multiplexing over ATM (IMA) was defined as a standard implementation of inverse multiplexing for ATM applications by the ATM Forum in 1997. The technique operates on a cell-by-cell basis across multiple discrete circuits through a cell-based control protocol that ensures that the cells are resequenced properly on receipt. An appealing attribute of IMAs is their ability to preserve the Quality of Service features that are inherent in ATM networks. Different traffic types can therefore be given different priority levels throughout the network. The solution is most often used by service providers to connect customers to fiber-based facilities, to enable switch-to-switch trunking, and to connect Digital Subscriber Line Access Multiplexers (DSLAMs) to a central office, as shown in Figure 8-8.

Multilink Frame Relay (MFR) refers to FRF.16, a multilink frame relay specification that has been adopted by the Frame Relay Forum. FRF.16 supports user-to-network and network-to-network interfaces. As with all approaches to imuxing, MFR bundles multiple

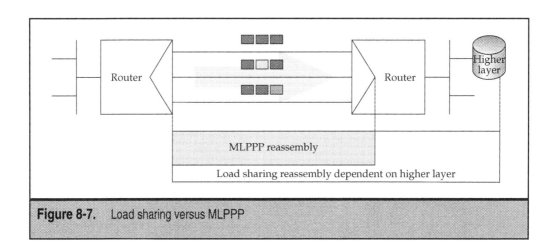

Figure 8-7. Load sharing versus MLPPP

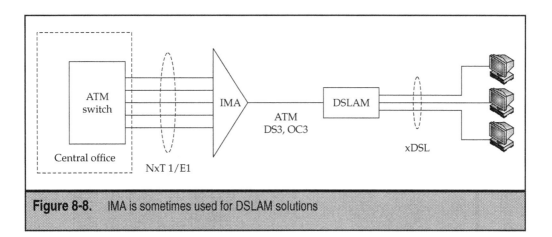

Figure 8-8.   IMA is sometimes used for DSLAM solutions

T1 or E1 links and presents them as a single physical interface to the Frame Relay data link layer. Frames are then distributed across each link and reconstructed at the far end of the connection.

The roots of NxT1 and NxE1 inverse multiplexing date back to the early 1990s, when Larscom and IBM were granted the original patent. The techniques employed back then were bit-based and were used to aggregate as many as eight T1s to form a megabit stream for different applications. The major difference between bit-based multiplexing and the ones we have discussed is that bit-based systems operate at the bit level instead of the cell, frame, or packet. The data stream is transmitted over the links one bit at a time in a round-robin fashion.

The predictable sequence of the bits makes for easy reassembly, as it is easy to anticipate which bit goes after which when a round-robin approach is being used for sending and receiving. The problem with bit-based solutions is that they are not governed by any standards, so different manufacturers employ different proprietary schemes that are not compatible. Table 8-4 compares the different inverse multiplexing techniques.

	Bit Based	Load Sharing	MLPPP	IMA	MFR
Number of T1/E1s per bundle (typical config)	2–8	2–6	2–8	2–32	2–28
CPU intensive	no	average	yes	yes	yes
Packet order preserved	yes	no	yes	yes	yes

Table 8-4.   Comparison of Various Imux Techniques

	Bit Based	Load Sharing	MLPPP	IMA	MFR
Support on channelized T3 interfaces	yes	yes	yes	yes	yes
Data link layer approach	no	no	no	yes	yes
Interoperability	no	yes	yes	yes	yes

Table 8-4. Comparison of Various Imux Techniques *(continued)*

Digital Timing, Clocking, and Synchronization

The dependence of digital links on TDM for higher data rates creates the need for an accurate time source on which both ends can rely. As one progresses further up the digital hierarchy to much higher speeds, issues of timing become paramount. Unlike asynchronous communication, which uses start and stop bits, synchronous communication is dependent on a reliable clock source to provide reference timing to the network,

Revisiting our analogy with the mailbox, if the receiving system does not have the same sense of time that was used by the sender, integrity of the data can be compromised. It would be highly probable that the receiving station would place the contents of each conveyor slot into the wrong receiving mailbox, if it were not synchronized to a common time.

Three different types of timing are used in network designs: *asynchronous, synchronous,* and *plesiochronous.* The difference among the systems lies in how they each distribute time.

Asynchronous systems are not dependent on the synchronization of clocks. Instead, they use bits in the data stream to tell when to start timing and bits that tell when to stop. The only thing that needs to be agreed upon between the two ends is when to start timing, how many bits make up the byte, when to stop timing, and how many bits should be used for parity checking. This method is suitable for slower links.

Synchronous systems distribute timing from a common clock. Each station or communication device calibrates its clock to a common primary time source. The telephone network uses a common primary clock as a *primary reference source (PRS).* Before the divestiture, AT&T used a cluster of cesium clocks located in Hillsboro, Missouri, as its PRS. The entire telephone network was designed in layers, called stratums, from the perspective of the distribution of timing signals, as depicted in Figure 8-9.

The layer with the PRS was called Stratum 1, and subsequent layers emanated outward toward the edge of the network, with each successive layer being less accurate than the preceding one. The key point was that each stratum derived its timing directly or indirectly from the PRS. Having a reference to a common time source meant that CSU/DSUs could operate within strict timing tolerances.

Plesiochronous systems use the same concept of a common clock, with the difference that there are many PRSs. This model makes sense especially when one considers the consequences of divestiture (see Figure 8-10). When there was a single telephone company, a single PRS made sense, although it was very expensive to implement. With many competing telephone companies, each maintains its own set of primary reference sources.

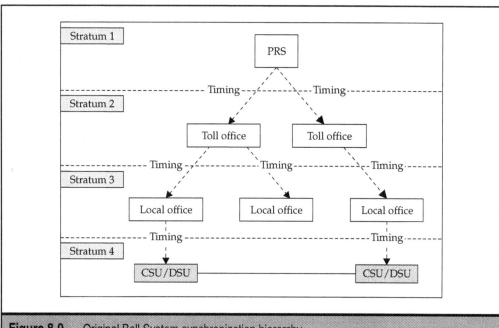

Figure 8-9. Original Bell System synchronization hierarchy

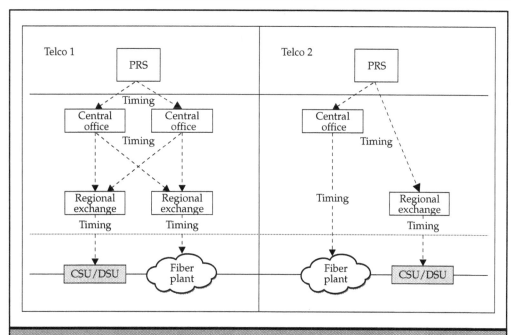

Figure 8-10. Plesiochronous designs

In this model, the CSU/DSU on both ends of a T1 circuit that spans multiple telephone networks will actually have different timing sources, each from a different telco. The timing between the two networks is finely tuned to ensure that both devices operate within precise tolerance. T-carrier systems are therefore really *plesiochronous* systems that use a *plesiochronous digital hierarchy (PDH)*—a derivative of a Greek word meaning *almost synchronous*—though we tend to view them as using a synchronous scheme.

Clocking and Timing   Performance standards for clocking are defined by ANSI/T1.101. The standards are based on Stratums numbered from 1 to 4, with Stratum 1 being the most accurate and the primary reference source for other Stratums. The accuracy of each Stratum decreases with its distance from the first Stratum. The higher the Stratum, the more errors or slips are expected.

A slip in time is considered to be a gain or loss of 125 microseconds, or one frame's worth of sampling time in a synchronous network. The accuracy of Stratum 1 clocks has a tolerance for one slip every 72 days. Table 8-5 provides the probability of slippage for each Stratum. Another term that is used in reference to clock sources and timing is *holdover*; this is the condition where the input synchronization source is no longer available and the clock drifts according to its own internal structure.

Stratum	Accuracy	Probability of Slip (Drift per Year)	Notes
Stratum 1	$\pm 1 \times 10^{-11}$	One slip every 72 days	Primary source
Stratum 2	$\pm 1.6 \times 10^{-8}$	One slip every 7 days	Can provide synchronization to other Stratum 2 clocks and Stratum 3 and 4 clocks
Stratum 3	$\pm 4.6 \times 10^{-6}$	One slip every 6 minutes (255 every 24 hours)	Can provide synchronization to other Stratum 3 devices and/or to Stratum 4 devices
Stratum 3E	$\pm 1 \times 10^{-8}$	One slip every 6 hours (4 every 24 hours)	Newer standard defined by Bellcore as a result of SONET requirements
Stratum 4	$\pm 3.2 \times 10^{-5}$	One slip every 4 seconds	
Stratum 4E			A proposed new standard for customer equipment

Table 8-5.   Stratum Hierarchy

Stratum 1 equipment is any that maintains a long-term frequency accuracy of 1×10^{-11} or better with optional verification to Coordinated Universal Time (UTC) and meets current industry standards. This equipment may be a Stratum 1 clock using a Cesium standard that defines an accuracy of $\pm1\times10^{-15}$, or it may be equipment directly controlled by standard UTC-derived frequency and time services, such as LORAN-C or Global Positioning System (GPS) radio receivers.

The LORAN-C and GPS signals themselves are controlled by Cesium standards. Because primary reference sources are Stratum 1 devices or are traceable to Stratum 1 devices, every digital synchronized network controlled by a PRS will have timing that is traceable back to a Stratum 1 source.

Stratum 2 nodes form the second level of the synchronization hierarchy. These clocks are synchronized to the PRS and can provide synchronization to:

▼   Other Stratum 2 devices

■   Stratum 3 devices, such as digital cross-connect systems (DCSs) or digital end offices

▲   Stratum 4 devices, such as channel banks or digital private branch exchanges (DPBXs)

Stratum 3 clocks can provide synchronization to other Stratum 3 devices and/or to Stratum 4 devices such as CSU/DSUs.

FRAME RELAY ACCESS SOLUTIONS

The alternative to building a solution is to buy it. These were the options available to companies that were considering building enterprise-wide networks. Today, buying a solution can mean fully divesting responsibility for the management, operation, and support of the network through some sort of outsourced arrangement. It can also mean maintaining some of these responsibilities—usually the local LAN, servers, and applications—while buying a packaged solution for the wide area portion.

Frame Relay fits into the latter category and is one of the most common access technologies among business customers. A Frame Relay solution may be a fully managed solution where the service provider provides the router and wide area connection as well as the management, operation, and support of the solution, or it may be a bearer service that encompasses the basic layer one and layer two functions of the transport.

Background

In the 1960s, mainframe computers were a shared resource, and users either had free access to them or had to buy time. In the campus environment, as well as in business, programmers accessed a shared computer through directly attached terminals. In the early days, these terminals were located in the same location as the computer and were directly attached via wires. This arrangement quickly evolved into one where a remote user no longer had to be on site but could connect from a single dumb terminal from a remote location over a dial-up connection.

Eventually, the requirement for one mainframe to connect to another became apparent. It was this need plus the inconvenience presented when a user had to connect to more than one mainframe—they had to either sign off from one and sign on to the other or use two terminals—that led to development of packet-based networks.

X.25, which operates at the first three layers of the OSI model, is a packet-based network solution that became popular during the 1970s and 80s. At the time of its development, the state of the telephone infrastructure was nowhere near the quality of present-day systems and as such had a great influence on the design features of X.25. The protocol, for instance, does error correction of every packet at every node in the network, which impacts performance. In contrast, modern packet-, frame-, or cell-based protocols rely on higher-layer protocols to check the integrity of the data. This change in philosophy was partially the result of infrastructure improvements made possible through the pervasive use of digital technology.

Frame Relay was developed to address the weaknesses of X.25 and to provide more functionality. The standards for the protocol are derived from the Integrated Services Digital Network (ISDN) Link Access Procedure for the D-channel (LAP-D)—later modifications are called LAP-F, Link Access Procedure for Frame Bearer Services.

An important characteristic of Frame Relay is that it exploits the advances that have been made in transmission technology. Earlier WAN protocols, such as X.25, were developed when copper and analog transmission systems were predominant. These links are much less reliable than the fiber media/digital transmission links available today. Over links such as these, link-layer protocols can forego time-consuming error correction algorithms, leaving these to be performed at higher protocol layers.

Greater performance and efficiency is therefore possible without sacrificing data integrity. Frame Relay is designed with this approach in mind. It includes a cyclical redundancy check (CRC) algorithm for detecting corrupted bits (so the data can be discarded), but it does not include any protocol mechanisms for correcting bad data.

The following is a summary of its improvements over X.25:

▼ Decreased protocol overhead through the elimination of error checking and limiting the protocol to the first two layers of the OSI model.

■ Digital switching through the development of interfaces for both T1 and E1 digital lines, with the original version being developed for use across ISDN interfaces. Digital circuits are less error prone, with less than one error in every 1Mb transmitted. With the improved infrastructure, Frame Relay was able to offload error checking to higher-layer protocols.

■ Higher speeds were made possible with Frame Relay. With the X.25 protocol, data was "piggybacked" on existing circuit-switched networks, which limited throughput to multiplexed DS0 rates of 64 Kbps. Frame Relay is able to operate at speeds in excess of T1 and E1 circuits and is more efficient in the use of bandwidth. Some applications of Frame Relay have it running on DS3 links.

▲ Support for other traffic types than just data. X.25 support only data, but today Frame Relay has built-in support for other traffic types like voice and video.

Table 8-6 lists some other key differences between X.25 and Frame Relay.

Feature	X.25	Frame Relay
Standards-producing organizations	CCITT, ISO, and others	ITU (CCITT), ANSI, and IEEE 802.2
Current typical speeds	9.6 Kbps to 64 Kbps	56 Kbps to 2 Mbps (and above)
Traffic type	Data	Data, voice, and video
Information unit type	Variable up to 4,096 bytes	Variable up to 4,096 bytes
Multicasting	No	Yes (optional)
Addressing	X.121 variable length (up to decimal 14 digits)	Fixed length (10-bit DLCI)
Explicit flow control per virtual circuit	Yes	No
Data link error correction	Yes	No
Topology	Star (typical)	Mesh/star/partial mesh

Table 8-6. Quick Comparison of X.25 and Frame Relay

The Technology

Frame Relay is a protocol for wide area access that operates at the physical and data link layers of the OSI model. It is a packet-switched technology that was originally designed for use across ISDN interfaces but is now almost exclusively used over non-ISDN interfaces. Using principles of statistical multiplexing, the technology allows the sharing of bandwidth between different applications. Variable-length packets are used for more efficient and flexible transfer of data through the network.

The Frame Relay specification defines the interface between the user and the network and the interface between two Frame Relay networks; it does not define the actual transport of the packets across the switching infrastructure. As a result, the infrastructure can utilize any switch technology, which may or may not be standards based. The only requirement placed on the network is for it to support the full suite of protocols and specifications defined in Frame Relay for access to the infrastructure.

Frame Relay employs virtual circuits that are uniquely identified by *data link connection identifiers (DLCI)* for communications between two end devices. Using this approach, multiple DLCIs can be bundled together across a single link, an approach that enables multiplexing different data streams destined for different destinations across the network.

The ability to use a single link for different destinations is a key differentia of using a packet-switched solution. In our previous discussions on circuit-based multiplexing, it emerged that different traffic streams could use the link but all had to have the same destination—the multiplexer on the other end of the link. Frame Relay has no such limitation, as the different virtual circuits can have different destinations that may be a few feet or thousands of miles apart.

Frame Relay Standards

Frame Relay was first proposed to the CCITT for adoption as a standard in 1984. The lack of standards had been a major stumbling block in its acceptance and deployment. It was hoped that published standards would lead manufacturers to adhere to a common set of specifications that would address interoperability issues. With interoperability issues addressed, it was believed that the way forward would be clear for faster global deployment.

It was not until 1990, however, that the technology gained some traction when Cisco, Digital Equipment Corporation (DEC), Northern Telecom, and Stratacom (eventually purchased by Cisco) formed a consortium, commonly called the Group of Four, to focus on Frame Relay technology improvements and deployment. The consortium focused on developing a set of specifications that conformed to, but extended the functionality of, those that were under consideration by the CCITT. The set of specifications, called the Frame Relay extensions, was developed for the support of more complex applications. They are collectively known as the *Local Management Interface (LMI)*, which has since been adopted by most manufacturers.

ANSI and CCITT (now the ITU) subsequently standardized a version of the LMI specifications that was originally developed by the consortium. Frame Relay is an international ITU standard that is defined in I.233, Q.922, and Q.933. In the U.S., it is an ANSI standard that is defined T.617 and T.618. In its definition, it is defined through three classifications: Frame Relay as an interface, signaling protocol, and as a network service.

The Interfaces

Frame Relay provides a packet-switching data communications capability that is used across the interface between user devices and frame relay network equipment (such as a router and switching nodes). User devices are often referred to as data terminal equipment (DTE) and network equipment as data circuit-terminating equipment (DCE). The network providing the Frame Relay interface can be either a carrier-provided public network or a network of privately owned equipment serving a single enterprise.

Frame relay consists of three interfaces: the *user-to-network interface (UNI)*, the *network-to-network interface* (see Figure 8-11), and the *local management interface (LMI)*, which is an optional extension.

User-to-Network Interface (UNI)   The UNI is based on the Link Access Procedure for Frame Bearer Services (LAP-F) and defines a set of procedures for communicating with a Frame Relay network port. It covers the entire link between the user port and the network port. The procedures for UNI are responsible for frame delimiting, alignment, and transparency; multiplexing; transparency; and error detection.

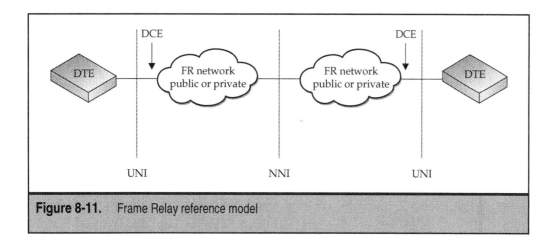

Figure 8-11.   Frame Relay reference model

Network-to-Network Interface (NNI)   The NNI set of procedures defines the interconnection between two Frame Relay networks. It has the basic set of procedures as defined by UNI plus additional features for congestion control, error handling, and link management.

Local Management Interface (LMI)   The LMI set of procedures is a set of extensions to the Frame Relay protocol that defines the process by which a user port and a network port exchange information regarding user status and configuration information for local UNI. It is responsible for getting the status of virtual circuits defined to the UNI, getting the status of connectivity between network ports and user ports, and storing address information. LMI will be discussed in more detail later in this chapter.

Permanent Virtual Circuits (PVCs)

Each FR UNI supports one or more *permanent virtual circuits (PVCs)*, which are logical connections that connect one user port to another user port (DTE to DTE via the network as shown in Figure 8-12). A PVC has characteristics similar to those of a leased-line connection: it is a fixed link that links two end points, it supports bidirectional traffic, any packets sent across the link will arrive in the order sent, and the end-to-end security is the same. Although a PVC behaves like a fixed point-to-point leased line, it operates a lot differently in ways that are transparent to a user.

In a leased-line connection, the actual link is a physical circuit that physically extends from one end of the network to another via switches. Packets using the link always take the same path through the same switches and the same wires. PVCs operate a lot differently. Once a packet arrives at the network port, it is switched from switch to switch across the network. Two packets, therefore, can take totally different paths through the network. It is the UNI's responsibility to provide features that makes all this transparent to the user. At the far end of the connection, the packets are presented to the user device in the order that they were sent.

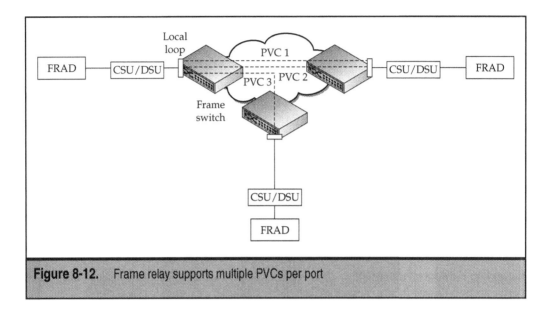

Figure 8-12.   Frame relay supports multiple PVCs per port

Because Frame Relay does not define the switching infrastructure, the actual method of getting a frame from one edge of the network to the other may vary by switch manufacturer. One vendor may use a scheme that always uses the same path through the network, whereas another may use a more dynamic switching scheme. Both approaches are valid because Frame Relay does not define the interior switching.

The data rate of the PVC is independent of the data rate of the access link. It is therefore possible to multiplex several PVCs across a single physical link (UNI). Taking this concept further, the PVCs can all be provisioned for different data rates, which may or may not add up to the data rate of the physical link. In fact, the sum of the PVCs can exceed the rate of the link; when this is done, the link is said to be *oversubscribed*. PVCs can also have an asymmetric data rate—one direction having a higher data rate than the other.

In our discussions on statistical multiplexing, we mentioned that statistical multiplexing techniques supported more users across the physical link than time-division multiplexing when traffic is predominantly bursty, as is LAN traffic. The reason for this is the statistical probability that all users will not be using the link at the same time. Using this probability, more users can share the link. If the traffic pattern is a constant bit stream instead of bursty, this becomes less true.

DLCI

To multiplex various data streams across different channels on a single link, the logical channels need to be defined at a layer higher than the physical layer. In Frame Relay, the data link layer is the layer at which the different logical channels are identified and managed. To do this, a *data link connection identifier (DLCI)* is defined for each logical channel as is evident in Figure 8-13. This identifier is used to distinguish one data stream from an-

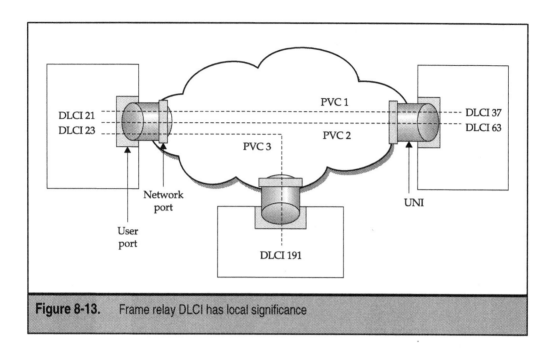

Figure 8-13.   Frame relay DLCI has local significance

other and has local significance. The consequence of this is that different switch ports can use the same DLCIs for each of their UNIs.

The PVC defines the actual end-to-end logical channel—user port to user port. Once the logical channel through the network has been established, it is further possible to assign other characteristics such as bandwidth. Multiple PVCs may be defined to a single circuit, with each PVC having its own bandwidth characteristic.

For example, a 56 Kbps circuit may have three PVCs defined, having respective bandwidths of 16 Kbps, 32 Kbps, and 32 Kbps. Notice, however, that the total of all three PVCs exceeds the capacity of the physical circuit. The circuit has a capacity of 56 Kbps, but the total bandwidth for all the logical channels is 80 Kbps.

To appreciate this concept, you need to remember that these are not physical links with physical speeds but logical links. In fact, it would be very difficult to enforce caps of 16 Kbps, 32 Kbps, and 32 Kbps for logical channels across the backbone. To the backbone, it is all data traveling at the speeds of the interswitch trunks. These caps are meaningful to the switches only when things are going wrong and are bandwidth that the service provider promises that the network will honor at any given time. The term used to describe these caps is *committed information rate (CIR)*.

Applying this to the three PVCs in our example, their respective CIRs are 16 Kbps, 32 Kbps, and 32 Kbps. This is the amount of data that the service provider guarantees to deliver across the network for each channel within a second of time. Implicit in this is the assumption that the service provider will not attempt to restrict the flow of the data to these caps; instead, the data will flow freely across the network, from UNI interface to UNI interface, at whatever speeds that the network was engineered to deliver.

The flip side to this is that the service provider does not guarantee anything above the CIRs. So if the network becomes busy, the probability of discarding the excess frames increases. The way the network differentiates packets that are within the CIR, and those that are not, is by setting a bit called a *discard eligible bit (DE bit)* in the headers of frames that exceed the agreed rate. Figure 8-14 shows the format of a frame and the different bits that can be manipulated by the switches.

The network will reliably transport all packets during times when the traffic patterns do not create congestion in the switches. During periods of congestion, the congested frame switch will begin to discard frames in order to relieve the congestion. The frames that are discarded are the ones that have the DE bit set; those that don't, receive priority treatment in order to honor the agreed limits.

A Glimpse in the Cloud   The only place in Frame Relay where connectivity from one switch to another is defined is the NNI. The normal interrouting between switches in a Frame Relay network is not covered. In an effort to tie together the elements that we have discussed so far, we will take a quick peek at how frames flow from switch to switch. It should be noted that this is a generic discussion that is not specific to a standard or vendor but is being presented as a means to complete the overall picture of the technology.

Table 8-7 follows the flow of a packet from a router to another router through the Frame Relay switching infrastructure. Figure 8-15 is provided as a reference that is to be used in context to the table.

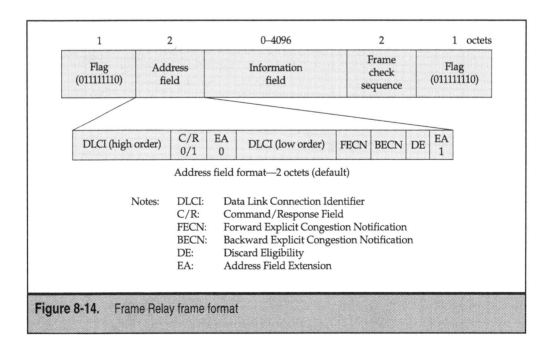

Figure 8-14.   Frame Relay frame format

Step	Action
1	Router A receives a packet destined to router B.
2	Router A appends a Frame Relay header with the proper DLCI of 101 and transmits it to network switch 1.
3	Switch 1 receives the packet on port 1.
4	Switch 1 uses a map table and replaces DLCI 101 with DLCI 100 and sends the packet to switch 2 via port 2.
5	Switch 2 receives the packet on port 1.
6	Switch 2 uses a map table and replaces DLCI 100 with DLCI 75 and sends the packet to router 2 via port 2.
7	The router strips the Frame Relay header and forwards it to the host.

Table 8-7.   Tracing a Packet Through a Frame Relay Network

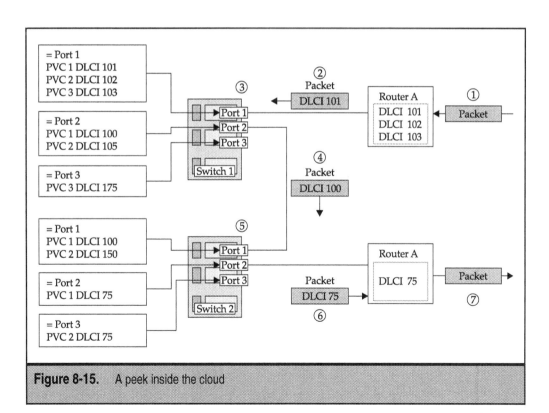

Figure 8-15.   A peek inside the cloud

Flow and Congestion Control

While Frame Relay does not employ an explicit method of flow control, it does contain mechanisms intended to notify users of congestion conditions. This is the extent of its notification; the end devices are left to decide how they choose to manage and respond to these notifications. The governing principle of its operation is to inform of a problem, not to attempt to correct it.

There are two methods used for congestion notification, *Forward Explicit Congestion Notification (FECN)* and *Backward Explicit Congestion Notification (BECN)*. Both methods rely on the setting of a bit in the address field of the frame header, as shown in Figure 8-16. When a packet travels across the network, a switch along the path that is experiencing congestion sets the FECN bit flag in the frame header. Packets on the return path have their BECN bit set.

Using this process, two things are accomplished. First, the FECN bit serves to inform the receiving end device that the packet passed through one or more congested nodes on the way to its destination. Using this information, the user device is able to inform a high- layer protocol, which may be able to use the information for pacing the rate at which it sends data. Second, for frames traveling in the opposite direction, the BECN bit

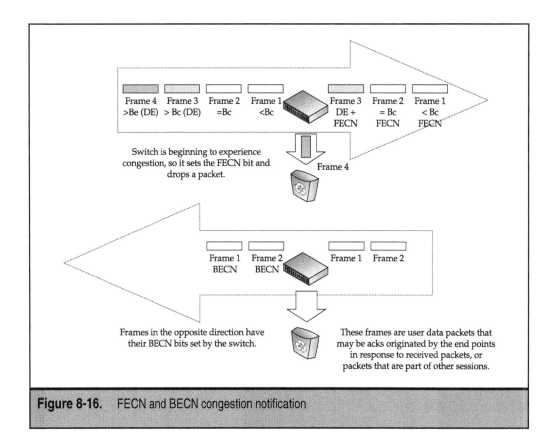

Figure 8-16. FECN and BECN congestion notification

serves to inform the receiving end device that any packets that it transmits may be subject to congestion on their way to their destination. Using this information, the application may or may not choose to pace the rate of transmission.

Notice that in both cases—forward and backward congestion notifications—Frame Relay relies on higher-layer protocols to respond to the congestion message and make the appropriate adjustments. It does not attempt to tell a switch to slow down. In fact, it takes the opposite approach: if the congestion gets chronic, it will begin dropping frames without sending any notification whatsoever that it has done so. This apparently cavalier approach to flow control and congestion control frees it to focus on basic functions such as delivering frames within a CIR at higher data rates than its predecessor.

Enforcing Subscription Rates

The foregoing explanation is a simplified version of how Frame Relay enforces CIR. This may be sufficient for most readers and is probably the depth of information that is necessary for most exposure to the technology. If you wish to dig a bit deeper, we will take a closer look at what actually happens.

The calculation and enforcement of CIR is a result of different parameters. These parameters are the CIR, the *committed burst size* (B_c), the *excess burst size* (B_e), and a *committed rate measurement interval* (T_c) that is not defined but is derived from these values. The parameters are explained in the following paragraphs, and Figure 8-17 presents an illustration of the concept.

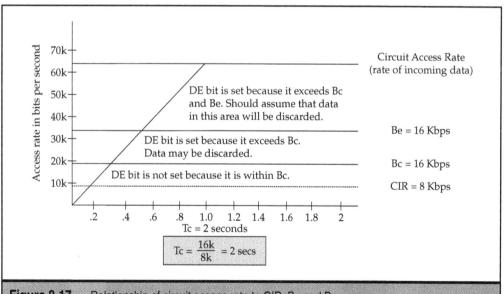

Figure 8-17. Relationship of circuit access rate to CIR, B_e, and B_a

Committed Information Rate (CIR)   CIR is the average throughput rate, in bits per second, that the user can expect from a PVC. In theory, the user should be able to transmit data continually without problems over a Frame Relay network, if a constant bit rate that is equal to CIR is attained. CIR is configurable for each PVC in the provider's Frame Relay switches.

Committed Burst Size (B_c)   The B_c is the maximum number of bits of user data (not bits per second) that the network commits to transfer across a given Frame Relay path over the committed rate measurement interval (T_c), which is a time interval. The time interval, T_c, is not configurable but is calculated as B_c / CIR in seconds and is the time interval over which the user may transfer B_c bits of committed data and $B_c + B_e$ bits of uncommitted data.

The ideal value for B_c for data PVCs is CIR / 8 so that T_c = 125 msec (obviously less is even better). For voice, a shorter delay is desired, so B_c = CIR / 100 is preferable so that the interval T_c = 10 msec.

Excess Burst Size (B_e)   The B_e is the maximum number of bits in excess of B_c (not bits per second) that the network will attempt to deliver on a given Frame Relay path over T_c. All frames in excess of B_c are sent with the DE bit set, which marks the frame as eligible for discard if congestion occurs. When congestion increases above a certain threshold, all DE frames are discarded.

Committed Rate Measurement Interval (T_c) T_c is the time period calculated by dividing B_c by CIR. It is used to determine the time period from which data will be measured in total bits, to determine if the user is within his agreement.

Each of the throughput parameters (CIR, B_c, B_e) has an effect on the others. For example, if the CIR = 1,000 bps and B_c = 2,000 bits and B_e = 0 bits, then T_c = 2,000 / 1,000 or 2 seconds. Note also that B_c does not have to equal CIR. Given this example, imagine that the user sends 2,000 bits of data in the first 0.1 second. The user should not send any more data until T_c expires, which is 1.9 seconds later, if the user is to be within the defined service parameters.

If more data is sent within the 1.9 seconds, it will be in excess of the defined service and may be discarded. The parameters given in the example therefore state that the user has the right to transmit 2,000 bits across the Frame Relay network in any two-second period. If the user exceeds the agreed 2,000 bits in two seconds, the excess data may be discarded by the network.

The relationships between the different parameters are explained by the following:

If CIR > 0, then the following applies:

▼ B_c = CIR

■ $0 = B_e$ = (access rate – B_c)

▲ $T_c = B_c$ / CIR

If CIR = 0, then the following applies:

- ▼ $B_c = 0$ (prevents divide by 0 error)
- ■ $0 < B_e$ = access rate in bits per second (If $B_e = 0$, then all frames would be dropped)
- ▲ $T_c = B_e$ / access rate in bits per second

Some service providers mandate that CIR must be equal to B_c, resulting in a T_c of one second. B_e may be configured for any value between 0 and the access rate minus B_c. Switches will treat data that is in excess of CIR in one of two ways. The first is to discard anything that is in excess of B_c. If B_e is configured for 0, then anything above B_c is eligible for discard. Some providers choose to use this option as a way to force customers to stay within the agreed service limits. This option is the least favorable to the customer, but it helps the service provider project bandwidth requirements across the infrastructure.

The second option is to always attempt a best effort at delivery and resort to discarding frames only when the switch begins to suffer from congestion. In this mode of operation, the bursts above B_c will be delivered if possible. However, if a burst above B_c is large enough to cause congestion, the switch will drop the packet.

In practice, the user should expect everything above B_e to be discarded. Because the PVC can pass through a number of Frame Relay switches, it is important that the CIR, B_c, and B_e configurations be the same on every switch. If not, the performance will be at the lowest defined throughput level. Some carriers offer a 0 CIR service, which means that the service provider will make a best effort for all packets but with no guarantee.

The Impact of Packet Size on Performance and Subscription Rates   An aspect of Frame Relay that is not often considered, is the impact of the frame size on subscription rates. The Frame Relay forum suggests a maximum size of 1,600 bytes. This value was chosen because it represents the size of a maximum Ethernet frame with the Frame Relay header added. This value when expressed in bits is 12,800 (1,600×8).

If the B_c is less than 12,800, any 1,600-byte packet may be flagged for DE. If the service provider strictly enforces his subscription policy, or if the network is highly utilized, the customer could be faced with a network that will only pass one 1,600-byte packet per second. If a low CIR is to be used, it would be best to ensure that B_e is high enough to provide acceptable levels of performance. However, any excess above B_c is subject to the enforcement policies of the service provider, so even this level of excess may be too much.

In reality, most packets will be less than 1,600 bytes, but it is a factor to consider when designing a Frame Relay connection or when determining the subscription rates of a service. The largest frame that a Frame Relay connection will handle is 4,096 bytes, the size of a token ring packet. For applications involving token ring LANs, the size of the packet makes this point even more relevant.

Link Management Interface (LMI)

The user-to-network interface and the network-to-network interface were the two original Frame Relay interfaces. The link management interface (LMI) was developed as a set of extensions to the protocol for the support of large and complex internetworks. Since UNI and NNI formed the basis of the network, the developers of the LMI extensions wanted to ensure that there was a way to properly exchange information with the myriad of devices that may attach to the network.

LMI was developed as a maintenance protocol that could be used between the user device and the network switch. The protocol enables proper control and supervision of the user interface and the possibility of checking the status of the subscriber connection. It is also used to inform the user about the addition, disconnection, or modification of a PVC. (For the structure and the fields within an LMI frame, see Figure 8-18.)

LMI is not a required protocol for the proper operation of Frame Relay. A device can connect and use all the features of the network, including FECN/BECN, CIR, PVCs, and DLCIs, without the use of LMI. However, when the feature is enabled, it provides additional features that can be incorporated into a more robust management strategy. It could, for instance, enable a network management server that is monitoring a router to generate an alert when an LMI response has not been received from the remote switch. By integrating these features into a management strategy, the user is better able to manage and respond to an outage.

Some LMI extensions are referred to as common and are usually implemented by every vendor who adopts the specifications. Other LMI specifications are referred to as optional. The common LMI extensions are based on the message formats of the ITU Q.931 recommendation.

Common LMI Extensions   Common LMI messages is usually fully implemented by vendors who choose to implement LMI. These common messages are classified under a single description: *virtual circuit status messages*. These messages provide communication and synchronization between the network and the user device, periodically reporting additions of new PVCs and deletions of those previously defined. These messages also provide information about the integrity of the PVCs.

Optional LMI Extensions   These extensions may optionally be implemented:

▼   **Global addressing**   Under normal conditions, connection identifiers have only a local significance and up to 1,024 DLCIs can be defined to a port. With this extension, the values inserted in the DLCI field of a frame are globally

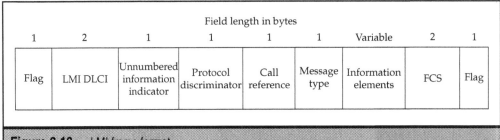

			Field length in bytes					
1	2	1	1	1	1	Variable	2	1
Flag	LMI DLCI	Unnumbered information indicator	Protocol discriminator	Call reference	Message type	Information elements	FCS	Flag

Figure 8-18.   LMI frame format

significant addresses of individual end-user devices (for example, routers). It allows a single port to be addressable by a single DLCI so that any user on the network is able to use the same DLCI for anything destined to the port. When global addressing is used, the entire network is limited to 1,024 DLCIs.

- ■ **Multicasting**   This feature allows the sender to send a single frame to a group of specific users. The receivers of the frame all belong to a common multicast group; any device that does not belong to the group does not receive a copy of the frame.

   Multicast groups are designated by a series of four reserved DLCI values (1,019 to 1,022). Frames sent by a device using one of these reserved DLCIs are replicated by the network and sent to all network ports in the designated set. The multicasting extension also defines LMI messages that notify user devices of the addition, deletion, and presence of multicast groups.

- ▲ **Flow control**   This is an enhancement to the existing congestion notification. It provides for a simple XON/OFF flow control mechanism that applies to the entire Frame Relay interface. It is intended for devices whose higher protocol layers cannot use the congestion notification bits.

Switched Virtual Circuits (SVC)

In addition to permanent virtual circuits that have characteristics similar to those of a leased-line circuit, Frame Relay defines a *switched virtual circuit (SVC)* that shares similarities with a dial-up connection. SVC is a specification that was added in later years of the technology. It is a temporary connection-oriented connection used in situations requiring sporadic data transfer between DTE devices across the Frame Relay network.

A communication session using SVC consists of the following states, which are also presented in Figure 8-19:

- ▼ **Call setup**   The virtual circuit is established between two Frame Relay DTEs on a temporary basis. The call setup process is similar to a user dialing into a network to establish a temporary connection with a remote device.

- ■ **Data transfer**   The virtual circuit, once established, is used to transfer data between the two user devices.

- ■ **Idle**   In this phase, the connection is still established but there is no activity across the link. If the connection remains in this state for longer than a predefined period, the connection can be terminated. The length of the idle period before termination is determined by a configuration parameter.

- ▲ **Call termination**   As with a dial-up connection, an SVC connection can also be terminated. Once the connection has been terminated, the call setup process needs to be repeated before any more data can flow.

The signaling protocols of ISDN are used for SVC connections. Few manufacturers of Frame Relay equipment support switched virtual circuits. As a result, there are few implementations of the technology.

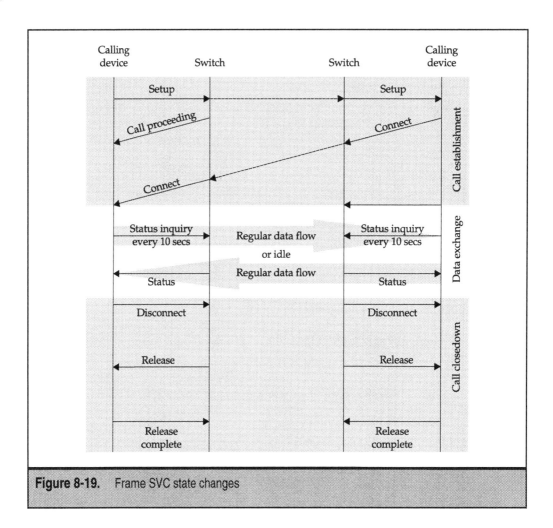

Figure 8-19.   Frame SVC state changes

SUMMARY

Corporate customers, universities, and industry have had a need for broadband solutions for as long as computers have been in existence. In the early years, their only option was to build a private enterprise network using traditional TDM and statistical multiplexing techniques.

Public X.25 networks, which were limited to data, eventually emerged and partially addressed their needs. Frame Relay has since replaced X.25 and has become a widely deployed technology for public and private networks. Advances in transmission technology have enabled Frame Relay to offload performance-limiting features like explicit flow control and error correction to higher protocol layers.

Frame Relay is a robust, well-understood, and widely deployed WAN protocol that operates at the physical and data link layers of the OSI reference model. Today it is used over a wide variety of network interfaces at speeds up to DS3. This chapter has focused on the common elements of the technology and the wider principles of digital transmission in an effort to provide a broad appreciation of the current state of Frame Relay and elements of its operation.

CHAPTER 9

Emerging Broadband
Access Technologies

In this chapter, we will explore some emerging access technologies. The term emerging technology has always been somewhat of a misnomer because the technology is almost always already in use by someone, somewhere—long before it is ever recognized as emerging. The curious thing about an emerging technology is that if it exists only in a lab—where one would expect it to emerge—it's hardly ever seen as emerging but more as a technology that has promise and one that may emerge sometime in the future. Because these technologies are already in existence, the term "emerging markets for new technologies" is probably more appropriate.

Looking back at all the access technologies that we have covered so far, DSL, cable, wireless, FTTx, and Frame Relay, one would think the market would be saturated with not much room for more. The reality is that broadband networking has such promise, and the market is so fertile, that there are always contenders seeking to dethrone incumbent technologies. Whether or not they succeed is dependent on many factors, some of which are performance, reliability, speed, recognition, and probably most important of all, cost.

EMERGING PASSIVE OPTICAL NETWORK TECHNOLOGIES

Today, the greatest market driver for consumer broadband access is the Internet. Maybe as we get closer to 2006, the impetus will change in the direction of digital television. Until then, the Internet will play a dominant role, which will in part contribute to the types of technologies that take root.

In Chapter 7, we reviewed fiber to the x as an access technology with a focus on ATM passive optical networks. Backed by established standards and major telecommunication companies around the world, the technology has the backing of an industry that is in the best position to push the technology. However, not everyone is convinced that ATM PON is the way to go. Some dissenters believe that the speeds are still too low and ATM is not a very good means of handling the types of traffic—IP and data—that are driving its growth.

For years, proponents of Ethernet have contended that ATM is not as efficient as Ethernet in handling IP. It has been their contention that 5 bytes for every 48 bytes of payload are a high price to pay for the transport of an IP packet. Considering an Ethernet packet of 1,518 bytes, this would take 38 cells with 158 bytes of overhead. It is not surprising, then, that one of the emerging access technology contenders that it is being positioned right alongside ATM PONs is Ethernet.

As digital television gets closer to becoming a reality, the benefits of ATM for transporting video may become a much more convincing argument. But until then, Ethernet should probably be taken very seriously as an access technology, for both business and residential consumers. This is especially true when data is the predominant market driver, as is evident in Table 9-1. The new application of Ethernet that is now challenging ATM PON is Ethernet passive optical networks, or EPON.

Application	Traffic Type	Data Size	Implications
Scientific modeling and engineering	Data files	100s of megabytes to gigabytes	3-D visualization of complex objects increases bandwidth requirement for desktops, servers, local network, and backbone.
Medical	Data files	100s of megabytes to gigabytes	Transmission of images between desktops and servers requires high bandwidth on local network and backbone.
Publishing	Data files	Up to 100MB	Full-color magazines and brochures prepared and transmitted directly from the desktop to digital-input printing facilities require high bandwidth for server and backbone.
Internet/intranet	Data files, audio and video streams	Large data files up to 100MB	Web browsers have evolved from straight text to a new generation of multimedia client/server applications that require high bandwidth for servers and backbone, low latency.
Data warehousing, SANs, and network backup	Data files	Gigabytes to terabytes	Disk mirroring, data warehousing, and backup requirements require high-bandwidth servers and backbone with low latency and secure systems. Warehousing may comprise gigabytes or terabytes of data distributed over hundreds of platforms and accessed by thousands of users, with the requirement for near-real-time data for critical business reports.
Desktop video conferencing and collaborative computing	Constant data stream	1.5 to 3.5 Mbps at the desktop	Collaborative workgroup applications that incorporate video conferencing are becoming commonplace. It requires higher bandwidth for servers and backbone, low latency, predictable latency, class of service reservation.

Table 9-1.   Traffic Types Driving Broadband Access

Ethernet Passive Optical Networks (EPONs)

EPONs are an emerging access network technology that provides a low-cost method for deploying optical access lines between the service provider and the customer. The technology should not be confused with Gigabit Ethernet, which is defined by the IEEE. EPONs build on the International Telecommunications Union (ITU) standard ITU G.983, which was defined for ATM PONs or APONs and developed by the Full Service Access Networks (FSAN).

A single event and two key transforming trends are aligning to redefine local and wide area networks as we once knew them. The first is the Telecommunications of Act of 1996, which opened up competition in the local loop. The second is the rapid decline in the cost of fiber optics and optical components, and the third is the growing demand for data and broadband services, which has placed a burden on carriers to upgrade existing facilities. These two trends and single event are aligning to reshape the networking industry.

As the cost of fiber falls, it has been slowly moving from the backbone outward into the access network market. Ethernet, the predominant LAN technology and an enabler of data and broadband services, is slowly creeping into the wide area. These two seemingly disparate trends have finally met, and a new breed of Ethernet technologies that are based on fiber has now emerged. In this section, we will examine one of those technologies, Ethernet PONs, and in the next, we will explore a parallel but more established development that is happening in Gigabit Ethernet. In both cases, Ethernet is now a viable contender for MAN and WAN networks.

Background

The operation of, and issues surrounding, FTTx solutions that use ATM PONs have already been addressed, and we will therefore assume a basic understanding of passive optical networks. As a result, our discussion of EPONs will be limited to points of the technology that differentiate it from APONs.

Passive optical networks are intended to address the speed gap that exists between the local area network and the backbone wide area network. Both have the potential for very high speeds, but the access to get there is greatly lacking. The LAN and the WAN backbone are like two great oceans, each carrying large ocean liners. The only way for a ship to get from one ocean to the other is via small tributaries of a river, which represents the access and metropolitan area network. The PON is meant to address this problem by creating large channels between these two great oceans.

Two companies are aggressively driving EPON, Alloptic and Salira Optical Network Systems. In May 2001, Alloptic announced that it was conducting field trials of EPON with 23 carriers in every major world market. The technology in trials is an IP EPON that incorporates burst-mode optical transceiver technology with data rates up to two times the downstream rate and five times the upstream rate of APONs. The Alloptic IP EPON achieves a downstream data rate of 1 Gbps and 800 Mbps upstream. APONs achieve rates of 622 Mbps downstream and 155 Mbps upstream. Salira Optical Network Systems intend to launch their products in 2002.

In November 2000, a group of Ethernet vendors, under the auspices of the IEEE, initiated the process for the development of a standard for Ethernet in the access network by forming a study group called Ethernet in the First Mile (EFM). The group, which consists of 3Com, Alloptic, Aura Networks, Cisco Systems, Intel, MCI/Worldcom, and others, aims to develop a standard that will apply the Ethernet protocol to the access market.

In addition to the IEEE study group, other standards organizations, including the IETF, the ITU, and ANSI, are also being explored for sponsorship of EPON standards. One approach that is being considered to speed the development of an EPON standard is to work cooperatively with FSAN, the developers of the APON technology. The approach would be to focus the effort on the development of a standard EPON MAC protocol and reference FSAN for everything else. This would be the quickest path to the development of an EPON standard.

Operational Features of EPONs

The main difference between an APON and an EPON is the way the data is transported. EPONs maintain the variable-length structure of Ethernet packets, with a maximum length of 1,518 bytes as defined by the IEEE 802.3 protocol. APONs use fixed-length 53-byte cells with 48 bytes of payload and 5 bytes of overhead as specified by the ATM protocol.

The topology and components of EPONs and APONs are the same. They both use an OLT (optical line terminator) at the CO, an optical splitter, and multiple ONUs or ONTs (optical network units or terminators). A single fiber connects the OLT to an optical splitter, which splits the fiber into multiple strands that connect different ONUs. The splitter makes multiple copies of the signal from the OLT and places them on the separate strands that connect the customer.

Because the network has a point-to-multipoint topology, the scheme for downstream broadcasting of a packet is different from the scheme used for upstream communication. In the downstream mode, the OLT is the sole contender for the channel. In the reverse, the channel is shared among several ONUs.

In the downstream mode, as shown in Figure 9-1, variable-length packets are broadcast to the connected ONUs, with each packet bearing a header that uniquely identifies it as belonging to a specific ONU. Each ONU accepts the packets that are intended for it and discards the rest. The term broadcast in this context describes the fact that the OLT sends a unicast packet to all ONUs but addressed to a specific ONU. Each ONU is able to see the packet because they all share the same network fiber infrastructure, but only the ONU that recognizes its address copies the packet from the network.

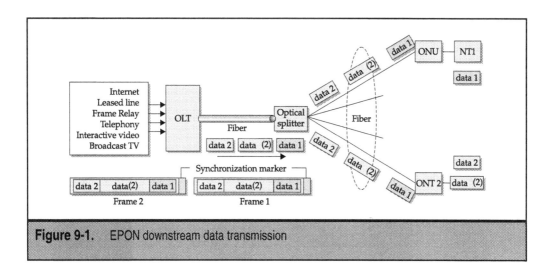

Figure 9-1.   EPON downstream data transmission

The downstream traffic is segmented into fixed-interval frames, with each frame carrying multiple variable-length packets. The header of the frame has a single-byte code, called a synchronization marker, that is used for timing synchronization between the OLT and the ONUs. The synchronization marker is transmitted every 2 ms.

In the upstream flow, shown in Figure 9-2, time-division multiplexing (TDM) is used to manage the way in which the ONUs access the channel. Each ONU is granted a time slot in which to transmit. The upstream traffic is also segmented into frames that are continuously transmitted every 2 ms, and each frame is segmented into smaller time slots.

Each ONU is assigned a specific time slot, which may be used to transport a single large packet or as many smaller packets as can fit within the slot. The time slot also includes some overhead, which consists of a guard band and a start of slot delimiter that contains timing indicators and signal power indicators (see Figure 9-3). When there is no traffic to transmit, the time slot is filled with an idle signal.

Benefits of EPON

The argument for EPONs over APONs is based on the growing requirement for IP-based services and the insatiable appetite for data. Besides, EPONs are a lot cheaper to implement because they can support more users per unit. The higher the ratio of ONUs to a single OLT, the lower the equipment cost on a per-user basis. With a higher ONU to OLT ratio, the performance is degraded, but that becomes less of a factor than with APONs because EPONs inherently operate at twice the speeds of APONs. Finally, the higher operating speeds of EPONs as compared to APONs is itself an advantage.

The strongest argument against EPONs is the fact that they are not backed by any standards. The technology, therefore, will potentially operate differently from vendor to vendor. The network described in the previous paragraphs is just one implementation of the technology; another system such as the Salira Optical Network Systems product may

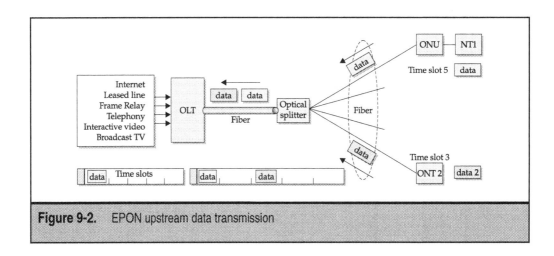

Figure 9-2. EPON upstream data transmission

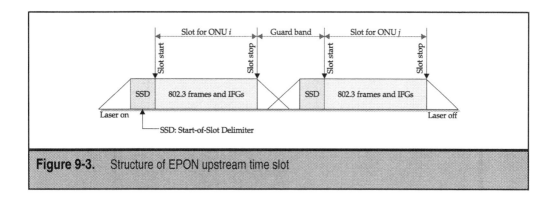

Figure 9-3. Structure of EPON upstream time slot

operate slightly differently from the one described. Standards address these differences by setting minimum guidelines on how a piece of equipment should implement a technology. This is not to say every equipment vendor will produce identical products with the same features; standards do not restrict vendors from adding features. At a minimum, each vendor must meet the requirements of the standard to ensure interoperability; beyond this, any number of enhancing features may be included as long as they do not prevent the equipment from operating within the established guidelines of the standard.

Another argument against EPONs is that they are still not fully defined and lack two major functionalities. One is the OAM&P—operational, administrative, maintenance, and provisioning—functionality, which is well defined (however complex) in APON systems but poorly defined in Ethernet systems. This feature is of key importance to operators in order to achieve effective network monitoring. The second missing functionality, which exists in APONs, is QoS.

APONs, in contrast, are based on established standards that are defined by ITU G.983. Because EPONs are evolving from the experienced gained from the deployment of APONs, it is highly probable that the two technologies will share a common development path. There is also the possibility that the standardization work that is being done on Ethernet will make reference to the FSAN document, which adequately addresses the issues of subscriber access networks. There is nothing in the FSAN document that precludes other protocols that are not ATM based.

EPON Standard: IEEE 802.ah Ethernet in the First Mile (EFM) Task Force

In July 2001, the IEEE announced the approval of a Project Authorization Request for Ethernet in the First Mile. The 802.3ah task force was authorized to carry out the work of drafting the standard pending approval by the IEEE Standards Association Standards Board. Ethernet in the subscriber access network is viewed as having several advantages over traditional first-mile technologies in terms of cost, network simplicity, packet-based efficiency, bandwidth, scaling, and provisioning.

The EFM Study Group has identified several key objectives that will be used to evaluate technical proposals brought before the 802.3ah Task Force. These key objectives include support of three subscriber access network topologies and physical layers:

▼ Point-to-point copper over the existing copper plant at speeds of at least 10 Mbps up to at least 750 m.

■ Point-to-point optical fiber over a single fiber at a speed of 1,000 Mbps up to at least 10 km.

▲ Point-to-multipoint fiber at a speed of 1,000 Mbps up to at least 10 km. This is the subscriber access that will define EPONs, as we know them today. To date, there has never been a version of Ethernet that works in a point-to-multipoint topology; this will be the first.

The project will also define operations, administration, and maintenance (OAM) for EFM, which includes remote failure indication, remote loopback, and link monitoring. The scope of the EFM working group is shown in Figure 9-4.

GIGABIT ETHERNET AS A BROADBAND ACCESS TECHNOLOGY

Over the years, Ethernet has proved to be a resilient technology that has become the de facto LAN technology in the business environment. The cost of components has fallen substantially, making it extremely affordable, and just when one thinks the technology

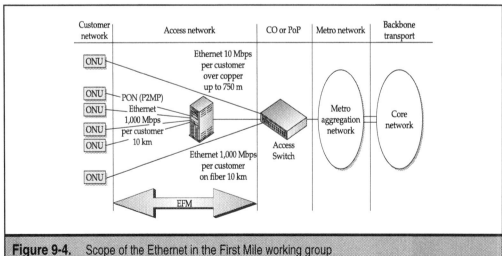

Figure 9-4. Scope of the Ethernet in the First Mile working group

has no room for further improvement, another, faster version appears. In fact, there was a time when ATM was being positioned as the technology that would extend from the desktop all the way across the network, supplanting Ethernet in the process. In response, 100Base-T emerged to be the favored technology that has broad applicability extending to the residential market.

As further proof of Ethernet's resiliency, *Long-Reach Ethernet (LRE)*, which we explored in Chapter 2, has evolved at a time when it is most needed. It supports broad- and narrowband communication across an infrastructure that would have been impossible in the closing years of the 1990s. Because of it, broadband access is now a possibility in buildings with substandard wiring that probably would have been too expensive to upgrade.

Gigabit Ethernet, as it is called at these speeds, still has room to run; equipment supporting a newer version of the technology that operates at 10 Gbps is already being introduced in the market. A new standard for this newer and faster version of Ethernet is to be published in 2002.

Based on its track record, its installed base, and the price per megabyte, many industry experts believe that Ethernet will ultimately be the final protocol that wins out over ATM in the wide area. Looking at its history, it would be wisest not to bet against it. Table 9-2 provides a quick comparison of the features and issues of Ethernet and ATM.

Feature	Ethernet	ATM
Speeds	10 Mbps, 100 Mbps, and Gigabit speeds	25 Mbps (desktop), 155 Mbps to 622 Mbps
Scalability	Scalable speeds use the same frame structure, allowing for backward compatibility.	Does not have the same granularity of speeds as Ethernet.
Troubleshooting	Same packet structure allows for easier troubleshooting between different Ethernet types.	Attaching an Ethernet segment to an ATM transport is a more complex environment to troubleshoot.
Management	Common management platform.	Has the potential to require different management platforms.
Cost	Cost per port is a lot lower mainly because of the rapid development of greater speeds.	ATM cost per port is a lot higher.
Strengths	Well-understood and pervasive technology for data applications. Less suited to applications that depend on a constant bit rate with low latency.	Much more suited than Ethernet to applications that require a constant bit rate with low latency.

Table 9-2.   Ethernet Versus ATM as an Access Technology

Feature	Ethernet	ATM
Integration	Because of the pervasive installed base of Ethernet LANs, it is logical to conclude that Ethernet as an access technology is easier to integrate than ATM.	ATM as an access technology requires the segmenting of Ethernet packets and would not be as easy to integrate as Ethernet into existing LANs. But it is more suited than Ethernet to a mixed application environment.
Migration	As a corollary to the points of integration, Ethernet provides an easier migration strategy for data applications and existing LANs. Not as suited to mixed environments as ATM.	The smaller cell sizes of ATM actually make it easier to adapt to applications that have a mix of frame sizes and latency requirements.
Security	About evenly weighted.	About evenly weighted.

Table 9-2. Ethernet Versus ATM as an Access Technology *(continued)*

Gigabit Ethernet

There seems to be no end to the voracious appetite for bandwidth. The edges of technology keep expanding as research and development efforts push the envelope of performance and redefine what is economically feasible. Ethernet, like the microprocessor, keeps being redefined through the extensions of the upper boundaries of its operation.

Interestingly, while the speed of microprocessors doubles every 18 months or so, each round of improvement of Ethernet is to the order of 10 times the previous speed. The growth incline of Ethernet has seen the speed go from 10 Mbps to 100 Mbps (there were earlier versions that operated at less than 10 Mbps) to 1 Gbps and now 10 Gbps. The good news is that each subsequent improvement keeps the original frame structure that is shown in Figure 9-5, allowing for backward compatibility. The operation of the protocol is also consistent across the speeds.

In 1995, the IEEE 802.3 committee formed a High-Speed Study Group to explore different means of obtaining Gigabit speeds while maintaining the existing Ethernet structure. As a result of these efforts, a new set of Ethernet standards have been issued that defines a new medium and transmission specification while retaining the CSMA/CD protocol of the 10/100 Mbps versions. Figure 9-6 shows the typical use of the new technology

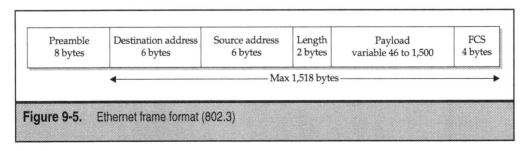

Preamble 8 bytes	Destination address 6 bytes	Source address 6 bytes	Length 2 bytes	Payload variable 46 to 1,500	FCS 4 bytes

◄─────────────────── Max 1,518 bytes ───────────────────►

Figure 9-5. Ethernet frame format (802.3)

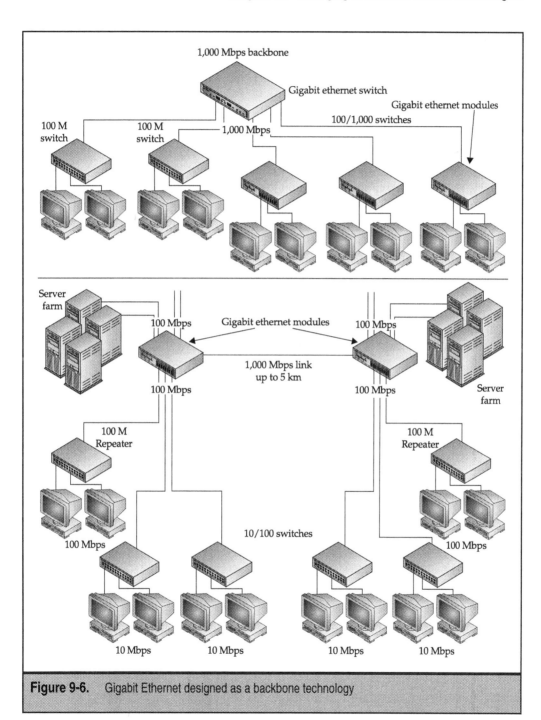

Figure 9-6.   Gigabit Ethernet designed as a backbone technology

as a backbone network for existing 10/100 Mbps LAN segments and servers. As a backbone technology, it is now evolving as a viable solution for metropolitan networks.

Protocol Architecture

The media access control (MAC) layer is an enhanced version of the basic 802.3 MAC algorithm. An optional *Gigabit Medium Independent Interface (GMII)* has been defined for all media except unshielded twisted pair (UTP) wires (see Figure 9-7). The GMII defines independent parallel eight-bit transmit and receives synchronous interfaces that are intended for chip-to-chip communication, thereby allowing for flexibility in mixing different MAC and PHY sublayer components.

Two signal-encoding schemes have been defined at the PHY layer. The 8B/10B scheme was defined for optical fiber and shielded copper media, and a five-level pulse amplitude modulation (PAM) is used for UTP.

MAC Layer   The improvements offered by GigE have a lot to do with improvements in timing, which enable the recognition of the preamble and the delimited fields at a much faster rate. The protocol supports both half- and full-duplex modes of operation, an enhancement that was introduced in the switched version of 100 Mbps Ethernet. In the half-duplex mode of operation, the protocol maintains the CSMA/CD feature of the original protocol. Figure 9-8 shows a flow chart of the CSMA/CD process.

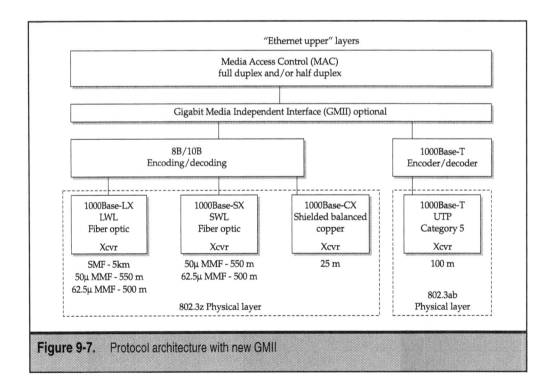

Figure 9-7.   Protocol architecture with new GMII

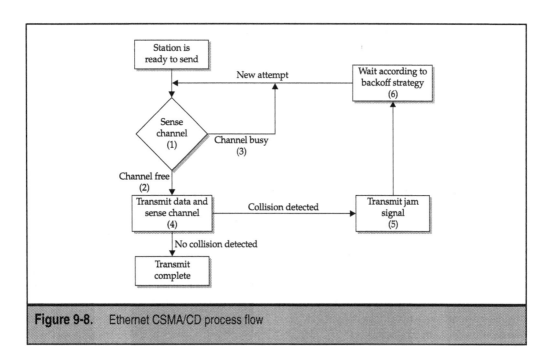

Figure 9-8.   Ethernet CSMA/CD process flow

At one billion bits per second, everything happens so quickly that the length of the frame and collision domain comes into play a lot more than in previous versions. At issue is the fact that at these speeds, Ethernet packets of the minimum size (64 bytes) could complete transmission before the transmitting station could sense that a collision has occurred, thereby violating the CSMA/CD rules.

To get around this problem, the slot time that was normally the size of the minimum Ethernet packet was increased to 512 bytes—eight times the original value. The minimum size of the packet remained at 64 bytes, but packets smaller than 512 bytes now have a carrier extension field following the CRC field (see Figure 9-9). The carrier extension

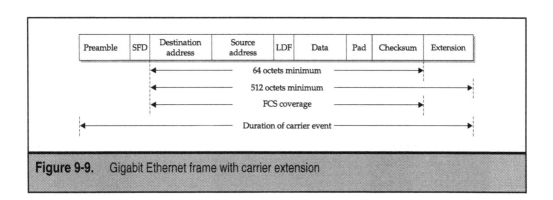

Figure 9-9.   Gigabit Ethernet frame with carrier extension

appends a set of symbols at the end of short MAC frames so that the frame is at least 512 bytes long. The extended frame length ensures that the frame length is longer than the propagation time at 1 Gbps.

By adding the carrier extension symbols after the data and FCS fields, the extra symbols are not calculated in the FCS CRC calculations. The receiver drops the extension bits once the frame arrives at its destination. The Logical Link Control layer (LLC) never sees the extension bits, so everything appears normal at this level. Packets of 512 bytes and above do not need, nor do they have this extension. With the change, the parameters for Gigabit Ethernet now look like those presented in Table 9-3.

The addition of the carrier extension field has the potential to impact the performance of smaller packets. Packet bursting, a new feature, has been introduced to offset any performance trade-offs created by the augmentation of packets less than 512 bytes. In order to fully utilize the bandwidth, the burst mode allows servers, switches, and other Ethernet-capable devices to send bursts of multiple small packets consecutively up to a limit without relinquishing control for CSMA/CD.

Devices that operate in full duplex are not subject to the carrier extension, slot time extension, or packet bursting changes because they utilize a dedicated medium rather than one that is shared. They continue to use the regular Ethernet 96-bit interframe gap and 64-byte minimum packet size. The new features are not needed because transmission and reception of packets occur independently of each other without interference or contention.

The full-duplex mode of operation is the predominant mode of operation for Gigabit Ethernet. The CSMA/CD tweaks—carrier extension and burst mode—are not generally used in most applications.

Parameter	10 Mbps	100 Mbps	Gigabit
slotTime	512 bit times	512 bit times	4,096 bit times
interFrameGap	9.6 microseconds	960 nanoseconds	96 nanoseconds
attemptLimit	16	16	16
backoffLimit	10	10	10
jamSize	32 bits	32 bits	32 bits
maxFrameSize	1,518 bytes	1518 bytes	1518 octets
minFrameSize	64 octets	64 octets	64 octets
addressSize	48 bits	48 bits	48 bits
extendSize	—	—	448 octets

Table 9-3. Parameter Values for Gigabit Ethernet

PHY Layer Support   Work on the PHY layers is addressed by two different working groups of the IEEE. The IEEE 802.3z working group handles work on the specification for a fiber optic interface and a shielded jumper cable assembly called "short-haul copper." They have produced several physical layer specifications. Initially, the specification for a UTP version was also being addressed by the 802.3z working group, but in March 1997, this was formally split off into a separate working group called 802.3ab.

The two working groups have defined four physical signaling methods: 1000BASE-SX (short wavelength fiber), 1000BASE-LX (long wavelength fiber), 1000BASE-CX (short-run copper), and 1000BASE-T (100-meter, four pair category 5 UTP). Table 9-4 provides a summary of the projects for each working group.

The encoding and transmission scheme used for the 802.3z PHY standard is 8B/10B, a scheme that was patented by IBM for its ESCON data links that are used to connect mainframes to front-end processors. With this scheme, which is also the scheme used in Fibre Channel, eight bits of data is sent in ten bits to provide the following benefits:

▼ Error detection (called disparity control)

■ Frame delimiting with data transparency, or in plain English, the ability to send any bit pattern and still be able to mark the beginning and end of a frame

▲ Clock recovery, signal transitions assist the receiver in clock recovery even if the sender has a slightly different transmission rate

The 8B/10B encoding scheme supports continuous transmission with a balanced number of ones and zeros in the code stream and detects single-bit transmission errors.

For 1000Base-T, IEEE802.3ab specification, a five-level pulse modulation technique is used.

Work Group	Project	Medium	Distance
IEEE 802.3z	1000BASE-SX	62.5-micron multimode fiber	220 m
		62.5-micron multimode fiber	275 m
		50-micron multimode fiber	500 m
		50-micron multimode fiber	550 m
IEEE 802.3z	1000BASE-LX	62.5-micron multimode fiber	550 m
		50-micron multimode fiber	550 m
		50-micron multimode fiber	550 m
		10-micron single mode fiber	5 km
IEEE 802.3z	1000BASE-CX	Shielded copper	25 m
IEEE 802.3ab	1000BASE-T	UTP	100 m

Table 9-4.   Summary Media and Distances

The Case for Gigabit Ethernet: The LAN WAN Missing Link

Historically, the local area network environment has been distinct and separate from the wide area network environment. Both network environments have taken distinctly different evolutionary paths that are rooted in the applications that drove the development of each environment.

The WAN infrastructure has traditionally been the domain of the telephone carrier and so had a development path that paralleled the development of the telco infrastructure, which was predominantly engineered for the delivery and trunking of voice circuits.

LANs, in contrast, evolved from the need to share expensive office resources such as printers and disk storage. Other than the technology differences, the significance of these two backgrounds is the fact that the LAN emerged from an environment that required the transfer of large data files at high speeds. The need for high speeds in the WAN was not as pressing as in the local environment.

To further complicate matters, the variety of applications that drive LAN technology has increased dramatically. With the emergence of applications such as multimedia applications that generate real-time video and audio, a high-performance end-to-end data communications infrastructure has become more critical than ever.

The requirements of these new applications are also impacting the WAN as it becomes a key component of end-to-end communication between remote offices. Historically, an 80:20 traffic split generally applied to local area networking: 80 percent of the traffic on the LAN remained within the same network segment, while 20 percent traveled off net. The changing trend in the way data is distributed and accessed suggests that the rule may also be changing and in the future, 80 percent of the traffic may be destined off net, with 20 percent remaining local. As a result, for the first time in their history these two evolutionary paths are now converging.

The Emergence of Switched LAN Technology

LANs began life as a shared medium, which was more than adequate for the applications and speed for which they were intended. As computer processing and storage have advanced and become commodities, computing power has been distributed to individuals rather than remaining a shared resource. This change also required a change in the networks that support these machines, one that reflected the growth of processing power on the individual's desktop machine rather than on a shared server.

Switched LAN technology emerged to fill this need. The introduction of the switched LAN created the potential to eliminate the contention scheme required for shared access and gave individual users the option of a dedicated channel that equaled the bandwidth that was previously shared. This was a development that was not limited to Ethernet but extended to other types of LANs as well. Among the LAN technologies, as shown in Table 9-5, the need for bandwidth was addressed by either creating more speed in the shared mode or creating a switched version that provided dedicated bandwidth or a combination of both.

Technology	Released	Speed	Description
Switched Ethernet	1991	10 Mbps	Improved performance by providing the option to dedicate 10 Mbps of bandwidth to each user or optionally share 10 Mbps.
Fast Ethernet	1994	100 Mbps	Improved performance of 10 Mbps Ethernet by a factor of 10. Instead of sharing 10 Mbps, users shared and contended for 100 Mbps of bandwidth.
Switched Fast Ethernet	1995	100 Mbps	Improved performance by providing the option to dedicate 100 Mbps of bandwidth to each user or optionally share 100 Mbps.
Gigabit Ethernet	1997	1 Gbps	Improved performance by providing the option to dedicate 1 Gbps of bandwidth to each user or optionally share 1 Gbps of bandwidth. This version also introduced Ethernet outside of the LAN into the MAN.
10 Gigabit Ethernet	2002 targeted for publication of standards	10 Gbps	Eliminated the contention scheme and limits operation to full duplex. Developed as a MAN and WAN technology.
FDDI	1987	100 Mbps	Shared media widely used as backbone technology for connecting workgroups.
OC3 ATM	1993	155 Mbps	Switched technology that supports LAN and WAN backbones and QoS.
OC12 ATM	1995	622 Mbps	Switched technology that supports LAN and WAN backbones and QoS.
Fast Token Ring (FasTR)	1998	155 Mbps	Software-configurable option on an IBM 2216 NWays Multiaccess Connector and the Multiprotocol Switched Services (MSS) Client Universal Feature Card (FC), designed to alleviate 16 Mbps bottleneck of 16 Mbps Token Ring. Token Ring is at the end of its development cycle.

Table 9-5. LAN Technologies Speed Improvements

As LANs and WANs converge, the features of switched LAN technology makes it more suitable for WAN use than the shared contention-based versions. This is not to say that a contention scheme is not appropriate for wide area applications; passive optical networks prove that there is a place for a shared medium. There is, however, a difference in the way LAN and EPON contention schemes operate, founded in the topologies of the applications. In a LAN environment, all devices are treated equally, and in contention-based Ethernet passive optical network, there is a governing device at the headend that controls access to the shared medium.

The encroachment of LAN technologies on the wide area network is directly related to need for greater speeds across the WAN, and the movement away from a shared medium to dedicated point-to-point connections, which looks a lot like wide area connections such as digital T1 and OC3. Full-duplex point-to-point switched LAN technology will be the main WAN application for LAN-based technologies.

Why Gigabit Ethernet in the WAN?

We have already seen that the development of point-to-point switched LAN technology aligns more closely with the technologies used in the WAN. The next obvious question is, why Ethernet for the WAN and not some other LAN technology? There are four—and potentially more—possible answers to this. First, other technologies such as ATM have already made it to the WAN—in fact, it was designed as a LAN and WAN solution. Second, consider the types of traffic that are driving the growth. The data presented in Table 9-1 shows that currently this is skewed in the direction of data and IP traffic, which is more suited to Ethernet.

The remaining answers, cost and installed base, are the two that present the most compelling argument for Ethernet in the WAN.

The cost model used by the publishers of the Ethernet standard is one that has been successfully tested with each revision of the standard. It is based on an expected increase of three times the cost for every ten-fold increase in speed, yielding a 3:1 improvement in performance versus cost. These are compelling numbers that remain unmatched by any other LAN technology. Cost, therefore, makes Gigabit Ethernet a compelling option for deployment across all segments of the network.

The fourth answer is the number of installed Ethernet devices—with an installed base of 600 million Ethernet nodes (an IEEE estimate), the implications are extensive. One implication of this number is that corporations have a significant investment in established Ethernet products that has to be protected. Any improvements in the technology and extensions of its applicability directly impact the return on their investment and the cost of supporting the network.

If the technology extends into the WAN, the existing operational and management platforms can be used to see beyond the local environment. With traditional WAN technology, a separate management platform is more likely required. Integration of the LAN and WAN technologies creates better economies of scale. And, seeing that the critical mass of the installed base is with Ethernet, it makes sense that it will be the protocol that encroaches on the WAN.

The Emergence of a New Breed of Service Providers

Gigabit Ethernet as a WAN technology is not just a theoretical debate. A new breed of service providers is emerging to make WAN Ethernet a reality. Service providers are emerging to address two markets that are best described as point-to-multipoint (e.g., EPON) and scalable point-to-point full-duplex applications.

One such provider of point-to-point WAN Gigabit Ethernet solutions is Yipes Communications. This company is an example of a new type of service provider that utilizes Ethernet technology in MAN and WAN solutions. Yipes provides an innovative suite of network solutions that is based on Gigabit Ethernet over fiber optic cable. These solutions are designed for LAN-to-LAN and LAN-to-Ethernet connectivity scalable up 1 Gbps in increments of 1 Mbps.

The interface to the Yipes network is a native Ethernet interface that logically extends the Enterprise LAN. This is a major change from traditional solutions that employ WAN equipment and interfaces like DSU/CSUs and V.35 interfaces, which require a conversion of the LAN traffic to a WAN protocol. Extending the LAN across the WAN allows native LAN protocols such as IP, IPX, and Appletalk and routing protocols such as OSPF and EiGRP to operate seamlessly across the entire wide area network.

An innovative feature of the service is its ability to deliver *bandwidth on demand*. In the traditional wide area model, speed upgrades of the circuit could take from days to weeks, sometimes extending into months. An upgrade from a 1.544 Mbps T1 circuit to 45 Mbps DS3 circuit requires the engineering of an entire circuit, which could take as much as one to two months or maybe more. Upgrades like these require careful planning and make sense only in cases where an ongoing need can be justified.

The Yipes solution is one where the bandwidth can be adjusted up or down on a moment's notice. The impact of this in business terms is that the provision of bandwidth is flexible and can be ordered for specific requirements rather than on a long-term basis. As an example, if a remote branch office does a once-a-week replication of disk images to a backup server at the headquarters, the bandwidth can be upgraded for these occasions and reduced once the job completes. This arrangement is very attractive to a business manager, as it allows more control over cost and better auditing of bandwidth usage.

The Yipes Gigabit Network   The Yipes network is based on a three-tiered ring architecture as shown in Figure 9-10.

The access layer consists of fiber optic rings that span a city along with customer premises equipment at customer distribution points. The fiber rings operate at 1 Gbps full duplex using Layer 2 switching or wire-speed layer 3 OSPF. The distribution points consist of Gigabit Ethernet switches that are located within customer buildings and are used to connect customer networks through 10Base-T or 100Base-T Cat 5 cable or 100Base-FX/1000Base-SX multimode fiber.

The distribution layer ties multiple fiber rings to a common Yipes point of presence, which hosts an aggregating gigabit router. Called Giga-PoPs, these are linked by a fiber ring to form a regional backbone. Giga-PoPs are able to route between each other at this layer.

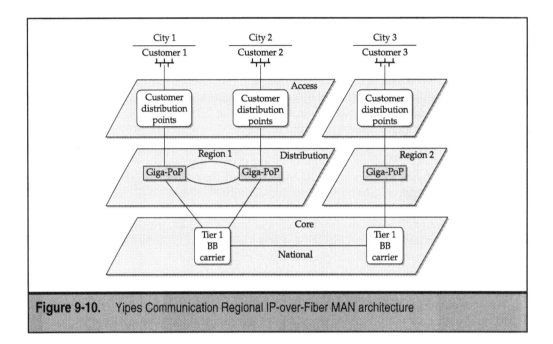

Figure 9-10. Yipes Communication Regional IP-over-Fiber MAN architecture

The core layer connects regional networks into a national network. This layer is implemented through multiple IP backbone carriers that are used to transport traffic between different regions. The connections to these tier-one backbone carriers are accomplished through the use of Gigabit Ethernet, OC-12, or OC-48 links at multiple peering points.

This model is being replicated throughout the U.S. in all major cities. The network is a scalable, redundant architecture that offers flexibility in traffic management, QoS, and bandwidth on demand. A comparison of traditional WAN solutions and Gigabit Ethernet WAN as viewed by Yipes is presented in Table 9-6.

10 Gigabit Ethernet: IEEE 802.3ae Emerging Standard

In December 2001, the 10 Gigabit Ethernet Alliance (10GEA) announced that all outstanding technical issues had been resolved and 10 Gigabit Ethernet (10GbE) was on track to final ratification by mid-2002. The Alliance, now over 100 members strong, was founded by 3Com, Cisco Systems, Extreme Networks, Intel, Nortel, Sun Microsystems, and World Wide Packets. Their mission is to facilitate and accelerate the introduction of 10GbE into the networking market.

The purpose of the proposed 10GbE standard is to extend the existing 802.3 protocols to an operating speed of 10 billion bits per second and to extend the application of Ethernet from a LAN technology to the WAN. The increase in speeds coupled with its ability to maintain compatibility with existing Ethernet standards provides for a wider applicability

Design Objective	Legacy Solution	Gigabit Ethernet Solution
Connectivity	Requires an estimate of future bandwidth needs and a decision on the most cost-effective service solution (Frame Relay, ATM, SONET). Each link has to be separately specified, designed, and configured.	All interfaces are Ethernet. Configuration of the interfaces is similar.
Scalability	Limited by protocols and service employed. Changing bandwidth has potential to impact interface and/or service.	1 Mbps to 1 Gbps on existing hardware and links.
Flexibility	Topology or service changes impact interfaces and necessitate hardware replacements that may not be useful elsewhere in the network.	Gigabit ports are all logically and physically interchangeable with like media.
Lead time to service upgrade	Lead times of one to three months.	Lead time of few minutes to a few hours.
Quality of Service	Dependent on line quality of copper circuits. Packet may be dropped by Frame Relay. Availability of 99.999% for SONET.	Latency comparable to Frame Relay or private circuit. Packet loss nears zero with no over-subscription. Availability of 99.99% or better.
Redundancy	Not available on copper-based links but is available on SONET circuits.	All-optical service provider uses only fully redundant fiber paths.
Support costs	Requires skilled knowledge of WAN protocols.	Requires knowledge of Ethernet.
Equipment costs[1]	SONET: $120,000 per gigabit.	Ethernet: $850 per Gigabit.

[1] Dell'Oro Group, Inc.

Table 9-6. Traditional WAN Solutions Versus Gigabit Ethernet WAN Solutions (Source: Spanning the Enterprise with Gigabit Ethernet, Yipes Communications, Inc.)

and better economies of scale for the growth and management of existing networks. Drivers for the new protocol include:

▼ An 83 percent cost savings per interface versus SONET

■ Tenfold better performance than 1GbE

■ Improved port density

■ A 30 percent annual price decline

■ Compatibility with previous Ethernet technology

■ A 40-kilometer reach

▲ Compatible with TDM and DWDM equipment

The draft standard for 10GbE is different in some respects from previous versions of Ethernet. The main area of change is the physical medium: 10GbE defines support only for an optical fiber infrastructure and operates only in full-duplex mode—see Table 9-7 for a comparison between 1000Base-X and 10GbE. Limiting operation to full-duplex mode means that the support for collision detection and resolution is not necessary, therefore eliminating the need for a CSMA/CD scheme. The protocol, however, is still Ethernet, and the packet structure remains unchanged.

The economic benefit of the protocol lies in the fact that the protocol and operation of the local environment can now be extended across the WAN. This ensures that existing investments in technology and infrastructure are not obsolesced by the new technology. The draft standard also attempts to address interoperability issues with other technologies such as SONET. Special attention was given to technology that enables an efficient transfer of Ethernet packets across a SONET link. The ultimate goal of the technology is to enable an end-to-end Ethernet solution or one that incorporates a SONET infrastructure. It is further hoped that the standard will help create a convergence between networks that are designed for voice and those for data.

Feature	1000Base-X	10GbE
Physical media	Optical fiber and copper	Optical fiber only
PMD	Uses fibre channel PMDs	Creates new PMD
PCS	8B/10B	64B/66B
Distance	Up to 5 km	Up to 40 km
SONET compatibility	No	Yes
MAC control	Half duplex (using CSMA/CD) plus full duplex	Full duplex only
Additions	Carrier extensions for half-duplex operation	XGMII, XAUI, SONET framing, SONET management information

Table 9-7. A 1GbE and 10GbE Comparison

IEEE 802.3ae Path to Standardization

In order for the technology to be adopted as a standard, the IEEE 802.3ae Task Force established five criteria that must be met by the new technology.

Market Potential   The first criterion imposed was that 10GbE must have a broad market potential and support a broad set of applications with the support of multiple vendors and multiple classes of customers.

As a measure of market potential, quantitative presentations were made to the task force indicating a significant market opportunity for higher speeds. At the 10 Gigabit call for interest, over 140 participants from 55 companies indicated their intent to participate in standardization of 10GbE. The level of commitment was a clear indication that there would be a widespread adoption of the new technology by many different vendors, thereby ensuring a broad range of equipment in support of a broad set of applications.

Compatibility with Other IEEE 802.3 Standards   The second criterion was that 10GbE must be compatible with existing Ethernet protocol standards, OSI, and Simple Network Management Protocol (SNMP) specifications.

In addition to the requirement for backward compatibility, the proposed standard had to define a set of management objects that are compatible with OSI and SNMP system management standards. It also had to conform to the full duplexing mode of the 802.3 MAC with the necessary tweaks to ensure proper operation at 10 Gbps speeds.

Distinct Identity   The third criterion was that 10GbE had to be substantially different from other 802.3 standards and solutions, thereby ensuring a distinct identity.

On the surface, this appears as a strange requirement. Why should 10GbE be substantially different from other forms of Ethernet? The logic behind this requirement was to ensure that the new technology satisfied a specific market rather than cannibalize existing ones. In other words, it had to be a unique solution for a problem, not an additional alternative to the problem.

As an upgrade path for exiting Ethernet users, which is the way it is intended to be perceived, the new technology maintains maximum compatibility with the installed base of over 600 million Ethernet nodes. The *management information base (MIB)* of the new protocol will be an extension of the existing 10, 100, and 1,000 Mbps Ethernet and so will present a consistent management model across all operating speeds.

Technical Feasibility   The fourth criterion was that 10GbE must have demonstrated technical feasibility before it can be ratified.

This point is why the announcement in December 2001 is so important. Stating that all outstanding technical issues had been resolved meant that the protocol had cleared a major hurdle in the path to being ratified. The technical challenges were not insignificant; to demonstrate technical feasibility the following points had to be proved.

▼   That it was feasible to use the 802.3 protocol in useful topologies that would address the target market at 10 Gigabit speeds.

■ That the protocol would scale successfully. Previous scaling of the 802.3 MAC to 1 Gbps had proved successful in research and actual deployment. The developers of the new 10 Gbps version hoped to build on the experience gained in scaling the protocol from 10 to 100 Mbps and from there to 1 Gbps. However, theory does not always result in an economical solution, so the previous successes were not a guarantee that efforts to scale to 10 billion bits per second would be successful.

■ To ensure compatibility, that the Layer 2 bridging function was compatible between the new and existing speeds. To bridge between two different speeds requires the bridge to do rate adaptation. The bigger the difference in speed between the two devices that are to be bridged, the more challenging the issue of rate adaptation becomes.

▲ That the PHY medium, the fiber optic cable, and components were reliable and met the requirements of different countries. The feasibility of the physical layer signaling at 10 Gbps on fiber optic was also an issue that had to be demonstrated. Feasibility also meant ensuring that the cost of doing so was reasonable.

Economic Feasibility   The final criterion was that 10GbE had to be economically feasible for customers to deploy, providing a cost benefit that includes the cost of installation and management for the expected level of performance increase.

The target cost of the solution was based on an expected increase of three times the cost of 1GbE components for a ten-fold increase in speed, resulting in an improvement of 3:1 in the performance versus cost ratio.

Are We There Yet?

Figure 9-11 shows a flow chart of the IEEE standards process. A "call for interest" is usually the first step in initiating a new standard. At this meeting, a "sponsor" is identified. A sponsor is a technical society or committee within IEEE that agrees to provide oversight during the standards development process. The next step is to gain approval of a Project Authorization Request from the New Standards Committee. Upon approval, a "working group" is formed, which is composed of interested parties. At this point, the countdown begins and the standard must be completed within four years.

If the work involved is very large, the working group will form a "task force" of people having specific expertise in the areas that the standard addresses. The task force will go off and work on various aspects of the standard. Once the work is complete, the working group votes on it, and if approved, the standard is passed on for "sponsor balloting," a process where the sponsor circulates the draft standard for a vote. Members of the sponsoring group or committee will vote.

It takes 75 percent of the ballots to be returned, and 75 percent of the returned ballot must approve the standard before it can be adopted. The IEEE Standards Association and its board have the final say in whether or not the draft should be approved. Once approved by this body, it may be published.

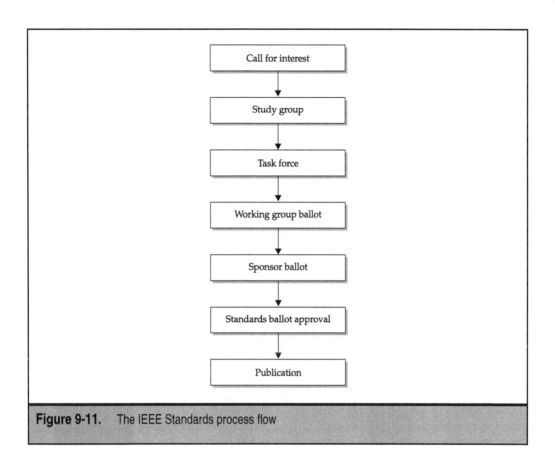

Figure 9-11. The IEEE Standards process flow

The actual structure of the IEEE is shown in Figure 9-12. The IEEE-SA (Standards Association) and its Standards Board oversees the process of standards formation through two committees: the New Standards Committee (NesCom) and the Review Committee (RevCom). The NesCom reviews each new request to ensure that it falls within the scope and recommends to the IEEE-SA whether the Project Authorizations and Request should be approved or denied. RevCom examines new and revised standards to ensure that they represent the consensus of the members of the IEEE Sponsor Balloting group, and advises the IEEE-SA on whether or not a standard should be approved.

The announcement that all technical issues have been resolved for the new protocol means that it has progressed to the stage of sponsor balloting. If approved, the draft goes before the Standards board for approval. The final ratification and publication should be sometime in 2002.

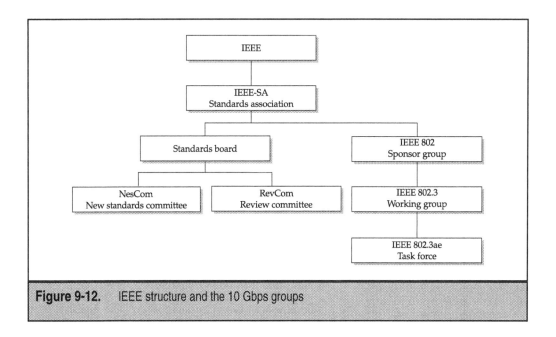

Figure 9-12. IEEE structure and the 10 Gbps groups

10GbE Architecture

All Ethernet protocols use the IEEE 802.3 Ethernet frame format, media access control (MAC), and a common minimum and maximum frame size. 10GbE, being a full-duplex fiber-based protocol, will not use the CSMA/CD protocol that is used by its predecessors. Other than this, its function is the same as Ethernet.

The architecture of Ethernet subdivides the physical layer into three sublayers: the physical medium dependent (PMD) sublayer, the physical medium attachment (PMA), and a physical coding sublayer (PCS). The PCS is responsible for the coding and multiplexing functions. 10GbE use the PCS to define two PHY types: The LAN PHY and the WAN PHY. The WAN PHY consists of an extended feature set in addition to those of a LAN PHY. Figure 9-13 illustrates the architectural components of the 10GbE standard.

The Physical Medium Dependent Sublayer The medium on which a signal travels has a direct impact on the speeds and distances that can be achieved using a particular technology. Each new generation of Ethernet has seen the development of new physical layer standards. 10GbE is no exception. While it is not unique in defining an interface for optical fiber—versions of 100 and 1,000 Mbps Ethernet also operate on fiber—it is the first to restrict the operation of the protocol to a single medium. This is not to say that future enhancements will not implement other types of media.

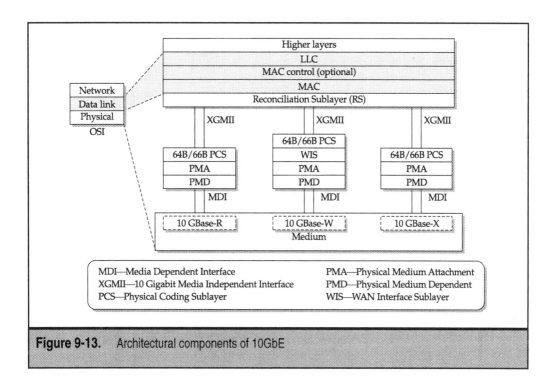

Figure 9-13. Architectural components of 10GbE

NOTE: *Single-mode fiber* is capable of achieving long distances because of the small diameter—eight to ten microns—of the fiber core through which the light travels. The small diameter—which is about the size of a wavelength—limits the passage to a single wavelength (or as close as possible) that is able to travel longer distances with little signal loss. Single-mode fiber is commonly used between buildings and in metro accesses. *Multimode fiber*, in contrast, uses a diameter that is significantly larger than single-mode fiber—50 and 62.5 microns is proposed for 10GbE. The larger diameter causes the light energy to be reflected within the core at multiple angles. This reflection creates multiple paths, each taking different times to traverse the cable, which creates distortion of the signal over long distances. Multimode fiber is the predominant type of fiber installed in LANs. The fundamentals of fiber optics are covered in Chapter 10.

The draft standard proposes a physical layer that supports different fiber optic media, as shown in Table 9-8, each having a different distance objective. Four different PMDs were chosen to meet these distance objectives. Figure 9-14 illustrates the likely applications in which each solution will be deployed.

PMD (Optical Transceiver)	Fiber Type	Diameter (microns)	Target Distance (meters)
850 nm serial	Multimode	50	65
1,310 nm WWDM (wide-wave division multiplexing)	Multimode	62.5	300
1,310 nm WWDM	Single mode	9	10,000
1,310 nm serial	Single mode	9	10,000
1,550 nm serial	Single mode	9	40,000

Table 9-8. 10GbE PMD Types and Distances

The Physical Medium Attachment (PMA) Sublayer   The PMA sublayer is responsible for the serialization and deserialization of the data that is to be transmitted. It supports multiple encoding schemes in support of the different PMDs that use an encoding that is specific to the medium it supports.

Physical Coding Sublayer (PCS)   The PCS sublayer consists of coding and a serializer or multiplexing functions. The structure of this layer is the only defining feature that differentiates the LAN and WAN PHYs. For the WAN PHY, the PCS operates in a serialized fashion. For the LAN PHY, it operates in two different modes: WWDM LAN PHY—acts

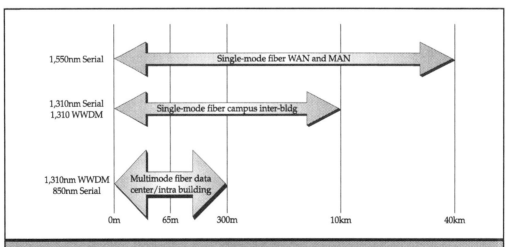

Figure 9-14. Likely application of 10GbE distances

as a multiplexer that multiplexes the data onto four 2.5 Gbps lanes; Serial LAN PHY—
acts as a serializer that serializes and deserializes the data to and from a single channel.

The WAN PHY operates exclusively in a serialized fashion. As can be seen in Figure 9-15,
the WIS is defined at this layer and is enabled whenever the interface is connected to a WAN,
and not used when connected to a LAN. The figure also serves to show the difference in
operation between the serial version of the LAN PHY and the Serial WAN PHY.

The LAN and WAN PHY   Both the LAN and WAN PHYs utilize the same PMD, so both
support the same distances. A WWDM version and a serial version further define the
LAN PHY. The WAN PHY operates only in serial mode.

The *WWDM LAN PHY* uses a physical coding sublayer that is based on four lanes of
8B10B—the same code used for 1GbE—with each lane operating at a data rate of 2.5 Gbps.
The WWDM PMD includes the following:

▼   Four laser transmitters

■   Laser drivers

■   An optical multiplexer

■   An optical demultiplexer

■   Four photo receivers

▲   Post amplifiers

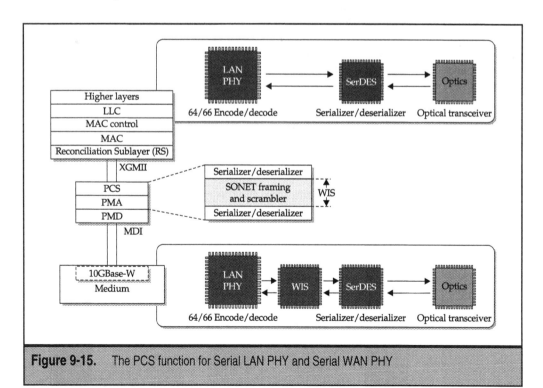

Figure 9-15.   The PCS function for Serial LAN PHY and Serial WAN PHY

The *Serial LAN PHY* uses a 64B/66B-coding scheme. It was originally intended to use the 8B/10B coding scheme for both the Serial LAN and WWDM versions, but this idea was scrapped as the 8B/10B version was found to be less cost effective for a serial implementation at 10 Gbps speeds. The committee instead adopted a new, more efficient and cost-effective 64B/66B code, which encodes 8 XGMII data octets to and from 66-bit blocks. The PCS transfers the encoded data to and from the PMA using a 16-bit-wide data path. An attraction of the code is the fact that it requires less complex hardware to implement, which has a direct impact on the cost effectiveness of the technology.

The serial PMD includes:

▼ A laser

■ Laser driver

▲ Optical receiver

For the *Serial WAN PHY*, an additional sublayer, the WAN Interface Sublayer (WIS), is required between the serial PCS and the serial PMA. The WIS sublayer ensures interoperability with SONET by including a simplified SONET/SDH framer in WIS.

The protocol is not designed for total compatibility with SONET but is considered "SONET friendly" as it adopts some of its features, the SONET framing being one those features and OC-192 link speed (9,953.281 Mbps) support being another. The sponsors of the protocol specifically shied away from the more expensive features of the SNET/SDH protocol. Instead, a more cost-conscious approach was taken in the development of the protocol, and common Ethernet PMDs were chosen to provide access to SONET infrastructures.

SONET, which was designed primarily for voice applications, is not as efficient with data. By integrating a 10GbE WAN PHY solution across a SONET infrastructure, packet-based IP can be transported more efficiently across that infrastructure. Finally, the WAN PHY provides most of the SONET/SDH management information, allowing the Ethernet WAN PHY links to be viewed as though they were SONET.

Reconciliation Sublayer (RS) and the 10 Gigabit Interfaces   The RS adapts the protocol of the Ethernet MAC into the parallel encoding of the 10 Gbps PCS. The *Extended Gigabit Media Independent Interface (XGMII)* is an optional interface that uses 32-bit data paths that are partitioned into four transmit and four receive lanes of 8 bits each. The RS is responsible for mapping the MAC data octets to and from the different data lanes of the XGMII in a round-robin manner. The RS also maps control signals to and from the XGMII when instructed to do so by the MAC or PHY.

An optional feature is the implementation of the XAUI interface (pronounced "zowie"; see Figure 9-16). The name is derived from the roman numeral 10 (X) and the Attachment Unit Interface of Ethernet (AUI) but has since been generally described as Extended AUI. This optional interface acts as an interface extender that extends the capabilities of the XGMII using the 10GBase-X, PCS, and PMA (see Figure 9-13). The XAUI may be used in place of, or to extend, the XGMII in chip-to-chip applications typical of most Ethernet MAC-to-PHY interconnects.

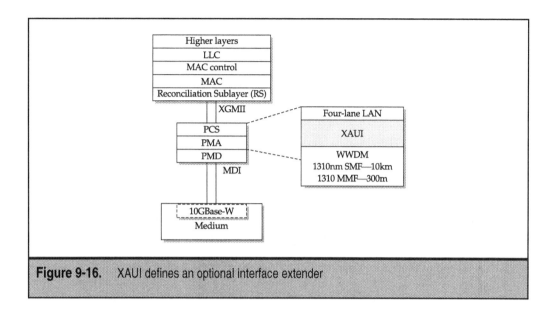

Figure 9-16. XAUI defines an optional interface extender

A limitation of the XMII interface is the fact that it is limited to a maximum distance of three inches. For applications that require greater distance, such as applications where the MAC and PHY need to be on separate boards, the XAUI interface extends the distance to 20 inches. The XAUI also require fewer pins and consumes less power. Finally, to confuse things even further, a third optional interface can be used as an alternative to XGMII and XAUI; called XSBI, it is based on the Optical Internetworking Forum SPI-4 interface. The bus is narrower than XGMII (16 lanes at 622–645 Mbps) and supports a longer reach.

SUMMARY

It is not a coincidence that this entire chapter of emerging technologies was spent discussing the emergence of Ethernet as a broadband access solution. The technology is well documented and well understood. It has the largest installed base of all LAN protocols with the lowest cost on a per-megabyte basis.

The proliferation of Ethernet nodes makes it a technology that will not disappear anytime soon. In fact, instead of diminishing, it is growing at a fast rate in multiple markets. Once a technology that was exclusive to the enterprise, the technology is now being deployed in the residential market. The protocol is also embedded in the operation of wireless networks, as it forms the basis for how IEEE 802.11b wireless access points communicate with each other.

The WAN is the last unconquered area of deployment, but the advent of 1- and 10-Gigabit Ethernet solutions along with innovative service providers like Yipes will probably change this quite quickly. The next few years will see a rapid convergence of

WAN and LAN solutions as companies clamber for solutions that support bandwidth-hungry applications like those mentioned in Table 9-9.

Ethernet is positioned to succeed in the WAN market as it has in others, and it could well become a ubiquitous standard for wireline subscriber access networks.

Application	Description
High-speed Internet access and Corporate VPNs for telecommuters and remote access	The increasing popularity of the Internet, its growing user base, and the emerging of applications such as streaming video all contribute to increased bandwidth requirements. The Internet is also being used as transport for corporate VPNs that places an even greater demand on bandwidth requirements.
Server connections	Server farms are becoming more commonplace. When servers are concentrated at a single location, the aggregate bandwidth required for the thousands of users accessing them drives the requirement to provision bandwidth for access and between servers.
Corporate LAN interconnectivity	As the speeds of local LANs increase and enterprises move more toward distributed workgroups, high-speed inter-LAN connectivity increases in importance. The wide area network has to be able to match the level of service that the local network provides in order to provide an infrastructure across which bandwidth-hungry and latency-sensitive applications can operate efficiently.
Growth of data	The amount of data that needs to be exchanged, accessed, or backed up is growing exponentially as different industries upgrade to the digital age. Telemedicine, distance learning, remote medical consultation, data warehousing, and workgroup integration of engineering CAD/CAM developers are just a few of the applications that are driving the growth of data.

Table 9-9.   More Bandwidth Driving the Need for 10GbE

PART III

Core Technologies

I n Part 1, we reviewed some of the technologies and trends that are driving the growth of broadband; we also looked at networking solutions employed within the home, multiunit buildings, and corporate offices.

Part 2 addressed the different types of access networks that are being deployed to provide broadband services and to bridge the bandwidth gap between the network core and the enterprise. In our review, we determined that the history and purpose of the network heavily influenced its architecture.

In Part 3, we will examine the technologies that make up the *metro and core network*. At the heart of this network are technologies that function as high-speed, high-performance bit pumps, pumping data through high-capacity pipes between different points in the network. The plumbing of the core network moves from the electrical to the optical domain, so instead of metallic wire, fiber optic is the predominant medium.

Chapter 10 provides a technical foundation on fiber optics and the technologies that are used to squeeze more bandwidth from a single fiber strand. In Chapter 11, we build on this foundation by exploring the protocols and technologies that are used within the core to move large volumes of data efficiently and quickly.

CHAPTER 10

Fiber Magic: Fiber Optics, WDM/DWDM

In this chapter, we explore fiber optics as a transmission medium. Before delving into the technical underpinnings of its operation, we explore the history of the technology. A historical perspective is important because it is a less understood and much more recent technology than wireline media. An appreciation of the evolution of light as a transmission method helps position fiber optics in a historical context alongside other, more popular technologies that receive much more publicity. The review of optical communication demonstrates that light as a transmission medium is not as new as it might first appear but has been around for centuries.

We next explore the technical details of the technology in enough depth to provide an appreciation of how it operates and its capabilities. The final sections of the chapter explore different technologies that are used to squeeze more capacity from a beam of light.

FIBER OPTIC AND THE MAGIC OF LIGHT

Fiber optics is the transmission of light through long fiber rods of either glass or plastic. The light is contained within the medium through a layer of cladding that surrounds the glass or plastic core. The core's being more reflective than the material surrounding it causes the light to keep being reflected back into the core, where it continues to be propagated down the fiber. Fiber optic cables are used to transmit voice, video, and data at close to the speed of light.

A fiber optic cable (see Figure 10-1) consists of a bundle of glass threads, each of which is capable of transmitting messages that are modulated onto light waves.

In the metro area and core network, the main communication medium is optical fiber. The use of fiber is a logical choice, as it is the only medium that can handle the amount, and mix, of traffic that current and emerging applications generate. It is also the only medium that can deliver the bandwidth that is required for these same applications.

From a cost perspective, fiber is also very attractive. With a metallic wire medium, the ongoing operational, network design, and maintenance are complicated by rusting components, temperature gradients, perforated insulation resulting in leakage, and loosened fittings, among other things. Fiber is not subject to these issues.

Fiber optics provides many benefits, a few of which are:

▼ Wide bandwidth

■ Higher speeds and performance

■ Low bit error rates

■ Immunity to electromagnetic interference

■ Low loss

■ Security

■ Light weight and small size

■ Safety

▲ Less expense and greater reliability than copper

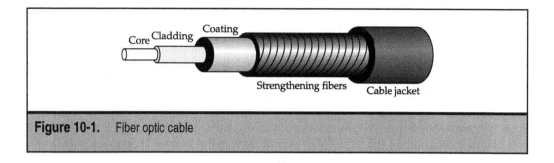

Figure 10-1.   Fiber optic cable

The applications of fiber optics govern the order of priority of these benefits: in some applications, a wide bandwidth and high speeds may be important; in others, the priority may be security rather than speed. The relevance of this observation is that the longer the list of benefits of a medium, the more applicable it becomes to a longer list of applications. Fiber optics' long list of benefits, which actually exceeds those just listed, makes it very appealing for a wide variety of applications. With these features, a strong case could be built for fiber optic as a replacement medium for wire in access and local networks.

A Historical Perspective

Optical communication has received a lot of focus in recent years. From the articles in the popular press and technology journals, to the success and problems of Global Crossing, a major player in the global fiber optic market, it could easily be mistaken for a technology breakthrough of the 1990s. Optical communication, however, emerged from the optical telegraph that predates the electric telegraph.

The first recorded experiment that led to use of optical communication was the "synchronous telegraph," an experiment that was conducted by Claude Chappe and his brother in the winter of 1790 and 1791. From the sketchy information that has been pieced together, it appears that this device operated through the use of synchronized clocks that were placed several hundred meters apart. When the pointer of one clock passed a certain number, a sound was made to indicate that the number that was being shown by the distant clock was significant. By referencing a dictionary of numbers and word associations, a message could be communicated.

The weakness of the system lay in its reliance on sound, which meant the communication was limited to the distance that the sound could travel. Not satisfied with the performance of the system, Claude Chappe began experimenting with other systems, a process that led to the panel telegraph, an optical means of communicating messages that never got off the ground. The panel telegraph, built in 1792, used a rectangular wooden frame with five movable panels that could be observed from great distances. Each of the five panels could be displayed or removed from view; in other words, it was capable of two binary positions. This crude invention was capable of communicating 32 possible combinations but was destroyed by an angry mob—who believed that it was a subversive means of sending messages to the nation's enemies—before it could be demonstrated.

Undaunted, Claude improved on the design by using a large horizontal beam called a *regulator*, with two smaller arms, called *indicators*, attached to either end. The entire structure, called the *semaphore telegraph*, is shown in Figure 10-2. The angles of the indicators along with the different positions of the regulator beam could be varied at 45 degree increments, yielding the possibility of encoding hundreds of symbols.

The legislative bodies, recognizing the importance of the technology for military purposes, agreed to fund the project to construct and test the system. After several messages were successfully exchanged using the test systems, the decision was made to establish a French state telegraph in July 1793. This invention was the first recorded successful public demonstration and adoption of the optical telegraph.

The optical telegraph was eventually superseded by the electric telegraph—a wire-based technology. It is something of a paradox that we have now come full circle and optical communication is now being acknowledged as the preferred medium for transmission, displacing wire-based systems in the process. So a type of system that wire transmission once replaced is now replacing wireline systems.

Figure 10-2. The semaphore telegraph, the first optical communication system

Guided Light Transmission

The advent of the telegraph did not stop work on the development of optical technology. Alexander Graham Bell, the inventor of the telephone, patented an optical telephone system that he called the Photophone in 1880. However, it was not as successful as his earlier wire-based system, the telephone, which is still in use today.

In our review of wireless systems, we determined that a transmission could be either guided through a medium like wire or unguided through free space. Optical transmission operates on the same principles; light can be either guided or unguided. The Photophone failed because it relied on unguided transmissions and air does not propagate light as reliably and predictably as wire conducts electricity.

The Photophone was able to transmit voice signals on beams of light by focusing sunlight on a mirror. The inventor then spoke into a device that vibrated the mirror. At the receiving end, a detector picked up the vibrating beam and decoded it back into a voice signal. The system was abandoned because the signals were too easily blocked or scattered. The invention, depicted in Figure 10-3, was donated to the Smithsonian Institute.

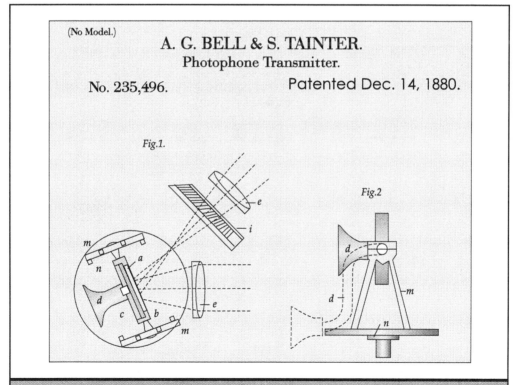

Figure 10-3. The Photophone, a telephone that uses light

Optical communication in contemporary fiber optic networks is based on the principle of guided light instead of unguided transmissions through the air. The path to modern fiber optic technology began in the 1840s, when Swiss physicist Daniel Colladon and French physicist Jacques Babinet observed that light could be made to travel along streams of water. As the water curved in an arc, the light beam followed the water's path, a property that became the basis for ornamental fountains made of brightly lit jets of water and the subject of further study in guided light, the basis to fiber optics.

Table 10-1 is a summary of the key contributors to research and development of technology that guides and bends light, and a chronology of significant events throughout history. The remaining paragraphs fast-forward to the years from 1950 onward, when fiber optics as we know it was developed.

Year	Contributor	Comment
1841	Daniel Colladon (Swiss)	Demonstrates light guiding in a jet of water in Geneva.
1842	Jacques Babinet (French)	Reports light guiding in water jets and bent glass rods in Paris.
1854	John Tyndall (Irish)	Demonstrates to the Royal Society that light could be conducted through a curved stream of water, proving that a light signal could be guided and bent.
1880	William Wheeler (American)	Invents a home lighting system that used a light source in the basement. A system of light pipes that were coated with a reflective coating was used to distribute the light energy from the source in the basement to the rest of the house.
1888	Dr. Roth and Prof. Reuss (Vienna)	Use bent glass rods to illuminate body cavities.
1895	Henry Saint-Rene (French)	An early attempt at television using a bent rod to guide light images (Crezancy, France).
1898	David Smith (American)	Applies for a patent on a bent glass rod device to be used as a surgical lamp.
1926	John Logie Baird (British)	Applies for a British patent on an array of parallel glass rods or hollow tubes to carry images in a mechanical television.

Table 10-1. Summary of Chronological Events Leading to Fiber Optics

Year	Contributor	Comment
	Clarence W. Hansell (American)	Outlines principles of the fiber optic imaging bundle in his notebook at the RCA Rocky Point Laboratory on Long Island. RCA files for U.S. patent Aug. 13, 1927, and later files for British patent.
1930	Heinrich Lamm (German)	As a medical student, assembles a bundle of optical fibers to carry an image. His motivation was to find a means by which he could view inaccessible parts of the human body.
1951	Brian O'Brien (American)	Suggests to Abraham C.S. Van Heel (Technical University of Delft) that applying a transparent cladding would improve transmission of fibers in his imaging bundle.
1952	Harold Horace Hopkins (British)	Applies for a grant from the Royal Society to develop bundles of glass fibers for use as an endoscope at the Imperial College of Science and Technology. Hires Narinder S. Kapany as an assistant when he receives grant.
1954	Abraham Van Heel, Harold H. Hopkins, and Narinder S. Kapany	Separately publish papers in the British Journal *Nature*. Kapany and Hopkins wrote papers on imaging bundles of unclad fibers. Van Heel wrote a paper on cladded fibers. The cladding described was transparent with a lower refractive index than the fiber strand, thus protecting the fiber from outside interference.
1960–1969	General interest	Decade of intense research and development.
1961	Elias Snitzer	Of American Optical, publishes a theoretical description of single-mode fibers. The fiber was so small that it allowed only a single wavelength to traverse it. The system had a light loss of one decibel per meter traveled.
1966	Dr. Charles K. Kao	Tells Institution of Electrical Engineers in London that fiber loss could be reduced below 20 decibels per kilometer for interoffice communications. Kao showed the need for a purer form of glass to help reduce signal loss.

Table 10-1. Summary of Chronological Events Leading to Fiber Optics *(continued)*

Year	Contributor	Comment
1970	Corning Glass scientists	Invent fiber optic wire called "Optical Waveguide Fibers" capable of carrying 65,000 times more information than copper.
1970 onward	General application	The use of fiber becomes more widespread. The military and the public telephone system as well as private enterprise begin using the technology.

Table 10-1. Summary of Chronological Events Leading to Fiber Optics *(continued)*

Fiber Optics: 1950s to Present

By the turn of the century, inventors realized that bent quartz rods could carry light and patented them as dental illuminators, and by the 1940s, many doctors used illuminated plexiglass tongue depressors. Optical fibers went a step further. Working with what were essentially transparent rods of glass or plastic stretched to be long and flexible, John Logie Baird and Clarence W. Hansell patented the idea of using arrays of hollow pipes or transparent rods to transmit images for television or facsimile systems.

The first person known to have demonstrated image transmission through a bundle of optical fibers was Heinrich Lamm, a German medical student. His goal was to look inside inaccessible parts of the body, and in a 1930 paper he reported transmitting the image of a light bulb filament through a short bundle. However, the unclad fibers transmitted images poorly, and the rise of the Nazis forced Lamm, a Jew, to move to America and abandon his dreams of becoming a professor of medicine.

Three reports on managing bundles were written by Abraham Van Heel, Harold H. Hopkins, and Narinder S. Kapany in 1954 (see Table 10-1) and published in the British Journal *Nature.* These reports proved to be a catalyst for a period of intense research and development of fiber optics technology.

In 1952, Harold Hopkins, a British subject, had applied for a grant from the Royal Society to develop bundles of glass fibers for use as an endoscope at the Imperial College of Science and Technology. Hopkins subsequently hired Niranda Kapany as an assistant once the grant was received.

Separately, Abraham Van Heel, a Dutch scientist of the Technical University of Delft, had also been working on imaging bundles. Through conversations with Brian O'Brien of the University of Rochester, he had embarked on a study of applying a transparent cladding to each fiber, a technique that O'Brien believed would improve the transmission of fiber in his imaging bundle.

The report in particular that had the biggest impact was Van Heel's, which proposed covering bare fibers with a transparent cladding of a lower refractive index than the core. This innovation protected the total reflection surface from contamination and reduced crosstalk between strands. Even then, the systems available at the time could not provide the performance necessary for long-distance communication.

It was not until the 1960s—a period of intense interest, research, and development of fiber optics—that the technology began to emerge as a viable solution for telecommunication. The systems up to that point had a signal loss of 1 decibel per meter, which was fine for an imaging device for peering inside the human body but was severely lacking in a means for long-distance communication.

In the early 1960s, Dr. Charles Kao, an engineer with Standard Telecommunications Laboratories, conducted experiments that investigated the properties of bulk glasses through comparison of samples of fibers from different fiber makers. His research led him to the conclusion that the lossiness of fiber was due to impurities in the silica and not the silica itself. He was convinced that reducing the impurities in the glass would result in the reduction of signal loss to less than 20 decibels per kilometer, thereby making optical fiber a viable medium for transmission of signals over long distances.

In 1966, he presented a paper at a meeting of the IEEE that forecasted that fiber loss could be reduced to below 20 dB/km. The bold and radical presentation that suggested that fiber loss could be reduced to such an extent, despite the fact that the best available low-loss fiber had a loss of 1,000 dB/km, attracted the interest of the British Post Office, which operated the British telephone network. The interest led to the funding of a study on ways to decrease fiber loss. In September 1970, the prediction became reality when Corning Glassware, now Corning, Inc., announced they had made a single-mode fiber with a signal loss of below 20 dB/km.

AT&T Bell Labs became interested in fiber optics in the mid-1960s, when it became obvious that light waves had an enormous capacity for carrying information and were immune to electromagnetic interference. Improvements in fiber components such as light emitting diodes, repeaters, connectors, and photodetectors led to the installation of the first light wave system in an operating telephone company in 1977.

The system was installed underground in downtown Chicago and ran for about 1.5 miles. Using glass fibers, the cable carried the equivalent of 672 voice channels and had the capability to provide a full range of telecommunications services that included voice, video, and data.

Technical Foundation

The historical account of optical communication and fiber optics shows that the breakthrough in fiber optics for long-haul telecommunication is a fairly recent occurrence. (It was not until as recently as the 1970s that the first fiber cable of sufficient quality to support long-distance transmission was developed.) It also shows that much research and development, from multiple industries, contributed to Corning's 1970s final breakthrough announcement that for the first time, they were able to develop a fiber that had an attenuation of less than 20 dB per kilometer.

The significance of the 20 dB/km figure was that it represented a minimum retention of one percent of the original light signal over a distance of one kilometer—the existing repeaterless distance for copper-based telephone systems. The work of the three Corning scientists, Drs. Robert Maurer, Donald Keck, and Peter Schulz, is recognized for laying the basis for the commercialization of optical fiber.

Contemporary fiber performance approaches the theoretical limits of silica-based glass systems and now achieves distances of over 100 km without amplification. From the perspective of the original specification, this represents a loss of 0.35 dB/km at 1,310 nanometers (nm)—one billionth of a meter—and 0.25 dB/km at 1,550 nm. The impressive improvements are a result of higher levels of glass purity and improved system electronics.

Light has an information-carrying capacity 10,000 times greater than the highest radio frequencies. After the viability of transmitting light over fiber had been established, the next step in the development of fiber optics was to find a sufficiently powerful and narrow light source. The light emitting diode (LED) and the laser diode were determined to be adequate for this function. The development of lasers went through many generations, culminating in the semiconductor types that are most widely used today.

After the Corning announcement, the telephone industry was cautious in embracing the technology. Innovation slowed somewhat, but eventually AT&T installed multimode fibers and standardized on a transmission speed of 45 Mbps—DS3 speeds. It was not long before single-mode fibers were shown to be capable of ten times the transmission rates of the older fiber types and of extending the range to 32 km as well. Single-mode fiber quickly became the standard in long-distance communication. We will be discussing single-mode and multimode fibers in more detail later in the chapter, but the main point to remember for now is that single-mode fiber has a much higher carrying capacity and supports longer distances than multimode fiber.

Finally, before we delve into the technology in more detail, we will introduce a final concept that will form the basis of further discussion later in the chapter. The development of fiber optics centers on four specific regions of the optical spectrum. These regions, called *windows*, lie between areas of high absorption, where the intrinsic properties of the material itself, impurities in the fiber, and defects in the glass absorb the optical energy, causing the light to become dimmer. These windows are at:

▼ **850 nm**   The wavelength at which the original systems were developed

■ **1,310 nm**   The S band, a short-wavelength band with much lower attenuation than the original window

■ **1,550 nm**   The C band, a conventional band with an even lower rate of attenuation

▲ **1,625 nm**   The L band, a long-wavelength band—a more recent addition that is currently being deployed

Basic Principles of the Operation of Light Through Optical Fiber

An important principle for the operation of optical fiber is that of *total internal reflection*. The principle is at the heart of the technology because it describes the way light behaves when penetrating a different medium. When a light source meets another medium of a different refractive index, it will either be *reflected* back into the medium from which it originated or alter its direction as it penetrates the other medium, through the process of *refraction*. The degree of reflection or refraction is directly related to the angle at which the light hits the second medium.

To better understand the concept, try to imagine what happens when you look into a pool of water. Looking down at a steep angle, you are able to see, below the surface, any objects that are on the bottom, but at a position that is offset from the actual position of the object. The position is somewhat distorted through the process of refraction. If you look out across the surface of the water at a flatter angle, you see the reflections of surrounding trees and other objects that are not within the water. The reason a flatter angle is more reflective than a steeper angle, which penetrates at a refractive angle, is that water and air have different indices of refraction and the angle at which the light strikes the other surface governs the way the light is treated.

It is important to understand this concept because it is at the very heart of how optical fiber works. The angle at which the light hits the outer walls of the glass has a direct impact on how much is reflected back into the medium, and how much escapes through refraction. To introduce some relevant terminology, we will slightly vary the analogy, this time interspersing it with terms that are relevant to the technology.

Assume there is a light source at the bottom of the same pool we just described. As the light rays from the source reach the surface of the pool, some of them continue through to the air at a refracted angle, while others are reflected back into the pool. As the *angle of incidence* increases, more rays are reflected. At some point as the angle of incidence increases, all light rays will be reflected back into the pool; the point where this happens forms the beginning of the *critical angle*, as shown in Figure 10-4. At the critical angle and beyond, all rays will be reflected back into the pool. Stated another way, increasing the angle of incidence above the critical angle results in total internal reflection.

The critical angle is defined by the ratio of the two indexes of refraction:

$\text{Sin } \theta_{\text{Crit}} = n_{\text{air}} / n_{\text{water}}$
Where n_{air} and n_{water} are indices of refraction of water and air.

Total internal reflection is the key to how well a beam of light travels through the fiber core. Controlling the angle at which the light waves are transmitted enables the efficiency of the transmission to be improved. The object is to stay above the critical angle. The difference in the indexes of refraction of the fiber core and its cladding causes most of the light to bounce off the cladding and stay within the core. Using this approach, the light is guided through the core, as illustrated in Figure 10-5.

Cable Construction For the efficient transmission of light through the fiber core, the core must have a higher refractive index than the cladding that surrounds it, just as, in the previous example, the refractive index of the water was greater than that of the air above it.

During the manufacturing process of optical fiber, the core of the fiber, which is made of glass (or sometimes plastic), is surrounded with a cladding that is also a type of glass,

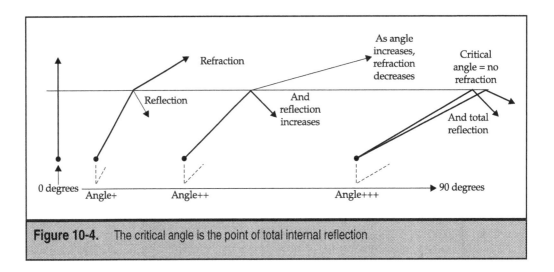

Figure 10-4.   The critical angle is the point of total internal reflection

but with a lower refractive index of the core. Both glass layers, the cladding and the core, are then enclosed in a protective acrylic buffer coating that is usually applied in two layers. Figure 10-6 illustrates the different layers of optical fiber.

The typical diameter of optical fiber including the cladding is 125 microns (micrometers, or millionths of a meter). The size may be doubled after the protective buffer coatings have been applied, but even at 250 microns it is still very small. To provide some perspective on the size, the diameter of a human hair is about 125 microns.

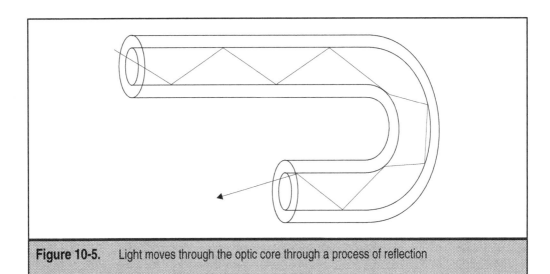

Figure 10-5.   Light moves through the optic core through a process of reflection

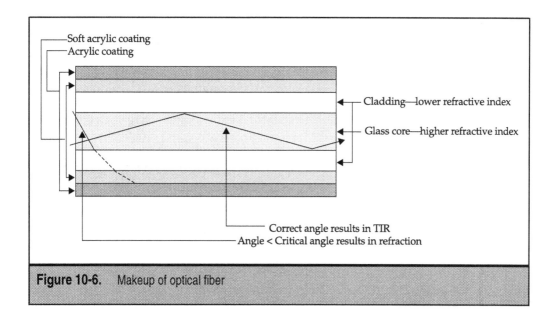

Figure 10-6.   Makeup of optical fiber

Figure 10-7 shows the typical sizes of optical fiber; note that the size of the core varies from 8 microns up to about 62.5 microns. At these sizes, the manufacturing process must be an exact science, as the purity of the silica has to be of a very high level and the size of the strands is extremely small. With a core size of 8 microns, the glass strand that constitutes the core of a single-mode fiber is about one-fifteenth the diameter of a human hair.

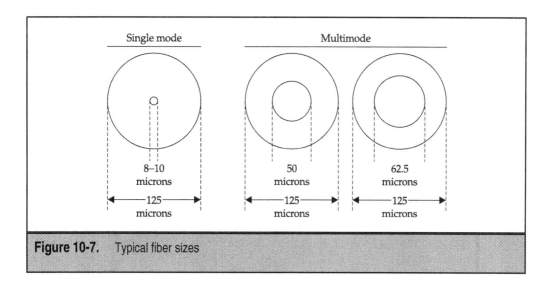

Figure 10-7.   Typical fiber sizes

The manufacturing process for optical fiber is a lot more involved than for copper or coax cables. The constitution of the fiber—its size and the purity of the material that composes it—has a direct bearing on the performance of the communication systems. This book is not about manufacturing, but it is well to appreciate the process of making optical fiber, as so much of its performance depends on the care taken in its manufacture. The process begins with the making of a *preform*, which is a solid rod of glass that has all the properties of the final product—core and cladding. The preform can be several inches in diameter and several feet in length. Different methods are used to make the preform:

- ▼ **OVD**   Outside vapor deposition, used in the U.S. and Europe
- ■ **VAD**   Vapor axial deposition, used mainly in Japan
- ■ **MCVD**   Modified chemical vapor disposition, also used in the U.S. and Europe
- ▲ **PCVD**   Plastic-activated chemical vapor deposition, which is an improved version of MCVD

In the first two processes, OVD and VAD, the glass preform—core and cladding—is built from the inside out. In the remaining two processes, the glass is built from the outside in. Rather than step through each process, we will quickly review the OVD process in order to give a general idea of how the process works. Beyond the two ways in which the preform is built, the remaining steps remain much the same for each process.

The glass used for optical fiber is actually made from the reaction of gases. In the laydown process—the step in which the preform is made—ultra-pure vapors of silicon chloride ($SiCl_4$), germanium chloride ($GeCl_4$), oxygen (O_2), and methane (CH_4) are passed through a burner, which causes a reaction that results in a fine soot of particles of silica and germania. These particles are deposited on the surface of a rotating target rod to form a preform, as shown in Figure 10-8. First the core material is deposited, and then the pure silica cladding is added.

Because both the core and the cladding are vapor-deposited, the preform becomes totally synthetic and pure. At the end of the process, the target rod is removed and the preform is placed in a furnace. The heat from the furnace removes any traces of water and transforms the preform into a solid, dense length of glass. The finished glass preform is then placed vertically in a draw tower and drawn into one continuous strand of glass fiber.

As the preform is lowered into the top of the draw furnace, the end begins to melt and forms a gob of molten glass. As the gob begins to fall, it creates a thin strand of glass that is threaded into a computer-controlled tractor assembly and drawn, after cutting off the gob. The diameter of the strand is monitored and controlled by speeding up or slowing down the process. Figure 10-9 illustrates the process.

Once the fiber has been created, it is tested for the following:

- ▼ **Tensile strength**   Must withstand a minimum of 100,000 lbs per square inch
- ■ **Refractive index profile**   Determine numerical aperture as well as screen for optical defects

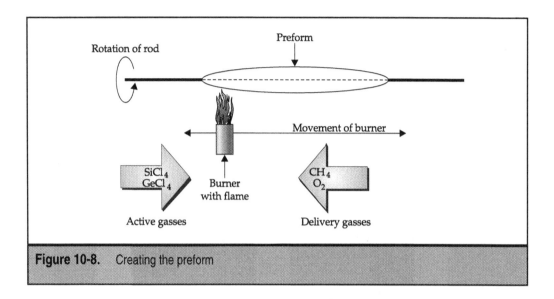

Figure 10-8. Creating the preform

- ■ **Fiber geometry**   Core diameter, cladding dimensions, and coating diameter must be uniform

- ■ **Attenuation**   Determine the extent that light signals of various wavelengths degrade over distance

- ■ **Information carrying capacity**   For multimode fiber, check the number of signals that can be carried at one time

- ■ **Chromatic dispersion**   Check the spread of various wavelengths of light through the core (important for bandwidth)

- ■ **Operating temperature/humidity range**   Test to ensure operation within predetermined ranges

- ▲ **Ability to conduct light underwater**   Test for undersea cables, where it is important

Types of Optical Fiber

There are two basic families of optical fiber, single-mode and multimode. Single-mode fiber has a very small core of about 8 to 10 microns, and multimode fiber has a core that is either 50 or 62.5 microns. They differ not only in the size of the core but also in the way they transport light.

Multimode fiber allows multiple light rays, called *modes*, to travel along the length of the strand. The different modes of light are created by directing light into the fiber at different angles. These different modes take different paths through the fiber as they reflect at different angles along the core. The different paths taken by the different modes means that they each travel different distances to arrive at their destination. The disparity

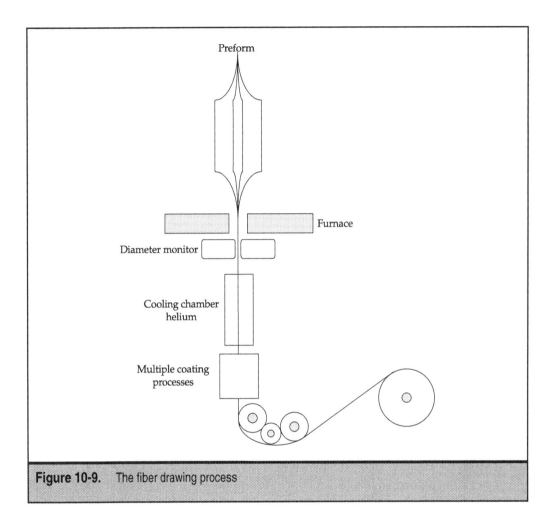

Figure 10-9. The fiber drawing process

among the different modes' times of arrival at the destination is called *modal dispersion*. This phenomenon has the effect of degrading the signal quality at the receiving end and ultimately limits the transmission distance. As a result, multimode fiber is not used in wide area applications, although it is common in local area networks.

Single-mode fiber allows a single mode of light to travel along the length of the strand. In this mode, the concept of modal dispersion does not apply, because there are no other light wavelengths to create problems. The single mode allows the light to travel greater distances at much higher speeds than multimode fiber. Because of the higher speeds and distances supported by single-mode fiber, it is typically used in long-haul networks that need to carry lots of traffic.

Both single-mode and multimode fiber can be further broken down into different fiber types:

▼ Multimode step index

■ Multimode graded index

■ Single-mode dispersion-shifted and dispersion-flattened

■ Single-mode non-dispersion-shifted

▲ Single-mode non-zero-dispersion-shifted fiber

Multimode Step Index Fiber   The number of possible paths through a optical fiber strand is determined by the difference in refractive indexes of the core and cladding, the wavelength, and the core diameter. *Step index* simply refers to the fact that the core has a higher refractive index than the cladding, or stated in another way, the refractive index steps up from the cladding to the core. Within the cladding, the refractive index is constant, and within the core, it is also constant.

As mentioned previously, dispersion is a negative effect in optical communication. Modal dispersion is a negative side effect of using multimode fiber in general. As is evident in Figure 10-10, the multiple paths through the fiber have different lengths, so the longer the fiber, the greater the difference in time it takes each mode to travel through the fiber.

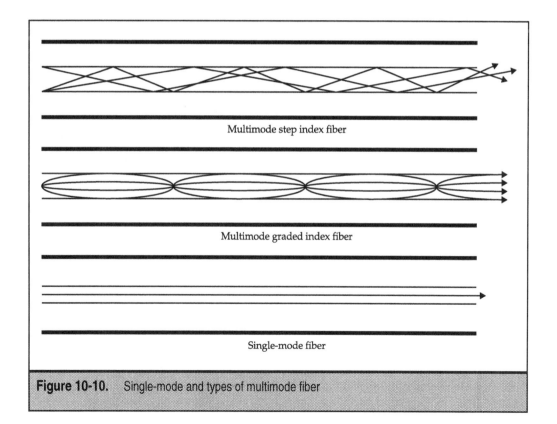

Multimode step index fiber

Multimode graded index fiber

Single-mode fiber

Figure 10-10.   Single-mode and types of multimode fiber

The different paths through the fiber cause the light signal to spread out over time as it travels toward the destination. So the longer the distance, the more distorted the signal, which translates to lower bandwidth. This is in essence the reason why multimode fibers serve shorter distances than single-mode fiber.

The relationship between bandwidth and distance is usually documented as the "bandwidth×distance" of the fiber. So a figure of 1,000 MHz×km means over a distance of 1 kilometer, 1,000 MHz of bandwidth can be used. It also follows that at 10 km, 100 MHz of bandwidth can be used.

Multimode Graded Index (GRIN)   The abrupt change in the refractive index between the core and the cladding is a contributor to modal dispersion. In multimode graded index (GRIN), the refractive index from the center of the core to the outer edge is graduated. The refractive index of the cladding remains constant. The graduation of the refractive index limits the amount of dispersion in the fiber and so limits the difference in time each mode takes through the fiber.

A graduated change in refractive index impacts light differently as the light approaches the edge of the core; the lower refractive index of the core causes light traveling closer to the core to travel faster than light traveling near the edge. The effect of this on the signal is a curving of light inward toward the core as it travels toward the edges of the core (see Figure 10-10). The objective is to keep each mode traveling at the same speed as it propagates through the fiber.

GRIN is a much better multimode fiber than step index. It has less modal dispersion, which results in greater speeds and higher bandwidth. However, it still does not approach the performance of signal-mode fiber for distance and bandwidth.

Single-Mode Fiber   Single-mode fiber is unaffected by modal dispersion because it uses a single mode. It is, however, affected by other types of dispersion, namely *waveguide dispersion* and *material dispersion*, the sum of the two being known as *chromatic dispersion*. In fact, these types of dispersion are not unique to single mode but are present in all optical fiber communication.

As the light is emitted into the fiber, some of the light enters and propagates through the cladding; waveguide dispersion is the result of this phenomenon. As we discovered in the previous paragraphs, the lower the refractive index of a medium, the faster light travels. The cladding has a lower refractive index to the core, so it follows that light that escapes into the cladding travels faster in the cladding than in the core, which means that the signal in the core arrives later at the end point of the fiber. The net effect of the different arrival times is dispersion.

The second type of dispersion, material dispersion, is the natural effect of the different speeds at which different wavelengths travel. Even the narrowest light beam will still have several wavelengths, and these wavelengths will travel at different speeds through the fiber, resulting in slight variations in arrival times.

Of relevance is the fact that these two types of dispersion tend to operate in opposite directions—one disperses outward and the other distorts inward. It is therefore possible to use one effect to counteract the other. During the manufacturing process of single-mode fiber, the fiber is designed so that these two dispersion effects cancel each other at a particular wavelength, called the *zero dispersion point*. In most single-mode fiber, the zero dispersion point happens at 1,310 nm and at this wavelength light will not disperse. However, the light will still be subject to attenuation as it traverses the fiber.

Commonly called standard single-mode fiber, *non-dispersion-shifted fiber* was designed for the 1,310 nm window. To optimize the performance of the optical fiber, this type of fiber is designed so that the zero dispersion point is near the 1,310 nm region.

As the use of optical fiber became more common, the C band window, 1,550 nm, was used to address the need for greater speeds and distance. The benefits of operating within this window are that it has a much lower attenuation rate than 1,310 nm and it operates within the frequency range of a type of amplifier called an erbium-doped fiber amplifier (EDFA), an in-fiber amplifier that will be addressed in more detail later in this chapter.

There is one drawback, however, with operating within the 1,550 nm window: the dispersion characteristics are severely limiting. To gain the benefits of the lower attenuation and greater operating distances within this window, a *dispersion-shifted fiber (DSF)* design was developed that shifted the zero-dispersion point to the 1,550 nm region.

The third type of single-mode fiber, *non-zero dispersion-shifted fiber (NZ-DSF)*, is designed specifically for DWDM applications. This design lowers the dispersion in the 1,550 nm region but does not lower it all the way to zero. By not totally eliminating dispersion, a controlled amount can be used to counteract nonlinear effects that can hinder the performance of the fiber.

NZ-DSF was invented to address a nonlinear effect called four-wave mixing. *Four-wave mixing* is caused by the nonlinear nature of the refractive index of the fiber. The nonlinear interaction between different DWDM channels creates sidebands that cause interchannel interference. These sidebands increase over distance and cannot be filtered out, which is the reason why dispersion-shifted fiber is not suitable for DWDM. It was found that small amounts of chromatic dispersion could be used to counteract four-wave mixing, hence the development of non-zero dispersion-shifted fiber.

The major types of single-mode fiber are summarized here:

▼ Non-dispersion-shifted fiber represents the most prolific fiber type deployed. It is suitable for TDM single-channel use in the 1,310 nm region or DWDM use in the 1,550 nm region when dispersion compensators are applied.

■ Dispersion-shifted fiber is suitable for TDM use in the 1,550 nm region but unsuitable for DWDM in this region.

■ Non-zero dispersion-shifted fiber is good for both TDM and DWDM use in the 1,550 nm region.

▲ Newer-generation fiber includes types that allow light to travel further into the cladding, creating a small amount of dispersion to counter four-wave mixing and dispersion-flattened fiber. Dispersion-flattened fiber permits the use of wavelengths further away from the optimum wavelength without the pulse spreading.

Light Sources and Detectors

Light emitters and detectors are active devices that reside on both sides of the fiber link. Light emitters perform transmit functions by taking an electrical signal and converting it to light pulses for transmission along the fiber. Light detectors receive these light signals and do the reverse function of light emitters. The optical signal is converted back into an electrical signal.

The light source, or light emitter, is one of the more expensive parts of the design of a fiber system. The characteristics of the light source can be a significant limiting factor in the performance of an optical link. These systems have to be able to emit a powerful and narrow light signal at a specific wavelength. Other key characteristics of these components are compactness and durability. Two basic types of light sources are used: light emitting diodes (LEDs) and semiconductor lasers (laser diodes).

LEDs are fairly inexpensive components that emit a wide light signal. Typically used on multimode fibers, LEDs are suitable for speeds of up to 1 Gbps but are not powerful enough nor are the beams narrow enough for single-mode fiber applications. Semiconductor lasers, or laser diodes as they are also called, are better suited to single-mode applications. They produce a much more powerful and narrow beam. These are the type of light emitters that are used in long-distance, single-mode fiber runs.

The process of conversion, or modulation, of an electrical signal into an optical signal can be accomplished by externally modulating a continuous wave of light or by using a device that can generate modulate light directly. In directly modulated lasers, the modulation of the light to represent the digital data is done internally. With external modulation, an external device does the modulation. Two types of semiconductor lasers are widely used, *monolithic Fabry-Perot* lasers and *distributed feedback (DFB)* lasers that are well suited for DWDM applications. DFB lasers have a very good signal to noise ratio, are capable of high speeds, and emit an almost monochromatic light.

Light detectors must be capable of receiving the different wavelengths used to represent the different digital data. Two types of light detectors are widely deployed, the *positive-intrinsic-negative (PIN)* photodiode and *the avalanche photodiode (APD)*. The PIN takes the different wavelengths and does a 1:1 conversion of photons to electrons, the exact opposite of an LED, which does a 1:1 conversion of electrons to photons. A PIN and an LED do nothing fancy with the signal other than absorb and convert, or convert and emit, respectively. APDs provide gain to the signal by applying amplification. APDs have a higher receive sensitivity and accuracy than PINs and are more expensive and temperature sensitive.

Challenges of Optical Fiber as a Transmission Medium

Just like wireline-based systems, optical systems have their own set of challenges that must be considered when building a fiber communication plant. A pulse of light traveling over a length of fiber will degrade over time. There is more than one factor contributing to this degradation, and different wavelength windows react differently to these factors. In this section, we will introduce you to some of these challenges.

Attenuation   As light signals travel along the length of a fiber strand, they weaken through the process of *attenuation*. Attenuation is caused by intrinsic factors of the fiber itself such as absorption and scattering of the light and external influences such as the stress on the fiber during the manufacturing process or bending.

Any imperfections in the glass during manufacturing contribute to the attenuation of light as it passes along the length of the fiber. If the fiber is not manufactured with precise uniformity, imperfections cause the light to scatter, which is the biggest contributor to attenuation in fibers today. The most common form of scattering, called *Rayleigh scattering*, is caused by small variations in the density of glass as it cools. These variations are smaller than the wavelengths used and so act as objects that scatter the wavelengths as they travel. Shorter wavelengths are more susceptible than longer wavelengths to scattering, a fact that limits the use of wavelengths below 800 nm.

A second intrinsic cause of attenuation is *absorption* of the signal by impurities within the fiber. These impurities cause the light to become dimmer as it traverses the fiber. Unlike Rayleigh scattering, absorption has more effect on longer wavelengths, becoming a significant factor for wavelengths above 1,700 nm.

The rate of attenuation in a fiber link has a direct correlation to the length of the fiber and the wavelengths being used. Figure 10-11 shows the loss in decibels per kilometer (dB/km) from scattering and absorption within the different wavelength windows. As is evident, the different wavelength windows are impacted differently by these factors.

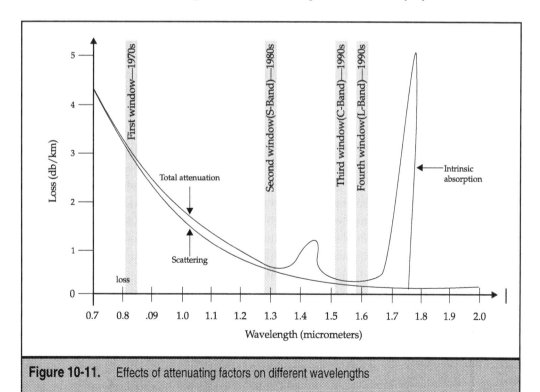

Figure 10-11.   Effects of attenuating factors on different wavelengths

Dispersion   *Dispersion* is the spreading of light pulses over time as they travel, resulting in distortion of the signal, thereby limiting the available bandwidth. In our review of the different types of fiber, we touched briefly on the different types of dispersions, modal and chromatic, the latter of which is the sum effect of waveguide and material dispersion. As we explore the different types of dispersions and their effects, we will take a closer look at modal and chromatic dispersion and introduce a third type of dispersion called *polarization mode dispersion (PMD)*.

Modal dispersion occurs in multimode fibers; it is the effect where it takes different length of times for different modes to travel through a length of fiber. The time difference in the different modal paths results in dispersion of the signal. Modal dispersion becomes more pronounced over longer fiber runs.

In single-mode fiber, chromatic dispersion has two components, material dispersion and waveguide dispersion. Material dispersion occurs as wavelengths travel at different speeds through the material. No matter how pure or narrow a light beam, it is composed of several wavelengths within a range. The different wavelengths travel at different speeds, and the difference in speeds among the wavelengths means they arrive at slightly different times at the receiver, resulting in dispersion.

Waveguide dispersion results from the difference in refractive index between the core and the cladding. The effects of the refractive index on different wavelengths are summarized as follows:

▼ Shorter wavelengths tend to be confined to the core; thus, the effective refractive index is close to the refractive index of the core.

■ Medium wavelengths tend to spread slightly into the cladding, thereby decreasing the effective refractive index.

▲ Longer wavelengths tend to travel even deeper into the cladding, and much of the light spreads into the cladding. This brings the refractive index closer to that of the cladding instead of the core.

So what does all this mean? The behavior of the different wavelengths as they travel the length of the fiber creates relative delays between wavelengths as they arrive at the receiver, and delays cause dispersion. Do not forget that the light traveling through the fiber is a composite signal of multiple wavelengths, each representing different digital data streams. Any delay in the component wavelengths causes distortion of the received signal.

Figure 10-12 shows the material and waveguide effects with the resulting total dispersion profile. Notice that the profile of waveguide dispersion is inverted with respect to that of material dispersion. This relationship allows the manufacturing process to vary both components in order to control the profile of total chromatic dispersion.

The core of the fiber should be perfectly round and symmetrical. The distortion of the roundness of the fiber or the ovality of the fiber shape causes *polarization mode dispersion (PMD)*. The ovality of the fiber results from defects in the manufacturing process or from any external factors that cause stress. PMD occurs in optical networks because an optical signal travels in two modes of polarization that oscillate at right angles to each other.

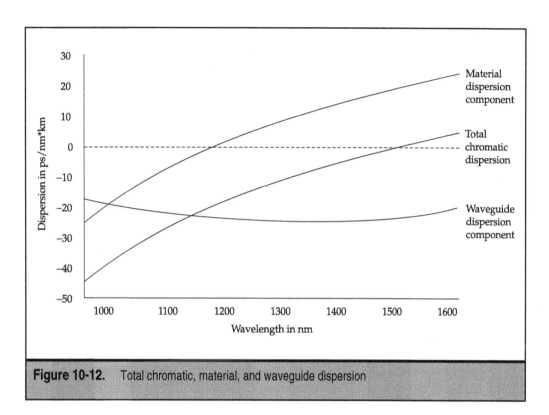

Figure 10-12. Total chromatic, material, and waveguide dispersion

These two modes may spread as they travel down a fiber, due to variation in a fiber's circularity. At the receiving end, the delay between the arrivals of these two modes of the optical signal is interpreted as dispersion, making it difficult or impossible to interpret the optical signal.

Unlike chromatic dispersion, PMD is subject to change over time as fiber is subjected to normal course of wear and tear. At speeds above 10 Gbps, PMD becomes a very big deal. It is considered by many system designers to be one of the biggest challenges in 40 Gbps systems.

Nonlinear Effects   We briefly touched on nonlinear effects in our discussion on NZ-DSF fiber, specifically as it relates to four-wave mixing, the phenomenon in which the interaction of different optical wavelengths creates self-inflicted interference. Unlike linear effects such as attenuation and dispersion that can be compensated, nonlinear effects accumulate and become more of a problem at higher power levels and greater distances, both being characteristics of DWDM applications. It follows, then, that nonlinear effects are deadly enemies of DWDM systems.

The challenge with nonlinear effects—sometimes called Kerr nonlinear effects—is that often they are not evident until the different component parts of an optical system are installed. These components may have tested fine during the isolated testing of the different subsystems, only to be a problem after they have been combined. Four-wave mixing is a very good example of a nonlinear effect that becomes evident only after the system has been installed. The use of a minimum amount of chromatic dispersion in NZ-DSF fiber is used to help mitigate this effect.

Four-wave mixing (FWM) occurs when two or more frequencies of light propagate through an optical fiber together. When these wavelengths get in phase with each other through a condition known as *phase matching*, light is generated at new frequencies using optical power from the original signals. Generation of light through four-wave mixing has serious implications for DWDM applications. The generation of new frequencies from two or three input signals is shown in Figure 10-13.

Other types of nonlinear effects include:

▼ **Stimulated Brillouin scattering (SBS)**   A nonlinear effect appearing in systems involving high-powered laser output. SBS is caused by light being back-reflected by an acoustic wave resulting from the interaction of the light with acoustic vibrations of silica molecules. The output power of an erbium-doped fiber amplifier, given sufficient input signal power, has the capability to produce SBS. Once the threshold power for SBS is achieved, all subsequent increases in power are completely back-reflected, causing system degradation.

■ **Stimulated Raman scattering (SRS)**   A nonlinear effect appearing in systems also involving high-power sources. SRS is caused by light scattered both backward and forward due to interaction between light and the vibrations of silica molecules. Molecules, the building blocks of matter, vibrate, and the warmer something is, the more the molecules vibrate. If light hits some of these vibrating molecules, a molecule can sometimes steal, or add to, some of the light's energy. When this happens, the wavelength of the light pulse can either increase or decrease.

▲ **Self-phase and cross-phase modulation**   Both of these effects, generically known as *phase modulation* effects, are caused by the refractive index of the fiber varying with the intensity of the light. Cross-phase modulation occurs between different wavelengths, whereas self-phase modulation happens within a single wavelength. These effects are compensated through adjustment of the power between channels in a DWDM system.

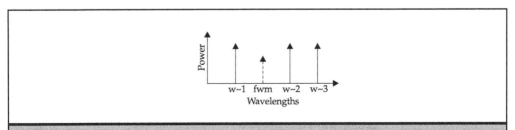

Figure 10-13.   Four-wave mixing, detrimental to DWDM

WDM AND DWDM

The two most commonly used wavelengths for optical communication are 1,310 nm and 1,550 nm. The 1,310 nm wavelength is popular for metropolitan and short-range applications because of the lower component cost and is the wavelength used for SONET/SDH systems. The 1,550 nm wavelength is capable of traveling longer distances and so is preferred for long-haul optical solutions. The components for a 1,550 nm system are more expensive than for a 1,310 nm system, but the benefits of increased distance—without the need for equipment to regenerate the signal—far outweigh the added costs.

The concept of *wavelength-division multiplexing (WDM)* is fairly simple to grasp: instead of a single signal, multiple optical signals are transmitted into a single optical fiber. It is the same concept as frequency-division multiplexing, in that different frequencies—wavelengths—are used to transmit multiple signals. The technology is illustrated in Figure 10-14.

WDM technology is comparatively new when compared with other multiplexing techniques such as time- and frequency-division multiplexing. Until the late 1980s, the predominant way of using optical fiber was to use a single optical channel across the fiber. This was due in part to the quality of the components that were in use up to that time. Improved fiber and fiber components paved the way for high-speed repeaterless transmissions, and so the stage was set for the next advancement, applying more wavelengths across the optical link.

At this juncture, it would be a fair question to ask, why introduce multiple wavelengths at all? If the technology had improved to the point that we can multiplex multiple optical signals across a single optical link, why not just use a single-wavelength system and keep raising the bit rate? This is a fair question that deserves some consideration.

First, the speed of technological advancements for a single-wavelength system could not keep up with the demand for bandwidth. When one considers that in March 2001, WorldCom and Siemens Information and Communication Networks, through its subsidiary Optisphere Networks, announced they had successfully boosted the transmission capacity of a single network fiber to 3.2 terabits per second, one understands the significance of WDM and dense wave-division multiplexing (DWDM). To achieve the 3.2 terabits

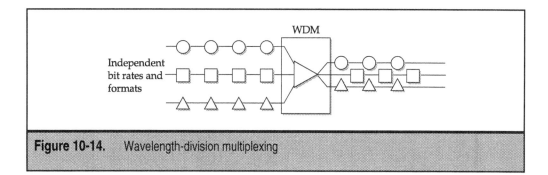

Figure 10-14.   Wavelength-division multiplexing

per second, 80 wavelengths of 40 gigabits per wavelength were used for a total capacity of 3.2 terabits per second. And we are only just beginning—Siemens already has demonstrated a lab version of 7 terabits per second.

For a single optical signal to do this would require the fiber and the components to operate at a serial bit time of 3.2 trillion bits per second. At this speed, the time slots for transmitters and receivers would have to operate at less than 0.001 nanoseconds per bit. The higher the speed, the more precise—and the more costly—the components must be to operate efficiently.

As we push the upper range of speed for a single signal, *jitter* becomes a factor in performance, a phenomenon where the timing gets out of sync and the receiver begins receiving bits in the wrong time slots (see Figure 10-15). If a single optical signal were to be used, instead of WDM, to achieve higher throughput and speeds, the speed would need to be limited to the point where jitter is not an issue. However, imposing limits on the speed means adding new fiber to make up for the lost capacity—an expensive option.

WDM

In a basic WDM system, multiple lasers are used to emit light at different wavelengths. The wavelengths are then combined, or multiplexed, onto a single optical fiber. At the far end of the fiber link, the multiplexed signal is demultiplexed back into the original streams through the use of a tunable optical filter that is tuned to the specific wavelength.

To understand the concept of a multiplexed and demultiplexed optical signals, consider a ray of sunlight. Under normal conditions, it appears to be a single color, white, but when directed through a prism, the single light breaks out into different light rays of varying colors, as shown in Figure 10-16. In this example, the prism acts as a demultiplexer, which breaks down the composite white light into a rainbow of different colors.

WDM was originally demonstrated in the early 1980s. Two light wavelengths at 850 nm and 1,300 nm were injected into an optical fiber at one end using a coupler that fused the two light sources. At the other end, another coupler was used to split the light into its original wavelengths through the use of two fibers, one made of silicon that was more sensitive to 850 nm and another made of germanium that was more sensitive to 1,300 nm. A filter was then used to remove any unwanted wavelengths. The end result was that the

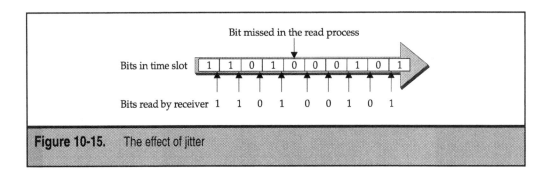

Figure 10-15.   The effect of jitter

Figure 10-16. A prism breaks out the component parts of a light beam

850 nm wavelength was separated onto the silicon fiber and the 1,310 nm, onto the germanium. This demonstration of wave-division multiplexing led to the development of wideband WDM that multiplexed two wavelengths at 1,310 nm and 1,550 nm.

In the late 1980s, with the emphasis shifting to the 1,550 nm transmission band for long-haul single-fiber runs, WDM systems were developed for the operating range of fiber optic amplifiers, 1,520–1,560 nm. The 1,550 nm window of an optical fiber could be considered the sweet spot of fiber; it has very low loss or low attenuation at this wavelength. In this range, WDM equipment was able to place four signals into a single fiber with an interwavelength spacing of 10 nm.

Well, we have since come a long way and are now able to place many more signals onto a single fiber. The number of wavelengths is usually a power of 2; the higher the number of wavelengths, the higher the speeds. Each wavelength operates at the speed at which a single wavelength would operate. Introducing an additional wavelength does not slow down the existing one; the additional wavelength is propagated along the fiber at the same speed as the first. If a single wavelength were used to produce a speed of 2.5 Gbps, then combining ten wavelengths would increase the capacity of the same link to 25.0 Gbps.

DWDM

Dense wavelength-division multiplexing (DWDM) is operationally similar to WDM. The difference is in the number of wavelengths and how densely they are packed. Whereas the early original WDM systems used a wavelength spacing of about 10 nm, newer DWDM technology increases the density of wavelengths to within 0.4 nm of each other and sometimes closer. With the introduction of DWDM, extremely high data rates have been achieved. Figure 10-17 shows the speed improvements that have been achieved as a result of DWDM and a projection of the speeds that should be possible in the near future.

WDM and DWDM are both techniques that use single-mode fiber to transmit multiple wavelengths of light. This should not be confused with multimode fiber, which directs light at different angles into a large fiber core to produce different modes of light.

DWDM systems are dependent on special fibers that use a non-zero dispersion-shifted fiber at the 1,550 nm wavelength. As previously discussed when we reviewed the different fiber types, NZ-DSF allows a limited amount of chromatic dispersion, which is used to compensate for nonlinear effects. Because of the higher speed and the density of the wavelengths, which may have wavelength spacing of less than 0.4 nm, these systems are dependent on precise engineering of the components of the optical link.

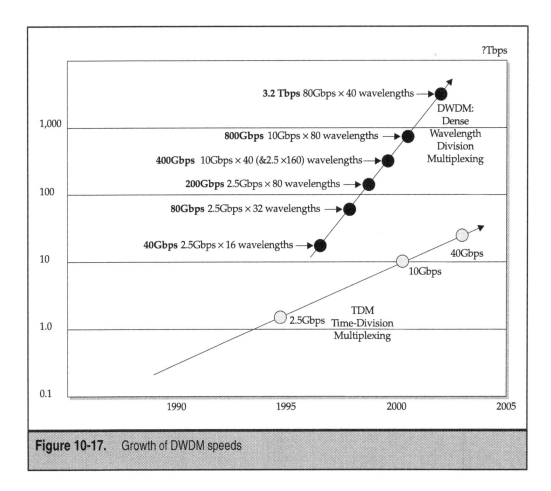

Figure 10-17. Growth of DWDM speeds

In addition to the use of specially developed NZ-DSF, DWDM systems are reliant on amplifiers such as the *erbium-doped fiber amplifier (EDFA)*, tunable laser diodes, and fiber Bragg gratings to isolate the individual wavelengths at the receiver.

Optical Amplifiers

There is a finite distance over which an optical signal can travel before it degrades to the point that it is undistinguishable by the receiver. To circumvent this problem, the signal has to be regenerated at different points along the path. Before optical amplifiers were used, regenerators were required every 60 to 100 km for each separate fiber carrying a single optical signal.

These regenerators were actually electrical regenerators that required the optical signal to be converted from the optical domain to the electrical and then back to the optical again. This was an expensive arrangement, as the cost of implementing these electrical regenerators extended beyond the regenerators themselves to the facilities in which to house them. Furthermore, the regenerators had to be in placed before the fiber facility could be lit, introducing the potential for delays.

Erbium-Doped Fiber Amplifier (EDFA)

A key DWDM enabling technology is the *erbium-doped fiber amplifier (EDFA)*. These amplifiers have made it possible to carry large amounts of data over very long distances. The amplifier itself, shown in Figure 10-18, is a fiber-based device that consists of an optical fiber having a core doped with the rare earth element erbium.

Light from one or more external semiconductor lasers in either of two pump bands, 980 nm and 1,480 nm, is coupled into the fiber, exciting the erbium atoms. Optical signals—at wavelengths between 1,530 nm and 1,620 nm—entering the fiber stimulate the excited erbium atoms to emit photons at the same wavelength as the incoming signal. The effect is the amplification of weak input signals to higher signal levels.

The essential characteristic of an optical amplifier is that it operates across a wide range of wavelengths, whereas the electro-optical repeaters used in SONET/SDH typically function only at the SONET/SDH wavelength of 1,310 nm. In electrical systems, amplifiers are a known source of noise that can interfere with the signals. In SONET/SDH applications, amplifiers were abandoned in favor of digital regeneration and repeater techniques that culminated in the SONET/SDH electro-optical repeaters.

DWDM systems have now come full circle, and amplifiers are now once again preferred, the key difference being that these amplifiers operate in the optical domain. Noise in optical systems is very low as compared to electrical transmission systems, in which electrical amplifiers have the tendency to amplify any preexisting noise along with the signal. With optical amplifiers, this is less of a concern because the noise in these systems is already so low that it is not usually a problem. However, given a long enough fiber run and the use of too many EDFAs, the cascade of amplifiers can make the noise a problem for the system. It would take several thousands of kilometers of fiber and over 80 EDFAs for this to become an issue. Table 10-2 gives a quick comparison of EDFAs and electro-optical repeaters.

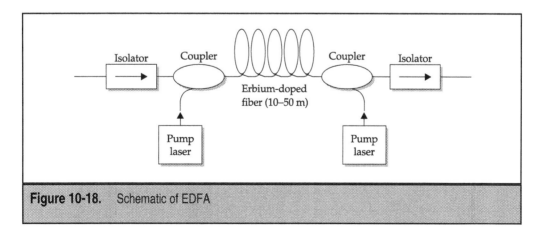

Figure 10-18.   Schematic of EDFA

EDFA	Electro-Optical Repeaters and Amplifiers
Not as complex and less prone to failure.	More complex and prone to failure. Require a receiver and transmitter circuitry, the optical-electrical- optical conversion, and amplification capabilities.
Built into the fiber and effective across a range of wavelengths.	Tuned to a specific wavelength and bit rate.
The composite signal is automatically amplified.	Require the demultiplexing of the signal, a conversion to an electrical signal, the amplification of these signals, and reconversion to an optical signal.
Lower per unit and operational cost.	Higher per unit and operational cost.

Table 10-2.   EDFA and Electro-Optical Repeater and Amplifier Comparison

NOTE: SONET systems, being systems that predate DWDM and cannot use EFDA, typically operate in the 1,310 nm range and rely on TDM. They rely on electrical regenerating amplifiers placed every 60 km or so along the path.

Multiplexing and Demultiplexing Techniques

In WDM and DWDM applications, multiple wavelengths representing different digital data streams must be combined into a single light beam for insertion into the optical fiber. This process of combining these signals is called *multiplexing*. At the receiving end, a single light beam, which is a composite signal of multiple individual signals or wavelengths, must be broken out into its component parts and placed onto separate fibers. This process, called demultiplexing, requires the separation of the individual signals before the light is detected by the photodetector, since the photodetector cannot selectively detect individual light wavelengths.

An optical fiber communication system can be either a unidirectional system that uses two different fibers for two-way communication or a bidirectional system that uses a single fiber for two-way communication. In unidirectional systems, a multiplexer is needed on the send side and a demultiplexer on the receive side of the connection. In this configuration, both a multiplexer and a demultiplexer are needed on both sides of the connection if two-way communication is to be achieved. Each fiber link has its own set of multiplexer and demultiplexer for communication in a particular direction.

In a bidirectional communication system that uses a single fiber, there is a multiplexer and a demultiplexer on either side of the fiber link, and different wavelengths are used for different directions. Multiplexers and demultiplexers can be passive or active devices. Passive devices are based on prisms, diffraction grates, or filters, and active devices combine passive devices with tunable filters.

Prisms Using the example of the prism that was given earlier in this chapter, we can see that the prism separates the different wavelength components of a composite light signal. It is also evident that these separate wavelengths are refracted at different angles from each other. By using a lens, these separate refracted wavelengths can be focused to a point where each needs to enter a separate fiber (see Figure 10-19). In the reverse direction, the same components can be used to multiplex different wavelengths into a fiber. This type of multiplexing/demultiplexing, called *prism refraction multiplexing*, uses passive components.

Diffraction Gratings Another approach that uses passive components is *waveguide-grating diffraction*, which uses a diffraction grating to separate the wavelength components. Based on principles of interference and diffraction, these systems use a diffraction grating to diffract the different wavelengths from a composite light beam to different points in space. A lens is then used to focus these wavelengths into other fibers. Prisms and diffraction gratings are now less common than the remaining systems that will be discussed.

Arrayed Waveguide Gratings (AWG) AWGs are based on the principle of refraction. In these systems, an optical waveguide router (another name for an AWG device) is used to separate the different wavelengths of a composite light beam. The optical waveguide router consists of an array of curved-channel waveguides—made of silica—of different path lengths. When light enters the system, it is diffracted and enters the waveguide array, whose members are all different lengths and are usually fixed into position upon some kind of base that forms optical circuits (in a similar way to electronic circuit boards).

The wavelengths travel down each waveguide, and the different lengths of the waveguide introduce phase delays, resulting in different wavelengths having maximal interference at different locations, resulting in turn in a separation of the wavelengths as they exit the output array.

There are several neat features of AWGs that make them quite a useful item for optical networks. The fact that they are based in so-called "optical integrated circuits" makes

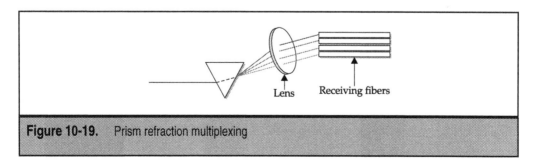

Figure 10-19. Prism refraction multiplexing

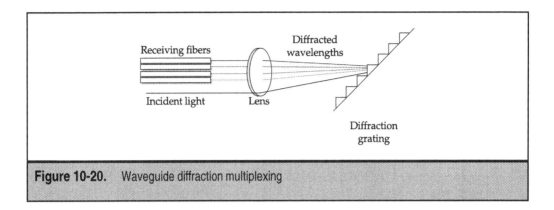

Figure 10-20. Waveguide diffraction multiplexing

them easy to integrate with other functions such as amplification. They can also be used as multiplexers by reversing the process, taking the separate wavelengths in from the right, as shown in Figure 10-21, and having them exit as a composite signal from the left. AWGs can also split or combine many wavelengths at a time, compared with fiber Bragg gratings, which require a separate section for each individual wavelength.

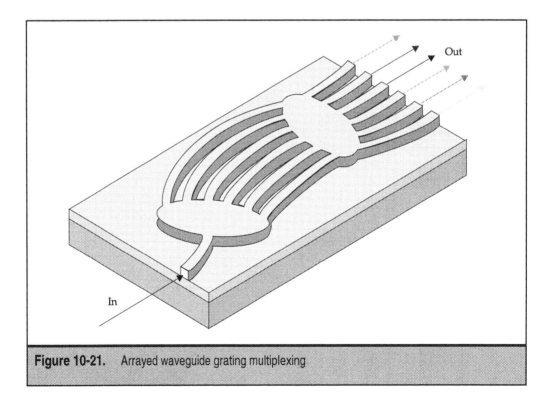

Figure 10-21. Arrayed waveguide grating multiplexing

Thin Film Filters   Another scheme uses thin filters that are placed in the path of the optical signal, as shown in Figure 10-22. The filters are designed to transmit a single signal and reflect the others. By placing multiple thin film filters, or multilayer interference filters as they are sometimes called, in a line, many wavelengths can be multiplexed.

Fiber Bragg Grating   A *fiber Bragg grating (FBG)* is made from a section of ordinary single-mode optical fiber—typically, a few millimeters to a few centimeters in length. Causing periodic variations in the index of refraction of the core lengthwise along the fiber forms the grating, as shown in Figure 10-23. This is accomplished by exposing the core to ultra-violet radiation in a regular pattern, which causes the refractive index of the fiber core to be altered in a corresponding pattern. The result is that light traveling through these refractive index changes is reflected back slightly, but the maximum reflection usually occurs at only one particular wavelength. The reflected wavelength—known as the *Bragg wavelength*—depends on the amount of refractive index change that has been applied and also on how distantly spaced these changes are. The light at the Bragg wavelength is selectively reflected while all other wavelengths are transmitted, essentially unperturbed by the presence of the grating.

By combining fiber gratings in various arrangements, many different wavelengths can be separated and coupled out. FBGs are used in a number of applications:

▼   DWDM filters

■   Dispersion compensation

■   Gain-flattening filters

▲   Fiber-tuned lasers

These devices have an important part to play in optical networks as a demultiplexing technology, as they allow discrete or multiple discrete wavelengths to be blocked or separated. FBGs are used in some advanced amplifier designs where they play the role of a filter in order to "flatten" the amplification so that signals at different wavelengths are amplified by equal amounts. For this application, fairly complex arrangements of different fiber gratings can be joined together to reflect just a small percentage of certain wavelengths, in order to reduce the signals to match the intensity of others.

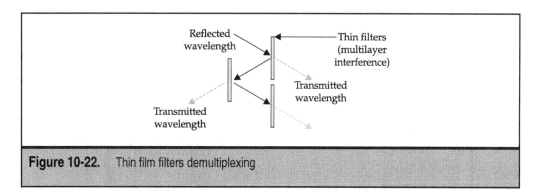

Figure 10-22.   Thin film filters demultiplexing

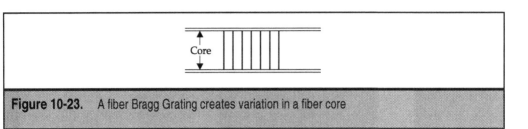

Figure 10-23. A fiber Bragg Grating creates variation in a fiber core

Optical Add/Drop Multiplexers (OADM)   It is not always convenient to have to multiplex or demultiplex an entire signal, especially when somewhere along the optical path a single wavelength needs to be extracted or inserted into the fiber. For these occasions, an *optical add/drop multiplexer (OADM)* is used. Rather than demultiplexing an entire signal, these systems are able to extract a single wavelength from the composite light beam. Conversely, they are able to insert a single wavelength into an existing composite optical signal. These systems play a key role in optical systems.

Two general types of OADMs exist. The first generation of the technology employed a system that was configured to drop or add a specific predetermined set of wavelengths. Once designed for the specific wavelengths, these systems spend their lives adding or dropping these wavelengths. The second generation of OADMs is more dynamic and can be reconfigured for different sets of wavelengths. This dynamic capability enables a more flexible approach and is more common in contemporary optical systems.

Tunable Laser Diodes

The tunable laser diode is also an important part of DWDM systems. It consists of a laser diode and a tunable component that is used to generate different wavelengths of light. In our previous discussion on light sources, it was determined that there are two basic types of light sources, light emitting diodes (LED) and laser diodes. We further determined that laser diodes, while being more expensive, were more suitable as a light source for single-fiber systems over longer distances.

In theory, lasers are a more efficient light source for optical communication because they produce a very narrow beam of a single wavelength. That is the theoretical description; in practice, however, no laser generates a single wavelength of light, although the light has a characteristically narrow linewidth. Another important aspect of lasers as compared to LEDs is the speed at which the light can be switched on or off, or modulated. Lasers can be turned on or off at a rate of one millionth of a nanosecond, or 10^{-15} seconds, and they produce higher powers than LEDs. A laser can inject 50 to 80 percent of its power into a fiber.

The fact that lasers actually emit an extremely narrow beam instead of a single wavelength is an important consideration for DWDM systems. The width of the beam, called the *spectral width*, is usually about 8 nm, about the size of the fiber core of a single fiber system. This narrow beam will favor a specific wavelength but will also have other wavelengths of varying strengths, resulting in a beam that resembles a bell curve. The reason the spectral width of the beam is important is that the wider the beam, the more prone it is to dispersion. Also, the narrower the spectral width, the more wavelengths can be packed into a single fiber.

Prior to tunable lasers, lasers of specific wavelengths were used. This meant that different lasers were needed for different wavelengths. Tunable lasers are capable of transmitting data on a wide range of wavelengths, rather than just on a single frequency. So instead of many laser components, a single laser is used, and the use of a single component has far-reaching cost implications, including less inventory to stock and track, and lower deployment costs.

An added significance of tunable lasers is their impact on *optical wavelength routing*, a means of optical cross-connecting. Tuning lasers creates the possibility of taking a wavelength that was input from one fiber and routing it out of another fiber strand. Intelligent optical devices that take a wavelength and determine which fiber to send it out on can make quick routing decisions. Using a tunable laser, these decisions can be implemented on the fly.

The ITU DWDM Grid

Table 10-3 shows the laser grid for point-to-point WDM systems based on 100 GHz wavelength spacing or about 0.8 nm, which has been published as part of the ITU Draft standard G.692. Manufacturers can vary the spacing between wavelengths, and a spacing of 50 GHz

Wavelength	Frequency	ITU Channel	Wavelength	Frequency	ITU Channel
1531.12	195.8	58	1547.72	193.7	37
1531.90	195.7	57	1548.51	193.6	36
1532.68	195.6	56	1549.32	193.5	35
1533.47	195.5	55	1550.12	193.4	34
1534.25	195.4	54	1550.92	193.3	33
1535.04	195.3	53	1551.72	193.2	32
1535.82	195.2	52	1552.52	193.1	31
1536.61	195.1	51	1553.33	193.0	30
1537.40	195.0	50	1554.13	192.9	29
1538.19	194.9	49	1554.94	192.8	28
1538.98	194.8	48	1555.75	192.7	27
1539.77	194.7	47	1556.55	192.6	26

Table 10-3. ITU Wavelength Grid

Wavelength	Frequency	ITU Channel	Wavelength	Frequency	ITU Channel
1540.56	194.6	46	1557.36	192.5	25
1541.35	194.5	45	1558.17	192.4	24
1542.14	194.4	44	1558.98	192.3	23
1542.94	194.3	43	1559.79	192.2	22
1543.73	194.2	42	1560.61	192.1	21
1544.53	194.1	41	1561.42	192.0	20
1545.32	194.0	40	1562.23	191.9	19
1546.12	193.9	39	1563.05	191.8	18
1546.92	193.8	38	1563.86	191.7	17

Table 10-3.   ITU Wavelength Grid *(continued)*

or 0.4 nm is not uncommon. Using this spacing, the number of channels is doubled, but it limits the rate per wavelength to about 10 Gbps. The smaller spacing also increases the potential for FWM and crosstalk between channels.

The problem with this arrangement is that the draft standard does little to assure interoperability, as the flexibility within the grid translates to the potential for differences in implementations that are incompatible. The main reason for this has been the speed at which DWDM technology has been developing. The ITU has been hard-pressed to keep up with standards that meet existing technology and needs.

SUMMARY

Optical networking represents the final chapter in the book of communication systems that rely on a physical connection between two points. There is nothing available beyond light, so this represents the final transmission system to be explored. We are nowhere near the end of the development cycle for light-based communication, however. Whether it is fiber-based or over-the-air, light as a transmission medium will see developments that have not yet even been conceived. Already, we have seen speeds of 7 terabits per second being demonstrated in labs, with working and deployed systems running at 3.2 terabits per second, the equivalent of 41 million concurrent telephone calls on a single fiber. This is indeed mind-boggling.

The exciting thing is that, while we have seen a century's worth of copper-based transmission systems from their humble beginnings to their current capabilities, with fiber-based systems, we have seen only 30 years worth of development, and it is highly probable that the best is yet to come.

CHAPTER 11

The Metro Area and Core Network

The technologies inside the metro area and long-haul (core) networks are a lot different from those found in the access network. The technologies in these networks are also very different from those used in the enterprise and homes. These networks being predominantly a fiber optic domain, the technologies are predominantly SONET/SDH and ATM, but other emerging technologies such as MPLS and optical solutions such as DWDM are also becoming more prominent.

Before we begin, this is probably a good time to position the different networks to which we will be referring. Figure 11-1 illustrates a network hierarchy that forms a point of reference for the networks we have been discussing. As is evident from the figure, the hierarchy is logically segmented into an access portion, a metropolitan area portion, and a long-haul core portion.

There is a natural tendency to regard the metro area network as a scaled-down version of the long-haul core network. In some respects this is true, as both are now predominantly fiber-based networks, with the metro area having shorter runs of fiber. However, an important differentiation is that the long-haul network tends to be a lot more stable in topology, whereas the metro area in contrast is an evolving entity and so the topology keeps changing. Another difference is that at the metro area level, the design and management are geared more toward the different types of traffic streams

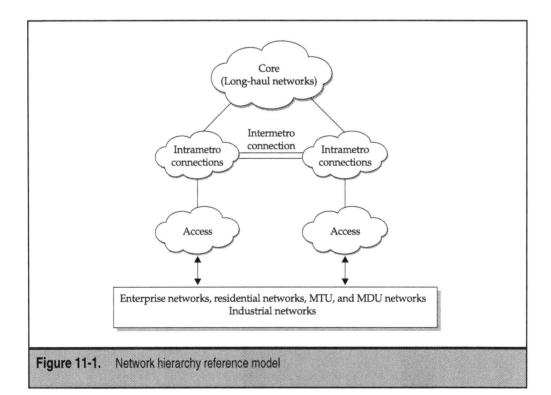

Figure 11-1. Network hierarchy reference model

and services. In the long-haul portion of the network, the emphasis is on big pipes and fast bit pumps that move data as quickly and efficiently as possible.

As we explore the different technologies in these two networks, we will generically refer to both as the core. In those cases where a distinct reference is needed, we will address each network separately as either the metro area network or the long-haul network.

SONET/SDH

SONET, short for *Synchronous Optical Network,* and SDH, an acronym for *Synchronous Digital Hierarchy,* are two standards that are very closely related. In the U.S., the standards for SONET are used; other places use SDH. Both standards specify interface parameters, data rate framing formats, multiplexing methods, and management methods for synchronous time-division multiplexing (TDM) over fiber.

Based on principles of TDM, multiple digital bit streams are input into a SONET/SDH system and are multiplexed into a composite signal over an optical fiber infrastructure. The input stream usually runs at a speed of 2.5 Gbps per second, with the resulting output signal being 2.5 Gbps times the number of input streams. So if four signals are multiplexed, as shown in Figure 11-2, the resulting output onto the fiber is 10 Gbps.

The arrival of SONET/SDH brought order to a somewhat chaotic approach to optical communication. Fiber optics being a new technology, equipment vendors recognizing its potential sought to capitalize on it by introducing proprietary equipment in the absence of standards. With the introduction of any new technology, the market for the technology emerges long before any standards. As a forerunner to standards, the market becomes the basis on which a decision can be made as to whether or not a standard is needed. If there isn't a market, then why bother with a standard?

So prior to the development of the SONET/SDH standard, many optical solutions existed that were mutually incompatible. If fact, there were no guarantees that a piece of equipment would be able to benefit from future technology improvements such as higher

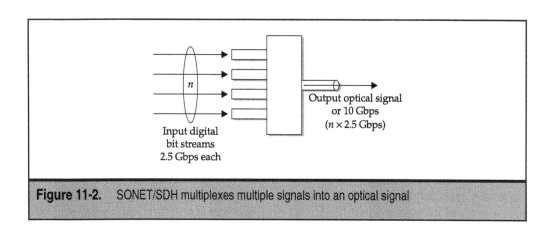

Figure 11-2. SONET/SDH multiplexes multiple signals into an optical signal

speeds, for example. With the introduction of SONET/SDH, telecommunications companies readily embraced the standard because it provided the means for interoperability between different vendor systems and it provided an upgrade path for any new technological enhancements.

At the turn of the 1990s, when the Internet became more prevalent and the requirement for more data grew, SONET was seen as a major enabler of the build-out of the infrastructure. The features of SONET that are now recognized as limitations were not a problem then. Its built-in reliance on a single wavelength of 1,310 nm and its top speed of approximately 10 Gbps appeared adequate at the time.

Standardizing on a single wavelength had its benefits, in that it limited the choice of components, in turn eliminating the complexity of different variables. Having a smaller number of choices made sense, as it simplified the life of the telecommunications operator. The only type of lasers that were required were those that operated at 1,310 nm. The number of fiber types was also limited, which created a more predictable operational environment with a single-wavelength system of a single-bit stream.

The upper limits of the technology also appeared to be more than adequate for applications that were available then. However, the speed limitation and the single-wavelength design of the technology bring it to a major crossroads today. Modern optical systems are capable of much higher speeds and utilize many more wavelengths, and so the question surrounding SONET/SDH is now, where does it fit into the current scheme of things and what is its future?

SONET/SDH Background and Architecture

It is important to note that the development of the first standards for optical communications was driven by telephone companies who looked at the technology as a way to upgrade the capacity for voice-based applications. With this in mind, many of the odd design features of SONET and SDH make more sense—odd from the viewpoint of how the technology approaches the handling of high-volume data traffic.

In 1962, AT&T began deploying T1-carrier circuits for the transport of voice channels between different central offices. Having a capacity of 24 voice channels, T1-carrier circuits worked well, as they provided a way of reducing the requirement for copper pairs and utilized digital technology to provide a virtual channel for analog voice signals.

In the 1970s when fiber optics became a possibility, telephone companies saw it as a solution for carrying even more voice circuits. The first commercial fiber circuits were deployed by AT&T in Chicago in 1977 and operated at 45 Mbps (DS3 rate). At that time, the plesiochronous digital hierarchy (PDH) of electrically based transmission systems was already defined and was the accepted standard by which different digital data rates were delivered. Based on a DS0 link of 64 Kbps, the PDH set fixed data rates that were multiples of the DS0 signal. A T1, for instance, was a multiplex of 24 DS0 signals, and a DS3, an even denser multiplex of the same signals.

Viewing fiber as a simple replacement for older copper-based systems and microwave communications, and in the absence of any standards, vendors designed and developed optical systems that used their own proprietary methods of operation. Once these systems were selected, the telecommunication company was locked into a specific optical solution from a specific vendor.

To open up competition and foster innovation, the drive toward a standard was soon completed. The emerging SONET/SDH optical standards, not surprisingly, were optimized for voice-based applications, as this was the main market driver for the technology at the time. The channel structure looked very similar to the legacy plesiochronous channels and data rates such as DS1s and DS3s, but instead of using a basic DS0 signal, the new standard was based on a basic signal rate of 51.84 Mbps. All higher line rates were multiplexes of this basic line rate, as shown in Table 11-1.

An important aspect of SONET/SDH to remember is that it is basically a channelized system—a throwback to the voice days and the PDH ways of doing things. So an STS-12 is actually made up of 12 ST-1s, and hence the reference to the overhead bits, which are the bits used for maintaining channel structure or the different multiplexes of the ST-1 signal. These overhead bits cannot be used for data, so the payload rate of an OC-12 line is actually the line rate less the overhead bits. The term *envelope* is used to describe the portion of the frame that is used to transport the payload or data. When SONET/SDH operates in an unchannelized mode, through a process known as *concatenation*, the extra overhead bits can be reused for data and so the full line rate of 622 Mbps for an OC-12 link becomes available.

Optical Level	Electrical Level	Line Rate (Mbps)	Payload Rate (Mbps)	Overhead Rate (Mbps)	SDH Equivalent
OC-1	STS-1	51.840	50.112	1.728	—
OC-3	STS-3	155.520	150.336	5.184	STM-1
OC-9	STS-9	466.560	451.008	15.552	STM-3
OC-12	STS-12	622.080	601.344	20.736	STM-4
OC-18	STS-18	933.120	902.016	31.104	STM-6
OC-24	STS-24	1244.160	1202.688	41.472	STM-8
OC-36	STS-36	1866.240	1804.032	62.208	STM-12
OC-48	STS-48	2488.320	2405.376	82.944	STM-16
OC-96	STS-96	4976.640	4810.752	165.888	STM-32
OC-192	STS-192	9953.280	9621.504	331.776	STM-64
OC-768 (Emerging)	STS-768	39813.12	38,486.016	1,327.016	STM-256

Table 11-1. SONET and SDH Data Rates and Digital Hierarchy

The difference between the OC- and the STS- designations is that OC- refers to *optical carrier* and describes the transmission characteristics of the transmission link, whereas the STS- refers to the *synchronous transport signal,* a reference to the SONET frame structure of the link. The n, the number after the OC-, refers to multiples of the basic line rate of 51.840 Mbps, so OC-12 or STS-12 refers to a multiple of 12×51.840 Mbps or 622.08 Mbps. The theoretical maximum value n is 255 as per the standard. However, as is evident in Table 11-1, an emerging version of SONET taps the technological achievements of current optical systems and has achieved a serial bit rate of 40 Mbps on a single wavelength. Thanks to the work being done on DWDM solutions and higher data rates, the upper limits of SONET have the potential to be redefined even further.

There is a deeper significance to the synchronous transport signal or STS in SONET. This designation actually refers to the electrical level. Unlike modern-day optical systems where the transmission of a digital data stream can remain in the optical domain as it is multiplexed, amplified, and passed through the system, SONET depends on the electrical domain for a lot of this work. In SONET/SDH, all frames must be created, multiplexed, managed, repeated at the electrical level. Herein lies another key difference from the technology of contemporary optical systems. The dependence on an electrical-optical-electrical conversion makes SONET/SDH less efficient and more expensive to deploy and maintain.

SONET and SDH Standards Work on the SONET standard was initiated in the U.S. by the ANSI T1 X1.5 committee and commenced in 1985. The CCITT (now known as the ITU) initiated their efforts in Europe on the development of standards for SDH in 1986. From the very beginning there was a difference in approach. The U.S. wanted a basic data rate that was close to 50 Mbps to carry DS1s (1.544 Mbps) and DS3s (44.736 Mbps), while the Europeans wanted a basic speed that would accommodate their standard data rates of E1 (2.048 Mbps), E3 (34.368 Mbps), and E4 (139.264 Mbps).

The U.S. standards organization wanted a basic bandwidth that was close to 50 Mbps, one that would carry DS3 signals. The European wanted one closer to 150 Mbps in order to carry the 139.264 Mbps signal. Eventually an agreement was reached and the U.S. data rates were accepted as a subset of the higher SDH data rate. This is the reason why the U.S. basic data rate for SONET is 51.840, or one-third of the SDH basic data rate. The SONET STS-3 data rate of 155.520 Mbps, a multiplex of 51.84, is equivalent to the basic SDH STM-1, which is the same as an OC-3 rate.

Both technologies have a lot in common, with the major difference being in the terminology and the allowances made for a larger frame size in SDH. Table 11-2 provides a reference for the relevant standards for both technologies.

Architecture and Terminology SONET makes use of specific terminology to describe the various components and makeup of the connection. The SONET/SDH nodes are called *add/drop multiplexers* or ADMs, the end-to-end connection is called the *path*, and the devices that are used at these end points are called *path terminating equipment (PTE).*

ANSI SONET Standards	Description
ANSI T1.105	Basic Description including Multiplex Structure, Rates, and Formats
ANSI T1.105.01	Automatic Protection Switching
ANSI T1.105.02	Payload Mappings
ANSI T1.105.03	Jitter at Network Interfaces
ANSI T1.105.03a	Jitter at Network Interfaces—DS1 Supplement
ANSI T1.105.03b	Jitter at Network Interfaces—DS3 Wander Supplement
ANSI T1.105.04	Data Communication Channel Protocol and Architectures
ANSI T1.105.05	Tandem Connection Maintenance
ANSI T1.105.06	Physical Layer Specifications
ANSI T1.105.07	Sub-STS-1 Interface Rates and Formats Specification
ANSI T1.105.09	Network Element Timing and Synchronization
ANSI T1.119	Operations, Administration, Maintenance, and Provisioning (OAM&P)—Communications
ANSI T1.119.01	OAM&P Communications Protection Switching Fragment
ITU SDH Standards	**Description**
ITU-T G.707	Network Node Interface for the Synchronous Digital Hierarchy (SDH)
ITU-T G.781	Structure of Recommendations on Equipment for SDH
ITU-T G.782	Types and Characteristics of SDH Equipment
ITU-T G.783	Characteristics of SDH Equipment Functional Blocks
ITU-T G.803	Architecture of Transport Networks Based on SDH

Table 11-2.   SONET and SDH Standards

The path through the network may pass through many nodes, or the path could be direct from one node to another. In either case, the connection between the nodes is called a *line*. A path therefore may consist of many ADMs and lines. Within a line—that is, a link between ADMs—there may be *regenerators* of the optical signal. These signal regenerators electrically regenerate the optical signal; the signal therefore has to be converted from an optical signal to an electrical signal, regenerated, and then converted back to an optical signal before it can be transmitted. Optical amplifiers had not yet been developed, so this was the only way to provide amplification and reshaping of the optical signal.

A link between regenerators or between an ADM and a regenerator is called a *section*. Thus, regenerators are considered *section terminating equipment (STE)*. Likewise, an ADM is considered *line terminating equipment (LTE)*, if it is being used as a line terminator. And as previously noted, in instances where an ADM is at the end point of a connection, it is called a PTE. Figure 11-3 graphically illustrates the different components and terminology of a SONET/SDH network.

What may not be immediately obvious from the foregoing descriptions of the component parts of the technology is that these components constitute an architectural model. Like most technology, SONET uses an architecture that consists of layers of functions. Layered functions, as in the case of the OSI, allow flexibility in designing and implementing equipment. By employing a layered model, manufacturers are left with enough latitude to add functions at each layer while maintaining interoperability between systems by ensuring that the interfaces between layers are consistent and meet published standards.

Layering also facilitates choice in the layer in which a manufacture chooses to operate. Without layers, equipment manufacturers would be required to produce equipment with the full feature sets and functionality of the entire architecture. By having a layered model, a manufacturer can concentrate on only certain functions, thereby tapping its ability to be more cost efficient and innovative. Figure 11-4 shows the components previously discussed in a layered architecture.

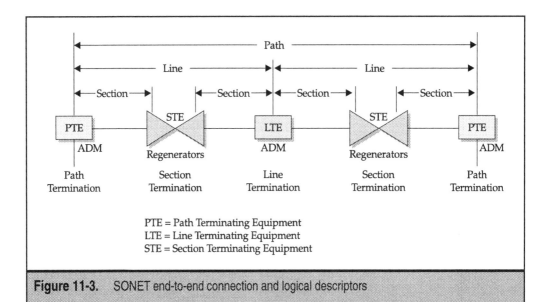

Figure 11-3. SONET end-to-end connection and logical descriptors

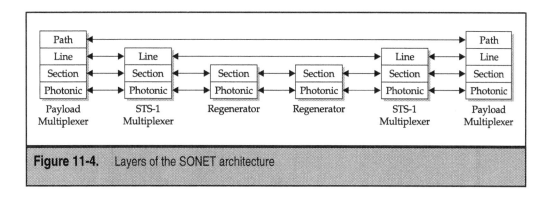

Figure 11-4.   Layers of the SONET architecture

Figure 11-5 shows how SONET relates to the OSI model. As noted, SONET operates at the physical layer, so even though it has a layered architecture, it is important to understand that it defines the operation within the scope of the physical layer of the OSI model.

SDH Layers   Like SONET, SDH defines different layers of operation. One of the more confusing elements of SONET and SDH is the difference in terminology. Conceptually, and for the most part technically, they are very similar. However, SDH uses different terms to describe the architectural components of the technology.

Common to both technologies is the description used to describe the end-to-end connection, or more precisely, the point where the SDH frames are originally assembled and the point where they are disassembled and exit the SDH network. In SDH, this is also called the path, and just as in SONET, the devices that sit at the ends of the path are called PTEs (path terminating equipment). Beyond these terms, things are different.

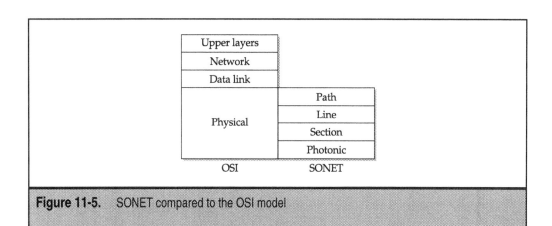

Figure 11-5.   SONET compared to the OSI model

In SDH, a *network element (NE)* is the generic name of any component that provides network-multiplexing functions for an SDH network. PTEs are therefore NEs as they multiplex and demultiplex the payload as it enters or exits the SDH network. The following NEs are typical PTEs in an SDH network:

▼ Low-order multiplexer (a low-speed multiplexer)

■ Wideband cross-connect system

▲ Subscriber loop-access system

Figure 11-6 shows an SDH end-to-end connection along with the names of the various components that make up such a link. Note that beyond the PTE and path descriptions, the remaining descriptions are different from those of SONET. However, the logical makeup of the link is similar to SONET. As shown in the figure, a *multiplex section* is defined as the transmission medium along with the associated equipment required for transporting a signal between two consecutive NEs. This is functionally similar to the line in SONET. The SDH *multiplex section terminating equipment (MSTE)* is a network element that originates or terminates STM signals. Examples of SDH MSTEs are:

▼ Optical line terminal

■ Radio terminal

■ High-order multiplexer (high-speed multiplexer)

▲ Broadband cross-connect system

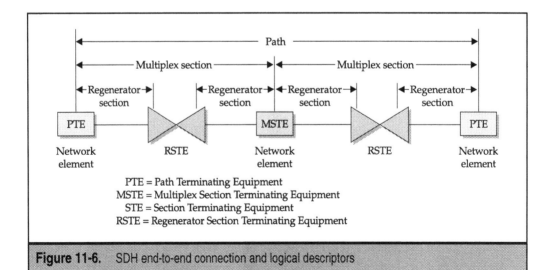

Figure 11-6. SDH end-to-end connection and logical descriptors

The final component of the end-to-end connection of an SDH network is the *regenerator section,* which is functionally equivalent to the *section* of SONET. The regenerator section of the network is the portion of the network that consists of the connection between—including the terminating points—a network element and a regenerator or between two regenerators. The regenerator section terminating equipment (RSTE) is simply the regenerator of the signal. The difference therefore between a multiplex section and a regenerator section is whether or not an RSTE is on either or both sides of the link; if it is, it is a regenerator section; if it is not and the two end points are multiplexers, then it is a multiplex section.

SONET Line Interfaces

SONET/SDH defines both electrical and optical interfaces; Figure 11-7 demonstrates that a SONET multiplexer is logically divided into an electrical component and an optical component.

An electrical interface is defined for intraoffice applications. It consists of a coaxial cable interface for distances up 900 feet for STS-1 signals and 450 feet for STS-3 signals.

Three basic optical interfaces for short, intermediate, and long reach were defined for SONET, and all use single-mode fiber.

▼   **Short Reach (SR)**   Applies to distances of up to 2 km with a signal loss of no more than 7 dB. For this interface, an LED or a *multilongitudinal mode (MLM)* laser is used at the 1,310 nanometer wavelength—a laser that emits light over a fairly large spectrum of several nanometers.

■   **Intermediate Reach (IR)**   Applies to fiber runs up to 15 km with a signal loss of no more than 12 dB. At this distance, a low-power (50 mW) single longitudinal mode (SLM) or MLM laser is used at either the 1,310 nm or 1,550 nm wavelength. An SLM laser emits a much narrower beam that significantly reduces chromatic dispersion.

▲   **Long Reach (LR)**   Applies to distances up to 40 km with a signal loss in the range of 10–28 dB. A high-power (500 mW) SLM or MLM laser is used at either the 1,310 nm or 1,550 nm wavelength.

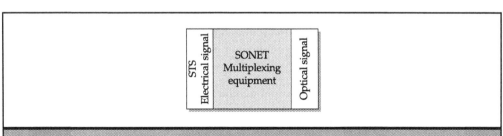

Figure 11-7.   SONET defines electrical and optical interfaces

The STS-1 Frame Structure

Across a fiber link, the transmission of digital data is achieved through the modulation of a light signal. By sensing that a light source is switched on or off, the receiver is able to determine a 1 or a 0 bit. SONET/SDH defines a structure in which these bits are transmitted by defining a low-level framing protocol for use across optical links. *Framing* is simply a term to describe the fact that a block of bits is transmitted with some type of structure that has a means for determining the boundaries of the structure. Parts of the block may be devoted to overhead for the network provider to manage the network, and other parts, for the payload. The part that carries the payload is called an *envelope*.

SONET/SDH uses fixed-length frames and so only needs to delineate the start of a frame. Had it used a variable-length frame, we would also have had a requirement to know the end of the frame, but this is not the case with SONET/SDH. Normally, when we think about a packet or frame structure such as Ethernet or HDLC, we tend to imagine it as a single-dimensional structure that has bits consecutively occurring from beginning to end, or left to right if we were viewing it as a part of a document. The single dimension assumes that different grouping of bits, from the beginning to the end of the frame, has a specific meaning: destination address, source address, etc. The SONET frame structure adds a second dimension to this concept. In addition to reading left to right, the frame structure reads top to bottom.

The basic SONET STS-1 frame is structured as shown in Figure 11-8, with nine rows of 90 octets called columns. The SDH STM-1 frame, being much larger—remember the basic rate in SDH is three times that of SONET—has nine rows of 270 columns and thus is three times the size of the SONET frame. In both cases, the frame is transmitted from left to right and top to bottom. The octet in the upper-left corner of the figure is transmitted first,

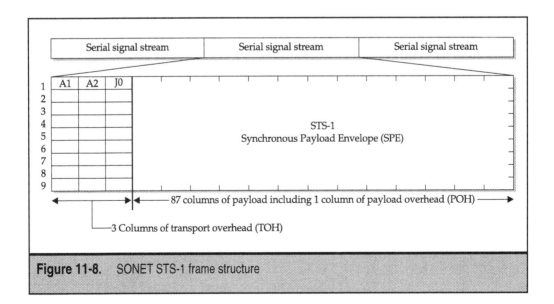

Figure 11-8. SONET STS-1 frame structure

followed by the second octet of the first row and so on. At the ninetieth octet, the process is repeated from the second row, and the first octet of the second row is transmitted until the entire row is completed. This repeats for all nine rows.

The concept of a two-dimensional frame—left to right and top to bottom among multiple rows—is just for the convenience of documentation and explanation. In reality, all 810 octets are transmitted serially. However, the concept of multiple rows does assist in remembering that every 90 octets, or each row, has 3 overhead octets.

Section and Line Overhead   The first three columns of the SONET frame are called the *transport overhead (TOH)*, and the 87 columns following the TOH, the *synchronous payload envelope (SPE)*. The SPE also contains a column overhead that is called the *Payload Overhead (POH)*. The actual number of columns available for the payload is therefore 86 columns by 9 rows (90 columns less 3 columns of TOH and 1 column of POH).

The transport overhead actually consists of *section overhead (SOH)* and *line overhead (LOH)* as shown in Figure 11-9. The first three rows of the transport overhead are the section overhead, and the remaining rows, the line overhead.

Framing is accomplished by the first two bytes, called the A1 and A2 octets as shown in the figure. When the frame is transmitted, all frames with the exception of the first three, A1, A2, and J0, are scrambled to ensure that octets within the frame do not duplicate the A1 and A2 bit sequence, thereby causing framing errors. The bit patterns of the A1/A2 octets are 1111 0110 0010 1000 (0xF628). The receiver searches through the stream for bit patterns that match the A1/A2 octets. Once found, they enable the receiver to gain bit and octet synchronization and once synchronized, everything is done from there on octet boundaries. SONET/SDH, therefore, is octet synchronous, not bit synchronous.

The STS-1 Signal   So far, we have accounted for only 810 octets (9 rows and 90 columns), or 6,480 bits of the 51.84 million bits that are transmitted in a second. Since SONET/SDH is a technology that was designed to transport legacy voice traffic, the specifications were closely aligned to the DS0 hierarchy. Now here is an important point that is critical to

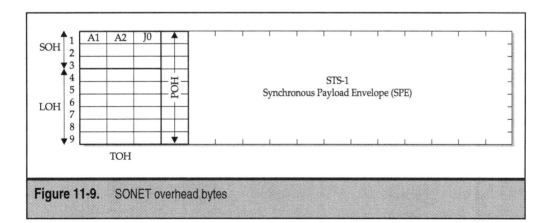

Figure 11-9.   SONET overhead bytes

understanding the mechanics of the operation of the technology: In a previous chapter, we explained the origins of the DS0, in which we explained that the DS0 signal is a 64 Kbps stream that was derived from a sample of 2 × a 4 kHz (that is, 8,000 Hz) voice pass band in accordance with Nyquist theorem.

The granularity of the voice sample is one octet (8 bits), and samples are taken 8,000 times per second or every 125 microseconds. The relevance of this information lies in the way it translates to the basic SONET data rate, in that a SONET frame is sent every 125 microseconds. Another way of viewing this is that SONET was optimized for voice traffic, as was the DS0 signal—the basis of the plesiochronous digital hierarchy.

This is important in understanding the mechanics of SONET's operation because a SONET frame is transmitted every 125 microsecond no matter how high the line rate gets. As the line rate gets higher, the SONET frame gets bigger by some amount of octets that is sufficient to maintain the frame rate of 8,000 frames per second. So, for an STS-1 signal, 8,120 octets or 6,480 bits (9 rows × 90 columns) are transmitted 8,000 times per second, yielding a data rate of 51.84 Mbps (6,480×8,000).

SONET/SDH Interleaving and Higher-Level Frames   In this section, the examples given are specific to SONET; the concepts, however, apply to both SONET and SDH. The higher multiples of the STS and STM signals in SONET and SDH are all derived from multiplexing the basic STS-1 or STM-1 frame. A SONET STS-N or STM-N frame basically contains N STS-1s or STM-1s, respectively.

STS-3 can be thought of as a multiplex of three STS-1 bit streams in a single channel so that the resulting channel is three times the rate of an STS-1 signal. When streams are multiplexed in this manner, the multiplexing is done at an octet level rather than a bit level (see Figure 11-10). The resulting channel is one where octet A1 of the first STS-1 stream is transmitted, followed by octet A1 of the second STS-1 stream, and finally A1 of the third signal. The process repeats with A2 of all three streams being transmitted

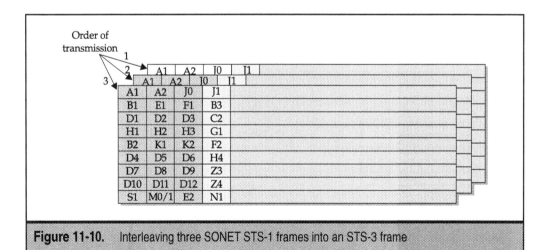

Figure 11-10.   Interleaving three SONET STS-1 frames into an STS-3 frame

consecutively, all the way up to the last octets of each STS-1 stream. This type of multiplexing is carried out for all data rates of SONET and SDH, including STS-192 and STS-768, the newest higher-rate version of SONET. Using this technique, SONET/SDH maintains the frame rate of 125 microseconds.

The resulting STS-3 frame maintains the same number of rows as those of an STS-1 frame (9 rows). The columns, however, are the sum of all three STS-1 frames, so the final structure of the STS-3 frame is 9 rows by 270 columns.

Concatenated Mode   Having a channelized structure for STS-3 and higher-level multi-plexes is useful only when one plans on utilizing these channels for specific applications, as is the case when the predominant traffic is voice. In many contemporary applications, however, the greater speeds offered by SONET/SDH are of more use if the entire bandwidth could be used as a single channel.

An example of one of these applications is ATM, a technology that is much better served by a single channel of the available bandwidth. Both SONET and SDH have the ability to *concatenate* a number of STS- or STM-1 payloads to create a single higher-speed payload. In such cases, a lower case *c* is used to indicate a concatenated mode of operation, so an STS-3c designation means that the STS-3 is being used in a concatenated mode with a single channel for a higher-speed payload. Thus, an STS-3c is a single payload (SPE) data stream of 150.336 Mbps and an STS-48c is an SPE of 2.405 Gbps.

Under normal operating conditions, the payload of an STS-1 frame would have a single column of payload overhead (POH). In the concatenated mode of operation for an STS-3c frame, there is only one column of payload overhead even though the signal is a concatenation of three STS-1 frames, each having its own POH. The fact that STS-3 is the first logical step up from an STS-1 frame means that a single column of payload overhead can be used to represent the three concatenated channels. However, higher levels of concatenation such as STS-12c use *stuff columns,* which are filler columns that are not customer data. Figure 11-11 shows these filler columns after the POH column and provides an equation—$(N/3) - 1$—that can be used to determine when these stuff columns are used and how many are used. Using the equation, an STS-3c frame would yield 0 stuff channels:

$3/3 = 1$
$1 - 1 = 0$

However, with an STS-12c frame:

$12/3 = 4$
$4 - 1 = 3$

so three stuff columns are used in an STS-12c frame. It also follows that stuff columns are present only when N (in STS-Nc) is greater than 3.

To ensure the foregoing is understood in the correct context, we will summarize the high points of the discussion. A concatenated channel has a single SPE, which cannot be used to transport DS1s or DS3s; that is, for what the channelized version—STS-3—was designed. The unchannelized structure, or the concatenated payload, was created for

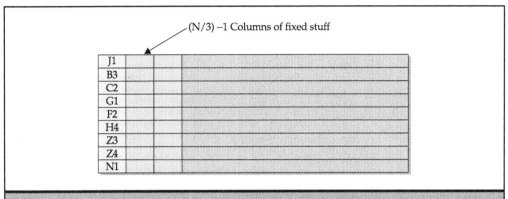

Figure 11-11. Concatenated payload with formula for when stuff columns are used

applications such as the transport of ATM cells. The SDH hierarchy starts with the STS-3 or STS-3c structure, but in this case they are called STM-1 in both instances (a lowercase *c* is not appended for concatenated frames). In SDH, more information is provided in the frame structure to tell the receiver that the payload is concatenated.

Virtual Tributaries The STS-1 signal was specifically designed to provide transport for a DS3 channel. The lower increments of DS3, such as the 1.544 Mbps of a DS1, or the 2.048 Mbps of a CEPT signal (European E-1 equivalent of the T1), are supported by designating a portion of the STS-1 frame as a *virtual tributary (VT)*. In the context of SONET/SDH, subrates or lower rates usually refer to data rates below DS3. A subrate payload is just a payload that does not map directly to an STS-1 but requires virtual tributaries.

The STS-1 frame can operate either without VTs or with seven VT groups (VTGs), but not both. This means that the frame can be divided into VTs of different sizes, but it cannot be divided into five VTs, for instance, with the remaining cell being used for the transport of ATM cells. It is all or nothing. Four VT sizes are used to accommodate four different inputs:

▼ **VT-1.5** Three columns of 9 octets each (27 bytes). At 8,000 frames per second, this provides capacity for 1.728 Mbps, which is sufficient for transporting a T1 (1.544 Mbps DS1 signal). Twenty-eight VT-1.5s can be multiplexed into the STS-1 signal.

■ **VT-2** Four columns of nine octets each (36 bytes). At 8,000 frames per second, this provides a capacity of 2.304 Mbps, which is sufficient for the 2.048 Mbps of CEPT. Twenty-one VT-2s can be multiplexed into the STS-1 signal.

■ **VT-3** Six columns of nine octets each (54 bytes). At 8,000 frames per second, this provides the capacity for 3.456 Mbps, which is sufficient for a DS1C signal of 3.088 Mbps. Fourteen VT-3s can be multiplexed into the STS-1 signal.

▲   **VT-6**   Twelve columns of nine octets each (108 bytes). At 8,000 frames per second, this provides the capacity for 6.912 Mbps, which is sufficient for a DS2 of 6.176 Mbps. Seven VT-3s can be multiplexed into the STS-1 signal.

The foregoing should be enough to provide a basic understanding of the function of a VT. To gain a better grasp of how this works in practice, we will take a brief, but more detailed, look at how this all fits together. Now remember, the SPE of an STS-1 frame consists of 9 rows of 87 columns, or stated another way (in case you have forgotten what an SPE is), the payload of an STS-1 frame is 9 rows of 87 columns. The actual usable portion of the SPE is 86 columns, because one column is dedicated to POH, or payload overhead.

The 86 remaining columns of the payload is broken into 7 groups of 12 columns, or VTGs. However, 7 groups of 12 yield 84 columns, leaving 2 extra columns. Using the POH as column 1, the extra two 2 are placed at columns 30 and 59 and are blocked from use. The number of columns that are used for virtual tributaries, therefore, is 84, so the structure of an STS-1 SPE is really 84 columns by 9 rows when VTs are used, as shown in Figure 11-12.

The resulting total payload data rate when VTs are used is 48.384 Mbps; 84 columns by 9 rows yields a payload of 756 bytes or 6,048 bits. With a frame rate of 8,000 frames per second, the resulting speed is 6,048×8,000, which is equal to 48,384,000 bits per second.

The seven VTGs are interleaved into the 84 columns in the same manner described for interleaving STS-1 frames into higher levels of SONET.

Taking a closer look at one of the VTGs, it is made up of 12 columns (and 9 rows), equaling a gross rate of 6.912 Mbps calculated as follows:

12×9 = 108 bytes
Multiply this by 8 for the bit equivalent = 864 bits
Multiply this by 8,000 per second = 6,912,000 bits per second

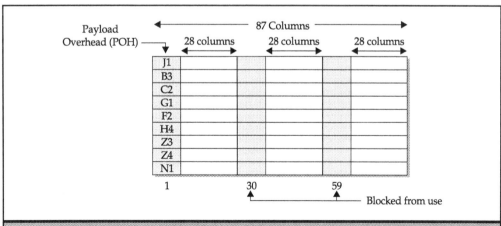

Figure 11-12.   SPE structure of STS-1 frame when virtual tributaries are used

Notice that this basic structure of 7 groups of 12 columns is the functional equivalent of VT-6. However, while 6.912 Mbps is fine for transporting a DS2, it is wasteful if it is used to transport a lower rate such T1, E1, or a DS1C. By subdividing the 12 columns, we are able to accommodate the different data rates of the plesiochronous digital hierarchy. So using a three-column group (instead of 12), we get a gross bit rate of 1.728 Mbps or a VT-1.5. Four columns yields a VT-2, and six columns a VT-3. Figure 11-13 provides a visual summary of the different groupings and interleaving.

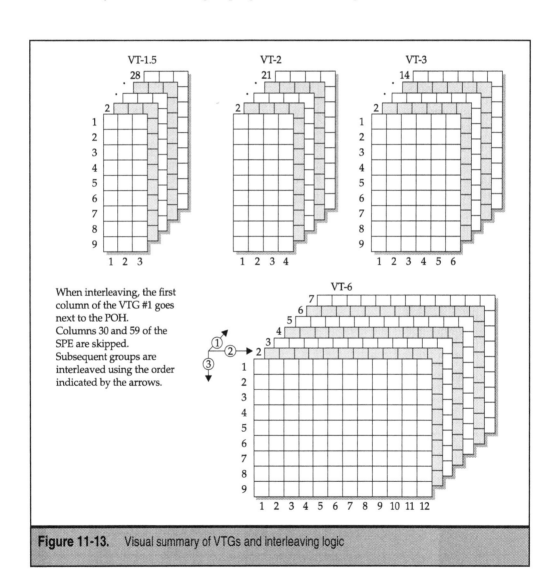

Figure 11-13. Visual summary of VTGs and interleaving logic

SONET/SDH Network Elements

Within a SONET/SDH network, there are many devices that perform different functions. An introduction to some of the more common elements of the network is appropriate, and so we will list the components and provide a brief description of the functions of each.

▼ **Terminal multiplexer** This device is an end-point device that is meant to reside at the customer premises but instead is mainly found in the central office. It functions as an entry-level path-terminating terminal multiplexer, acting as a concentrator for DS1s as well as other tributary signals.

■ **Regenerator** Used to redefine and amplify a fiber signal that has degraded because of attenuation. The optical signal is converted to the electrical domain for processing and then sent back to optical for forwarding. A regenerator clocks itself off the received signal and replaces the section overhead bytes before retransmitting the signal. The line overhead, payload, and POH are not altered.

■ **Add/drop multiplexer (ADM)** Provides interfaces between the different network signals and SONET signals and allows signals to be added or dropped off at any point in the network. At an add/drop site, only those signals that need to be accessed are dropped or inserted. The remaining traffic continues through the network element without requiring special pass-through units or other signal processing.

■ **Drop and repeat (D+R)** Also known as drop and continue—a key capability in both telephony and cable TV applications. With drop and repeat, a signal terminates at one node, is duplicated (repeated), and is then sent to the next and subsequent nodes.

■ **Digital cross-connect system (DCS)** A more sophisticated version of the ADM, it is used to access the STS-1 signals, accept various optical carrier rates, and perform switching functions. It is ideally used at a SONET hub. One major difference between a cross-connect and an add/drop multiplexer is that a cross-connect may be used to interconnect a much larger number of STS-1s.

■ **Digital loop carrier (DLC)** Used in what is known as a carrier serving area (CSA) to connect a high number of customers using ordinary copper wire. Chapter 4 provides a more detailed description of CSA and digital loop carriers. Figure 11-14 provides a visual summary of its application.

▲ **Matched nodes (MN)** Used to interconnect SONET rings, these devices provide an alternate path for the SONET signals in case of equipment failure. This feature is also known as signal protection.

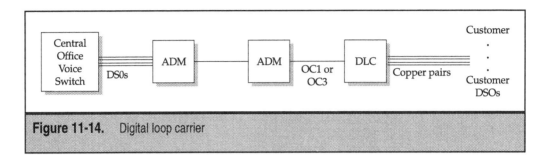

Figure 11-14.   Digital loop carrier

SONET/SDH Topologies

A SONET/SDH network can be deployed in a linear fashion, as depicted in Figure 11-15, illustration A; or as a ring, as depicted in illustration B of the same figure. When configured in a linear fashion, with the ADMs connected in a straight line, the fiber links between them may consist of either two fibers or four. With four fibers, one set can serve as a protection or backup pair, in the event of a failure. However, the weakness of this design lies in the fact that all four fibers can be cut, causing a disruption to the network. Of course, special care could be taken with the fiber runs to ensure path diversity, but this can become quite challenging and expensive very quickly.

By far, the ring topology is the most common SONET/SDH topology, especially in metro area networks. Linear topology is more common in the long-haul core network. The immediate benefit of a ring topology is its robustness. What makes it more robust than the linear topology is the fact that no single failure isolates the ring, as each node has

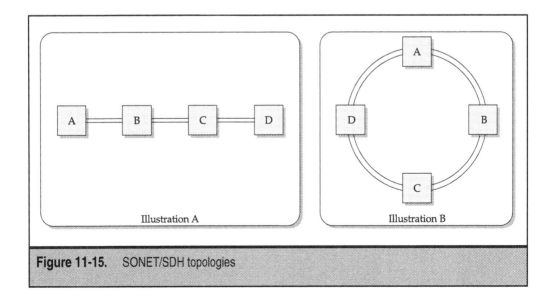

Figure 11-15.   SONET/SDH topologies

an implied second path from it to the other nodes. In other words, in a linear topology if all the fibers between nodes B and C of illustration A of Figure 11-15 are cut, there is no way for B to get to C. If, however, this happens using the ring topology depicted in illustration B of the same figure, B still has a path to C via nodes A and D. Ring systems are therefore relatively immune to interruption because of the multiple access path that is implied in the ring itself.

Types of Fiber Rings

Fiber rings are generally deployed as either a two-fiber ring or a four-fiber ring. A two-fiber ring can be operated as a unidirectional ring or a bidirectional ring. In the unidirectional mode of operation, a single fiber is used for the active ring with the second fiber being used for backup, as what is called the *protection fiber,* if the active fiber fails. During normal operation, all communication flows over the active fiber in a single direction as illustrated in illustration A of Figure 11-16.

The weakness of a unidirectional ring is the propagation delays between transmitting and receiving traffic between two nodes. Using Figure 11-16 as an example, node B is able to send traffic to node A directly. However, traffic flowing from node A to B must be routed via nodes C and D, introducing propagation delays.

With a bidirectional two-fiber ring, as shown in illustration B of Figure 11-16, both fibers are active, each being used for sending traffic in a different direction. On the outer first ring, traffic is sent counterclockwise, and on the inner ring, clockwise. In this configuration, there is no concept of an active ring and a protection ring, because the two fibers are working fibers. Using this configuration, propagation delay between transmit and receive traffic is not as much an issue, as communication between A and B flows over the

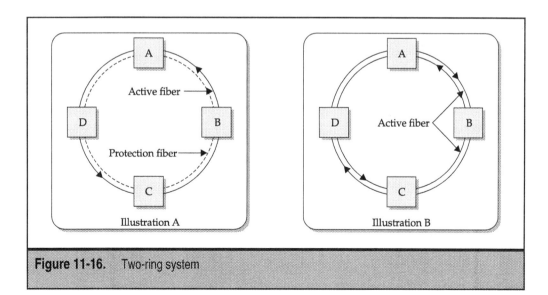

Figure 11-16.   Two-ring system

two fibers connecting them. In this configuration, each fiber in either direction can be used at only half its capacity because the second half is reserved for backup.

Four-fiber rings are always operated as bidirectional rings. In this configuration, one pair of fibers is active and the other pair is used as protection fibers. Unlike in two-fiber bidirectional systems that can use only half the capacity of the fiber, the full capacity of the working pair in a four-fiber system can be used. With four fibers, full link recovery is possible if one or two or even four fibers are cut.

Two backup systems, or ring protection methods, are used in fiber: path switching and line switching.

Path Switching Systems

Path switching can be implemented in either a unidirectional or bidirectional ring. Path switching is always implemented when using a unidirectional ring, where it is known as a *unidirectional path-switched ring (UPSR)*. In this system, traffic is transmitted in both directions on the ring by using one direction on the active fiber and the other direction on the protection fiber.

Within the network, ADMs are the entry and exit points for traffic entering and leaving the ring. As is shown in Figure 11-17, the two fibers are connected to the receive and transmit portions of the ADM. The traffic that enters or leaves the ring can be any signal ranging from a DS1 up to a much higher PDH or SONET signal. If the signal is a DS1, the ADM would put it into a virtual tributary, which would then be put into a virtual tributary group, which would be placed in an STS-1 SPE, and this STS-1 might be further multiplexed into a higher-level signal such as an STS-12. The DS1 would be transported to an ADM at the point where the DS1 signal is meant to exit the ring. At this point, the DS1 would be removed from the ring and the VT carrying the DS1 traffic would be the path of the connection.

The path could also be another type of signal. For instance, if the traffic is ATM cells that have been placed into a concatenated STS-3c for presentation to the ADM, the ADM could conceivably multiplex the STS-3c into an STS-48. In this case, the STS-3c is considered the path as far as the SONET/SDH ring is concerned. The point of this is that for path switching to work, both the entry and the exit nodes for a path must operate at the same level.

In UPSR, the path is sent over both counter-rotating rings (active and protection fibers) from the transmit side of the ADM, as shown in Figure 11-17. On the receive side, both fibers are monitored and the best traffic is selected. If an error is detected, switching between fibers is immediate, since the receiver is monitoring both paths on both fibers. Additionally, the receiver, without any assistance from the transmitter, accomplishes restoration of the ring. Because no assistance is needed from the receiver, no APS *(automatic protect switch)* communication channel is needed in UPSR.

The foregoing description is valid for a fiber cut, one of the most common failures of SONET/SDH rings. If an ADM fails, or if multiple fiber cuts remove an ADM from the ring, the loss of the ADM would be detected by the ADMs immediately connected to it. These ADMs must put out an *alarm indication signal (AIS)* on each path that originates from, or terminates at, the failed node. This is an immediate indication to the device at the other end of the path that something is wrong.

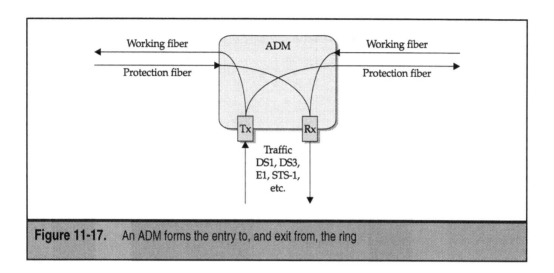

Figure 11-17.   An ADM forms the entry to, and exit from, the ring

Line Switching Systems

As previously stated, bidirectional rings can be either two-fiber or four-fiber rings. We also determined that in a two-fiber bidirectional ring, only half the capacity of the fibers can be used, because the other half is set aside for protection. In contrast, a four-fiber ring has no such limitation and the full capacity of the fiber pair may be used, because the remaining pair is reserved for protection.

In a four-fiber ring, two methods of recovery are possible. If the failure is limited to a single fiber, the intelligent SONET equipment recognizes the failure and switches traffic from the working pair to the protect pair between the affected SONET nodes only. This type of protection is referred to as a *span switch,* and it occurs in less than 50 milliseconds. If multiple fibers fail, software resident in the SONET equipment nearest the failure recognizes the loss of signal and switches traffic automatically from the working pair to the protect pair and sends the signal in the opposite direction around the ring to the terminating end. This type of switch, referred to as a *line switch,* occurs in less than 200 milliseconds.

The only recovery method in a two-fiber ring is a line switch (also called a *ring switch*), sending data in the opposite direction over the two fibers. As a result, this recovery mechanism is called a *bidirectional line-switched ring,* or BLSR.

Figure 11-18 shows a four-fiber ring that has a failure of multiple fibers between nodes A and B. The failure of the fibers is detected by both nodes A and B through the loss of a signal on both working and protection fibers. These two nodes will send a signal backward using bytes K1 and K2 of the line overhead. These two bytes are called the *automatic protection switching (APS) channel.* ADMs C and D pass these octets on, since they are not addressed to them. When nodes A and B receive the signals, they bridge the signals as shown in Figure 11-18.

Once this bridge has been formed, the signals arriving on fiber 1 at ADM B would ordinarily be transmitted to ADM A on fiber 1. Now, the signal is put on fiber 4, which

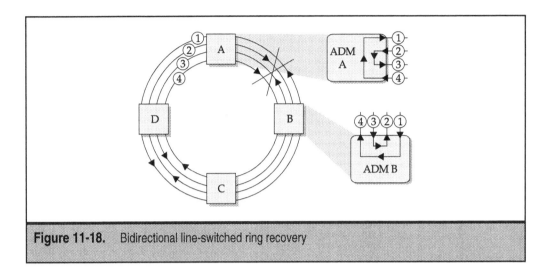

Figure 11-18. Bidirectional line-switched ring recovery

carries it in the reverse direction to ADM A. In turn, ADM A bridges the signal from fiber 4 onto fiber 1. The end result is that the signal from ADM B ends up on fiber 1 at ADM A, the fiber across which it normally arrives.

The reverse is also true: signals from ADM A would normally arrive at ADM B on fiber 2. Using the same process just described, the signal is transported to fiber 2 on ADM B after it arrives on fiber 3.

ASYNCHRONOUS TRANSFER MODE

The telephone network was developed for, and has evolved from, a single fundamental requirement: the transport of voice traffic and the switching of these voice calls. As a circuit-switched network, the PSTN is inefficient at transporting data, as it has a different traffic profile than voice. The development of packet switching improved on these inefficiencies by introducing features that were more suited to data applications.

The major inefficiency of the PSTN network as it relates to the transfer of data lies in the fact that the channel capacity is dedicated to a connection even when no data is being transferred. The development of packet switching addressed this and allowed communication between ports of different data rates through the technique of packaging the data into packets and combining it with extra information. By dividing the data into packets, the channel could be shared by multiple applications as long as each packet contained the appropriate information to identify the application to which it belongs and the sequence in which it should be reassembled. However, the drawback with this approach, in addition to those already mentioned, is that the packet has to contain information that can be used to route it through the network and to assure that errors were not introduced along the way. All this contributes to a decrease in the throughput of the network.

Asynchronous Transfer Mode (ATM) was developed to bridge the gap between the requirements of applications that benefit from a circuit-switched approach and those that benefit from packet switching. Universally, ATM has been accepted as a technology that handles any kind of information—voice, data, image, text, and video—in an integrated manner. Many of the broadband access technologies discussed throughout the chapters of this book demonstrate that it has found a place in many different platforms, technologies, and protocols. A connection-oriented, packet-switching technique, ATM uses a fixed-length 53-byte cell with 48 bytes of payload and 5 bytes of overhead. The following are the requirements that ATM was designed to address:

▼ Support all existing devices as well as emerging services

■ Utilize network resources efficiently

■ Minimize the switching complexity

■ Minimize the processing and buffer requirements at intermediate nodes

■ Support very high transmission speeds

▲ Guarantee performance requirements of existing and emerging applications

As a high-speed packet-switching technology, ATM was designed to be the final network technology that we would ever need. It was built to function equally in a LAN, MAN, or WAN environment regardless of the application or traffic type. In reality, the deployment of the technology has not been as pervasive as it was once billed. Today, it is mainly found in wide area networks, and whereas it was once positioned as the ultimate networking solution, it is now being positioned as a technology that complements competing technologies.

Architecture and Basic Functions

The advent of fiber as a communications medium created the need for standards that eventually led to the development of SONET/SDH. During the same time frame, the concept of an intelligent fiber-based network that could efficiently transport different services was also being explored; it became known as a Broadband Integrated Services Digital Network (B-ISDN). After a period of study that lasted from 1984 to 1988, the ITU decided that the promise of B-ISDN could be realized through the use of SONET/SDH and ATM.

The standards for SONET/SDH defined how information could be packaged, multiplexed, and transmitted over the network, and they addressed issues of interoperability between different vendor equipment. This very limited scope of the SONET/SDH standards—limited from the perspective of not requiring any knowledge of the traffic that is being transported—made SONET/SDH ideal for transmitting existing and emerging traffic types. Missing, however, was a switching mechanism to complement SONET/SDH in the B-ISDN model—an intelligent switching fabric that had the capability to switch all types of traffic at very high speeds. ATM was determined to be that solution and was developed as the switching structure for B-ISDN.

The development of the ATM technology is based on the efforts of the International Telecommunication Union Telecommunication Standardization Sector (ITU-T) Study Group XVIII to develop the Broadband Integrated Services Digital Network (B-ISDN) for the high-speed transfer of voice, video, and data through public networks. The ITU I.121 recommendation, entitled *Broadband Aspects of ISDN,* defined the operation of ATM as it relates to the B-ISDN protocol reference model. Table 11-3 provides a reference for some of the more significant standards and specifications for B-ISDN and ATM.

ITU Standards	Description
I.211	B-ISDN Services
I.361	B-ISDN ATM layer cell transfer performance
I.362	B-ISDN ATM adaptation layer (AAL) functional description
I.363	B-ISDN ATM adaptation layer (AAL) specification
I.2100	B-ISDN signaling ATM adaptation layer (SAAL) overview description
I.2110	B-ISDN ATM Adaptation Layer—Service Specific Connection Oriented Protocol (SSCOP)
I.2130	B-ISDN signaling ATM adaptation layer—Service specific coordination function for support of signaling at the user-network interface (SSCF at UNI)
I.2140	B-ISDN ATM adaptation layer—Service specific coordination function for signaling at the network node interface (SSCF AT NNI)
Q.2610	Broadband Integrated Services Digital Network (B-ISDN) Usage of Cause and Location in B-ISDN User Part and DSS 2
Q.2650	DSS2/ B-ISUP Interworking Recommendation defines the relationship between DSS2 and Signaling System No. 7
Q.2931	B-ISDN (ATM) Signaling
Q.293x	Generic Concepts for the Support of Multipoint and Multiconnection Calls
Q.2961	Broadband Integrated Services Digital Network (B-ISDN) Digital Subscriber Signaling System No. 2 (DSS 2) Support of Additional Traffic and QoS Parameters
Q.2963	Broadband Integrated Services Digital Network (B-ISDN) Digital Subscriber Signaling System No. 2 (DSS 2) Connection Modification
Q.2964	B-ISDN Look-Ahead
Q.297x	Point-to-Multipoint

Table 11-3. ATM Standards

ITU Standards	Description
Q.298x	Broadband Integrated Services Digital Network (B-ISDN) Digital Subscriber Signaling No. 2 (DSS2) User Network Interface Layer 3 specification for Point-Point Multiconnection Call Control

ANSI	Description
T1.403.01-1999	ISDN Primary Rate—Customer Installation Metallic Interfaces, Layer 1 Specification
T1.511-1997	B-ISDN ATM Layer Cell Transfer-Performance Parameter
T1.627-1993 (R1999)	B-ISDN ATM Layer Functionality and Specification
T1.629-1999	B-ISDN ATM Adaptation Layer 3/4 Common Part Functions and Specification
T1.630-1999	B-ISDN ATM Adaptation Layer Constant Bit Rate Service Functionality and Specification
T1.634-1993	Frame Relay Service Specific Convergence Sublayer
T1.635-1999	B-ISDN ATM Adaptation Layer Type 5 Common Part Functions and Specification
T1.636-1999	B-ISDN Signaling ATM Adaptation Layer Overview Description
T1.637-1999	B-ISDN ATM Adaptation Layer—Service Specific Connection Oriented Protocol (SSCOP)
T1.638-1999	B-ISDN Signaling ATM Adaptation Layer—Service Specific Coordination Function for Support of Signaling at the User-to-Network Interface (SSCF at the UNI)
T1.640-1996	B-ISDN Network Node Interfaces and Inter Network Interfaces—Rates and Formats Specifications
T1.644-1995	B-ISDN—Meta-Signaling Protocol
T1.645-1995	B-ISDN Signaling ATM Adaptation Layer—Service Specific Coordination Function for Support of Signaling at the Network Node Interface (SSCF at the NNI)
T1.646-1995	B-ISDN Physical Layer Specification for User-Network Interfaces Including DS1/ATM
T1.648-1995	B-ISDN User Part (B-ISUP)
T1.652-1996	B-ISDN Signaling ATM Adaptation Layer—Layer Management for the SAAL at the NNI

Table 11-3.   ATM Standards *(continued)*

ANSI	Description
T1.654-1996	B-ISDN Operations and Maintenance Principles and Functions
T1.656-1996	B-ISDN Interworking between SS7 Broadband ISDN User Part (B-ISUP) and ISDN User Part (ISUP)
T1.657-1996	B-ISDN Interworking between SS7 Broadband ISDN User Part (B-ISUP) and Digital Subscriber Signaling System No. 2 (DSS2)
T1.658-1996	B-ISDN Extensions to the SS7 B-ISDN User Part, Additional Traffic Parameters for Sustainable Cell Rate (SCR) and Quality of Service (QOS)
T1.662-1996	B-ISDN ATM End System Address for Calling and Called Party
T1.663-1996	B-ISDN Network Call Correlation Identifier
T1.664-1997	B-ISDN Point-to-Multipoint Call/Connection Control
T1.665-1997	B-ISDN Overview of B-ISDN NNI Signaling Capability Set 2, Step 1
IETF	**Description**
RFC 1483	Multiprotocol Encapsulation over ATM Adaptation Layer 5 specifies two encapsulation methods for providing TCP/IP protocol suite support over ATM virtual circuits
RFC 1577	Classical IP and ARP over ATM specifies an initial application of IP and ARP in an ATM environment as simply replacing LAN "wires" with an ATM virtual circuit
RFC 1626	Default IP MTU for use over ATM AAL5 specifies (and provides the reasons for) the IP MTU over ATM/AAL5 virtual circuits as 9,180 bytes
RFC 1754	IP over ATM Working Group's Recommendations for the ATM Forum's Multiprotocol BOF Version 1 represents the initial requirements given to the ATM Forum's Multiprotocol BOF for operating the TCP/IP protocol suite over ATM networks
RFC 1755	ATM Signaling Support for IP over ATM specifies the ATM call-control needed in order to support the TCP/IP protocol suite over ATM networks
(IPATM Draft)	IP over ATM: A Framework Document
(IPATM Draft)	Support for Multicast over UNI 3.1 based ATM Networks

Table 11-3. ATM Standards *(continued)*

ATM Reference Model

Figure 11-19 shows the B-ISDN reference model, which is now recognized as the ATM reference model. This model depicts the logical functionality that is supported by the technology. The functionality of ATM spans the physical layer and part of the data link layer of the OSI model.

As is evident from the figure, the ATM model is composed of multiple layers and planes. The layers of the reference model consist of the following:

▼ **Physical layer**   The ATM physical layer manages the medium-dependent functions for transmission and is logically divided into two sublayers: the physical media dependent (PMD) sublayer and the transmission convergence (TC) sublayer.

■ **ATM layer**   This layer is responsible for establishing connections, cell multiplexing, and switching functions, enabled through the data contained in the cell header.

▲ **ATM adaptation layer (AAL)**   This layer is responsible for isolating higher-layer protocols from the details of the ATM processes and is not required on ATM end systems or needed on the nodes that are doing a pure routing function, such as ATM switching nodes. This layer is further subdivided into two logical sublayers called the segmentation and reassembly (SAR) sublayer and the convergence sublayer (CS). As the name implies, the SAR is responsible for the segmentation of the data units into cells and the reassembly of the cells into their original data unit structure. The CS is responsible for ensuring that different traffic types receive the classes of service that they require. The combination of the ATM layer and the AAL is roughly the same as the data link layer of the OSI model.

The higher layers represent the different B-ISDN services.
The reference model is also divided into planes that span all layers:

▼ **Control plane**   Responsible for generating and managing signaling requests.

■ **User plane**   Responsible for management and transfer of data.

▲ **Management plane**   Consists of two components: *layer management,* which is responsible for the management of layer-specific functions such as detection of failures and issues resulting from problems with the protocol, and *plane management,* which is responsible for managing and coordinating functions of the entire ATM system.

In 1991, the ATM Forum—jointly founded by Cisco Systems, NET/ADAPTIVE, Northern Telecom, and Sprint—was established to accelerate the use of ATM and to promote awareness of its capabilities. By 1996, a full set of specifications, called the Anchorage

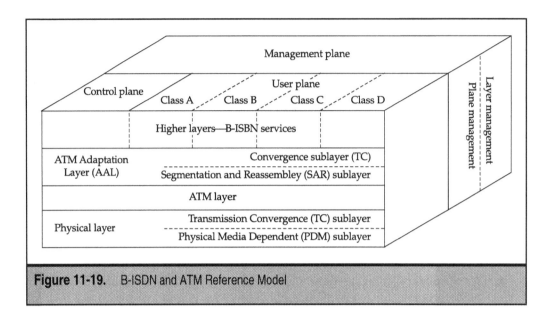

Figure 11-19.   B-ISDN and ATM Reference Model

Accord, was developed to build and manage an ATM infrastructure and ensure backward compatibility.

As shown in Figure 11-20, terminal systems or end points use all layers of the model. ATM switches, those that provide intermediate switching functions, use only the PHY and the ATM layers.

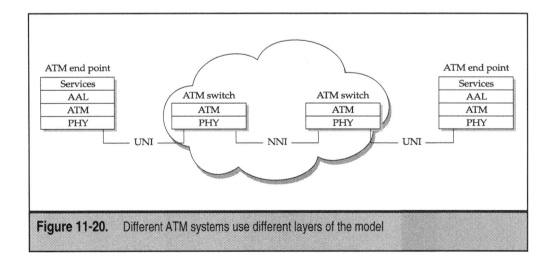

Figure 11-20.   Different ATM systems use different layers of the model

Physical Layer

The *physical media dependent sublayer (PMD)* is responsible for receiving and sending the transmission and for synchronizing a continuous flow of bits. Different media are supported: fiber with SONET/SDH, DS3, and E3, to name a few. The PMD is responsible for meeting the unique specifications of each.

The transmission convergence layer is responsible for the following:

▼ **Cell delineation**   The ability to determine and maintain cell boundaries

■ **Header error control sequence generation and verification**   Ensures the validity of the header by verifying the header error control

■ **Cell rate decoupling**   Inserts or suppresses unassigned ATM cells in order to adapt the rate of valid ATM cells to the payload capacity of the transmission system

■ **Transmission frame adaptation**   Packages ATM cells into frame formats that are acceptable to the particular physical medium

▲ **Transmission frame generation and recovery**   Generates and maintains the frame structure that is appropriate for the physical medium frame structure

ATM Layer

This layer is responsible for establishing connections and for multiplexing and switching cells through the network. We will cover the essentials of switching in a later section.

ATM Adaptation Layer

The AAL is responsible for taking the data that it receives and breaking them into the 48-byte payload of an ATM cell. The user data that it receives, voice, video, IP, Appletalk, etc., are called *service data units (SDU)* and are usually a lot larger than the required size of an ATM cell. Several types of ATM adaptation layers are specified, and each maps to a different service class as defined by B-SDN. Another way of viewing ATM adaptation layer is as the layer that adapts the needs of different applications to what the ATM layer actually offers.

B-ISDN defines four service classes, called Classes A through D. Each service class defines different types of services, which map to the different adaptation layers. The following paragraphs provide an additional level of detail of each of the adaptation levels, and Table 11-4 shows how each adaptation service maps to the different service classes. Finally, Table 11-5 provides a quick summary.

ATM Adaptation Layer 1 (AAL1)   ATM adaptation layer 1 operates in circuit emulation mode, providing a connection-oriented service that is appropriate for applications like voice or video that require a constant bit rate . Because of the nature of the traffic supported by this mode, AAL1 requires timing synchronization between the source and the destination. As a result, this mode of operation is dependent on a medium that supports clocking, such as SONET/SDH.

Class	Adaptation	Service
Class A	AAL1	Constant bit rate circuit emulation, e.g., legacy telephony
Class B	AAL2	Variable bit rate, e.g., packetized audio/video
Class C	AAL3/4 and 5	Connection-oriented data services
Class D	AAL3/4 and 5	Connectionless data services

Table 11-4. Service Classes and Adaptation Levels

Cells are prepared for transmission in three steps. In the first step, synchronous samples are inserted into the payload. The second step involves adding a sequence number (SN) field and a sequence number protection (SNP) field to provide information that can be used to verify and reassemble the data stream by the receiving AAL1 process. The third and final stage is to fill the remainder of the payload with enough single bytes to reach the maximum capacity, 48 bytes. This process is illustrated in Figure 11-21; notice that the SN and SNP are placed into the payload immediately after the header

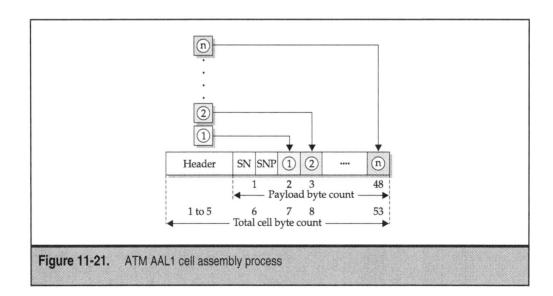

Figure 11-21. ATM AAL1 cell assembly process

information. The SN and the SNP occupy a single byte, and the remaining 47 are dedicated for use by the payload.

ATM Adaptation Layer 2 (AAL2)   AAL2 defines a traffic type that is similar to AAL1 constant bit rate but tends to be more bursty in nature. This traffic type is called variable bit rate (VBR) traffic. This includes services that are characterized as packetized voice and video and that are not dependent on a strict constant rate but do require a constant bit rate. In other words, services require a constant bit rate but are more tolerant than the strict bit rate required for circuit emulation.

To further qualify, AAL2 was designed to adapt the capabilities of ATM to the traffic requirements of low and variable bit rate applications such as compressed voice used in cellular environments. In this application, packetization delay and efficiency in the use of bandwidth resources are two important considerations. AAL2 achieves low packetization delay by supporting variable length packets, thereby allowing the application to choose the packet size that is most convenient for its delay requirements. The lower the size, the lower the delay.

High bandwidth efficiency is achieved through the use of multiplexing. Up to 248 AAL2 connections may be multiplexed in a single ATM virtual channel connection. Instead of having partially filled ATM cells, AAL2 fills the cell with AAL2 packets from several active AAL2 connections. The AAL2 process uses 44 bytes for payload and reserves 4 bytes for the AAL2 process.

VBR traffic has the further distinction of being either real-time (VBR-RT) or non-real-time (VBR-NRT). Both types are supported by AAL2.

ATM Adaptation Layers 3/4 (AAL3/4)   AAL3/4 supports both connection-oriented and connectionless data traffic. Designed for network providers, it is closely aligned with the Switched Multimegabit Data Service (SMDS) and is used to transmit SMDS packets across an ATM network. The following describes the four steps that are used in preparing a cell for transmission:

1. The convergence sublayer creates a protocol data unit (PDU) by prepending a beginning/end tag header to the cell and appending a length field as a trailer.

2. The segmentation and reassembly sublayer fragments the PDU and prepends a header to it.

3. A trailer that consists of a CRC is also appended to each PDU fragment.

4. The final step is to use the SAR PDU as the ATM cell payload. Figure 11-22 illustrates the process.

The SAR PDU consists of type, sequence number, and multiplexing identifier fields. The type field identifies whether or not the cell is the beginning, continuation, or end of a message. The sequence number fields are used to reassemble the data unit, and the multiplexing identifier identifies different cells from different traffic sources interleaved

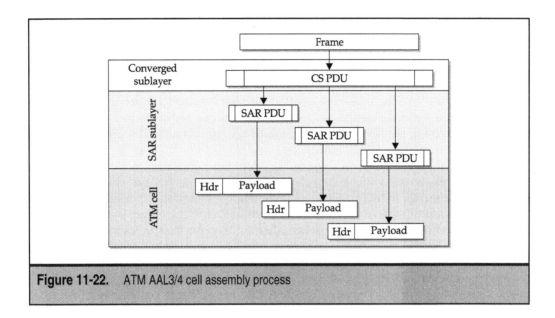

Figure 11-22. ATM AAL3/4 cell assembly process

within the same virtual circuit connection so that the correct cells are reassembled at the destination.

ATM Adaptation Layer 5 (AAL5)   AAL5 supports both connection-oriented and connectionless traffic and is the main AAL for data. This is the adaptation layer that is used to transport IP over ATM and ATM LAN Emulation. It is also known as the simple and efficient adaptation layer, or SEAL, because the SAR simply accepts the PDU from the CS and segments it in 48-byte SAR-PDUs without reserving any bytes in each cell.

Cells are prepared for transmission in three steps. The CS sublayer appends a variable-length pad and an 8-byte trailer to a frame. The pad is used to ensure that the resulting PDU falls on a 48-byte boundary. The trailer consists of a 32-bit CRC that is calculated on the entire PDU and the length of the frame. The SAR sublayer then segments the CS-PDU into 48-byte blocks that form the payload of the ATM cell. Messages cannot be interleaved as is possible in AAL3/4 because additional headers and trailers are not added at this step to identify different cells from different sources. In the final step, each block is placed into the payload of the ATM cell. Setting a bit in the PT field to 1 indicates the last cell that represents the end of the original frame. All other frames have this bit set to 0.

Table 11-5, provides a quick reference of the different ATM Adaptation Layers and Classes.

Service Class	Adaptation	Characteristics	Traffic Profile	End-End Time Synchronization	Traffic Example
Class A	AAL1	Suitable for traffic that is sensitive to both cell loss and delay. Intended to replace fractional and full T1 and T3 services but has a far greater range of available speeds. Uses one payload data byte for sequence numbering (leaving only 47 per cell for the data).	Isochronous constant bit rate services	Required	Digitized audio and video
Class B	AAL2	Similar to AAL1 constant bit rate but for more bursty traffic that includes services such as packetized voice and video. One main user of AAL2 is compressed voice (cellular systems). Uses 44 bytes for payload and reserves 4 bytes for the AAL2 process. Traffic type has the further distinction of being either real-time (VBR-RT) or non-real-time (VBR-NRT).	Isochronous variable bit rate services	Required	Packetized voice and compressed video
Class C and D	AAL3/4	Originally intended as two layers: one for connection-oriented services (such as Frame Relay) and one for connectionless services (such as SMDS). It was later decided that a single layer could do both. Intended for traffic that can tolerate delay but not cell loss. Uses four payload data bytes for error detection. Supports multiplexing of ATM cells.	Variable bit rate	Not required	Data LAN file transfers

Table 11-5.   Service Class Feature Summary

Figure 11-23 shows the relationship of the convergence sublayers to the types of traffic handled by ATM.

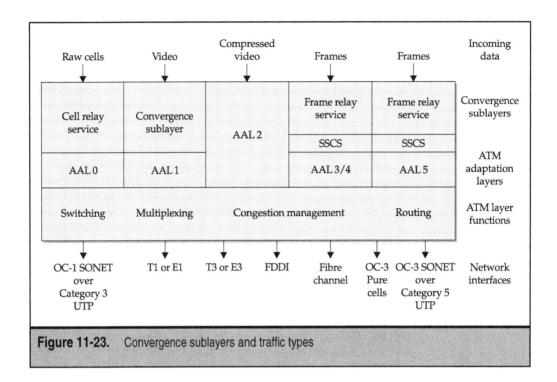

Figure 11-23. Convergence sublayers and traffic types

Basic Concepts

The constant transmission delay and guaranteed capacity of circuit switching are maintained in ATM. These features are combined with the flexibility and efficiency of packet switching networks, which are more suited to intermittent and bursty traffic. The ATM cell consists of 5 bytes of header and 48 bytes of payload for a total size of 53 bytes. Using the information contained in the header, ATM switches are able to determine the correct output port that connects it to the next switch in the path to the cell's ultimate destination.

Based on asynchronous time division multiplexing, ATM uses time slots that are allocated and are available on demand to user data. This is an important distinction of ATM. In our review of TDM in a previous chapter, we concluded that it was less efficient than statistical multiplexing because it results in wasted bandwidth. In our discussion, we determined that if there is nothing to send, the time slots in TDM are transmitted empty and cannot be used by another application. The flip side of this are those instances where an application has a lot to send but is limited by the number of time slots to which it has been allocated. In this case, the application has to transmit at the pace at which its time slots become available even though the remaining time slots may be empty because of no activity from the other applications that are sharing the link.

With ATM, time slots are used but are made available on demand. This means that if that same application has a lot to send, it will not be limited to the time slots to which it is assigned but will be allowed to use any available time slots.

The decision to use a fixed-length cell, 53 bytes long, was based on a compromise between the requirements of voice traffic and those of data. In fact, the requirements of voice and data in themselves have nothing to do with the packet sizes, which instead are based on the effects the characteristics of one traffic type have on the performance of the other. Voice samples are short and of a fixed length, while data has variable lengths that ranges from short to relatively very large when compared to voice samples. Under normal circumstances, this is not really a problem, but when one looks at the latency requirements of voice, it becomes a lot clearer why mixing the two on an unchannelized link could be a problem.

If voice were not sensitive to delay and did not have a requirement for packets to arrive at the destination within a predictable time frame, it really would not matter if it experienced delays along the way as a result of a much longer data packet. Imagine the unchannelized link as a major highway, the voice packet, a car, and the data packet, a train of variable length. Now this is a stretch, because in reality trains do not travel on highways, but for the purpose of the analogy, let's pretend they do. Let us assume that both the car and the train are traveling at the same speed, that both use the same entry ramp to enter the highway, and that the train arrives at the entry ramp ahead of the car. The length of time it takes the car to gain entry to the highway would be dependent on the length of the train. The shorter the train, the quicker the car gains access; the longer the train, the more delay the car experiences. This flies in the face of what is required for legacy voice traffic, that each packet must arrive at its destination within a predictable time frame.

However, if the trains were actually designed to be the same length as the car, the length of delay at the entry ramp would be a lot more predictable. This is the logic behind using smaller cells for ATM; the smaller cell size provides a more predictable delay for legacy voice or any other isochronous (time dependent) applications. The downside to this approach is that it can be inefficient for data transfers, especially at lower speeds, but this is the trade-off of compromise.

It should be noted that there is another set of reasoning that could be applied to the car and train analogy. The problem could be addressed by speed. If the train ran faster, that would have the same effect in lowering delays as decreasing the size at the lower speed. This is an argument that is used in support of using speed to address the problem. If technologies like DWDM can address the problem by using speed, it is argued that there will be no need for technologies like ATM.

The ATM Interfaces and Cell Structure

Like Frame Relay, ATM defines two interfaces, the User-Network Interface (UNI) and the Network-Network Interface (NNI). The UNI interface defines the interface between a user device or end system such as a host or a router and the ATM network. The NNI connects two ATM switches. The UNI and NNI concepts involve a further distinction that differentiates between a private ATM network, owned and operated by an enterprise, for instance, and a public ATM network operated by a public carrier.

A private UNI connects an ATM end device and a private ATM switch. A public UNI connects an end device or a switch, which may be privately owned, to a public ATM switch. A private NNI connects two ATM switches that are privately owned by the same organization, and a public NNI is one that connects two publicly owned switches.

Figure 11-24 shows the different interfaces and how they relate to each other. It also shows an additional interface called the *broadband intercarrier interface (B-ICI)*. This interface describes the connection between two public switches that are owned and operated by two different service providers.

The ATM standards define two cell structures for UNI and NNI connections. The fields of both UNI and NNI cell headers are shown in Figure 11-25. The difference between the two is the absence of the GFC in the NNI cell header. The following describes the fields and their meanings:

▼ **Generic Flow Control (GFC)**   Used to provide local functions such as the identification of multiple stations that share a single ATM interface. This field, which is usually not used, was envisioned as providing contention resolution and simple flow control for shared medium-access arrangements. This field is present only in the UNI cell and is absent in the NNI cell. In the NNI cell, the bits normally used by this field are allocated to the VPI field.

■ **Virtual Path Indicator (VPI)**   Used in conjunction with the VCI to identify the destination of the cell as it passes through the ATM network. In an NNI cell, 4 bits (taken from the GFC field in a UNI cell) are added to this field, which allows ATM switches to assign larger VPI values.

■ **Virtual Channel Identifier (VCI)**   Used in conjunction with the VPI to identify the destination of the cell as it passes through the ATM network. The switching function of the ATM network is dependent on these two fields.

■ **Payload Type (PT)**   Used to indicate multiple things. The first bit is used to indicate whether or not the cell contains user data or control information. If the

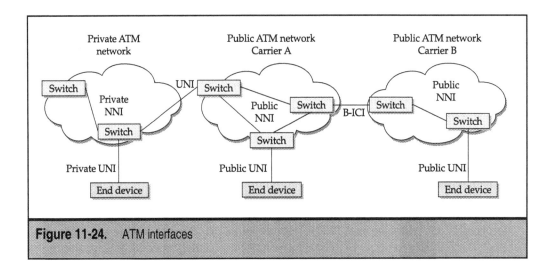

Figure 11-24.   ATM interfaces

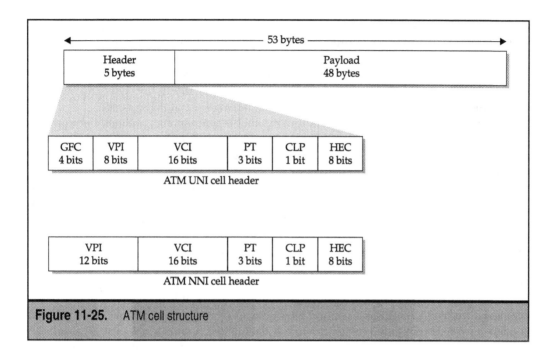

Figure 11-25.   ATM cell structure

cell contains user data, the second bit indicates congestion and the third bit indicates whether or not the cell is the last cell in a series of cells that represent a single AAL5 frame.

■ **Cell Loss Priority (CLP)**   Used to indicate if the cell should be discarded if network experiences extreme congestion. This bit is akin to the DE bit of Frame Relay.

▲ **Header Error Control (HEC)**   A checksum that is calculated on the first four bytes of the header.

The ATM Network and Switching Functions

ATM networks use a *virtual channel (VC)* between two points across which data flows. This connection must be established before any data can flow between the end points. Using the concept of virtual channels, ATM defines three different types of service: *permanent virtual circuit (PVC), switched virtual circuit (SVC),* and a *connectionless* service.

Like a PVC in Frame Relay, a PVC in ATM circuit is akin to a leased-line connection where the two end points are permanently linked. The advantages of a PVC include the guarantee of a connection between the two end points without the need for a process to set up and tear down the connection. However, the disadvantage is the fact that a PVC requires that every piece of equipment between the source and the destination be provisioned, which sometimes can be a manual process. Furthermore, no network resilience is available with a PVC in ATM. This may appear to contradict what was stated about Frame Relay PVCs in a previous chapter. Frame Relay defines the interface between a user and the network (UNI) and between a network and another network (NNI). Note the subtle difference; an ATM NNI defines the interface between ATM switches. The implication of this is that Frame Relay does not actually define how the switches communicate and how the data flows across the network—but ATM does.

By having a logical end-to-end relationship that is permanently defined between the end points, any cells placed on the logical connection (the PVC) have only a single option for a destination, the site that was predefined at the end of the logical channel. Call setup, therefore, is not required.

SVCs are functionally equivalent to a telephone call and unlike a PVC require some type of call setup. Whereas the end point of a PVC has the option of a single destination, an SVC has the capability to send a cell to any other end point. This approach is inherently more flexible than a PVC, but ATM, being fundamentally a connection-oriented service, places a requirement for a call to be set up to establish a connection between any two end points wishing to communicate. This connection remains established for the period of the session. This is not to say that the connection has to be active for it to remain in place.

Just as it is possible to place a phone call and not speak, it is possible to have an SVC established without the transfer of any user data. Once the end points decide to end the session, a process is initiated to tear down the established connection, a process similar to hanging up the phone. Once the call has been terminated, the logical path between the end points ceases to exist.

The connectionless service eliminates the need for call setups. With this type of service, the end points have a single connection to a *connectionless server (CLS)* in the ATM network. With a single connection, the end points are able to direct cells to any number of destinations. The deployment of connectionless servers throughout the network forms a virtual connectionless network that is overlayed on top of the connection-oriented ATM network. These CLSs, which are strategically connected to, or incorporated in, a subset of ATM switches, act as packet routers routing packets to other CLSs until the destination is reached. The topology of the virtual connectionless network may take the form of a ring, mesh, or star depending on the desired efficiency and robustness.

Switching Versus Routing

Before we get too far into the mechanics of switching within an ATM network, it is probably useful to review the basic difference between routing and switching. In a routed network, the routing information is carried in the header of each packet, and as result, the routing decision can be made on the fly. As a packet enters a router, the router is able to read the content of the header and, given the knowledge it has of the network at that moment in time, choose the path that is most appropriate. In a switched environment, the decision on the path taken through the network must be predetermined. The content of the packet, or the cell in this case, has a limited subset of the information that is carried in a packet that traverses a network that is designed for dynamic routing. This subset of information, five bytes in the case of ATM, carries the bare minimum information required for the switch to determine the appropriate interface to send the cell on.

So to summarize, in a network that is based on routing, the decision on the path through the network is done on the fly, but in one based on switching, it is predefined through means that may be manual or dynamic in nature. The net difference between these two approaches on the order in which packets arrive at the destination is significant.

In routing, packets entering the network in a given sequence are not guaranteed to arrive at the destination in the sequence in which they were sent. This fact, which is a feature of connectionless systems, adds overhead to the process of receiving these packets, as additional buffer space and processing are required to compensate for the packets being delivered out of sequence. Additionally, the time-out value of the receiving application must be balanced between the time it takes to recognize that a packet has been lost in the network and the time it takes to request a retransmission of the packet.

In switching, the transmission sequence of the packets is preserved. Because the path is predefined, packets entering the network in a particular sequence traverse and exit the network in the same order in which they entered. If a packet, or cell, is lost in the network, the receiving application is immediately aware because it expects to receive the packets in a given sequence. So if it receives packets 1, 2, and 4, it immediately knows that packet 3 is missing and may send an immediate request for retransmission.

ATM Switching

ATM, being fundamentally a connection-oriented technology, requires a logical path through the network to be established before data can flow. The logical path that it defines is called a *virtual channel (VC)*, functionally equivalent to a virtual circuit.

Within an ATM network, two types of connection exist, a *virtual path (VP)*, which is identified through a *virtual path identifier (VPI)*, and a virtual channel that is a combination of a VPI and a *virtual channel identifier (VCI)*.

A bundle of VCIs that share a common VPI creates a virtual path that is switched across the ATM network. The VPI and VCI are only locally significant to a specific link and are remapped as they pass through various switches within the path to the destination. The bundling of different VPs is called a *transmission path (TP)*, which describes the physical media that transport the VPs and VCs. Figure 11-26 illustrates the relationship of VCs, VPs, and a TP.

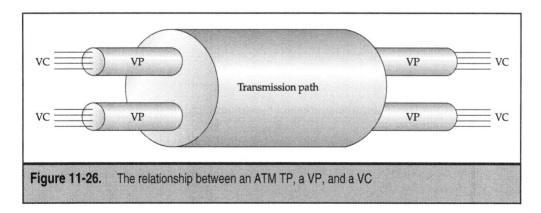

Figure 11-26.   The relationship between an ATM TP, a VP, and a VC

Conceptually, the VCI can be considered as the element that identifies different end-point devices in a particular site in the same way a room number identifies a specific room within a specific building. The VPI can be considered as the element that identifies the specific site in the same way the building number identifies a specific building. Within the building (VPI), there are many rooms, each having its own room number (VCI).

As a cell arrives at a switch, the switch looks at the VPI and the VCI that are contained within the header of the cell. Using a translation table that defines the appropriate outgoing ports for a particular VPI/VCI pair, the switch does two things. After determining the correct output interface as defined in the translation table, the switch replaces the VPI/VCI information with a new set that corresponds to the outgoing link and then sends the cell out the appropriate interface. The next switch, on the other side of the outgoing link, receives the cell with the new VPI/VCI value and uses that to determine its own predefined outgoing interface, and the process is repeated.

To better describe the process, we will track a cell as it enters and traverses an ATM network, but before we do that, we need to consider another key point about ATM. Before the traffic enters the network, it must have a contract for delivery through the network. The following section describes this concept.

The Traffic Contract

Each connection through the network has an associated contract for delivery of traffic through the network. The traffic contract specifies the negotiated characteristics of the connection. The terms of the contract that must be addressed are as follows:

▼   The traffic must have, at a minimum, a description that describes the maximum cell rate, known as the PCR, or peak cell rate, that can be sent over the connection in certain intervals or bursts. The maximum burst size (MBS) refers to how long a burst of cells at the PCR can last. Optionally, the sustainable cell rate (SCR) and burst tolerance (BT) may be defined. The SCR defines the maximum number of cells that can be sent continually over the connection and provides an upper bound on the conforming average of the connection. The BT limits the time a source is allowed to send traffic at its peak cell rate.

■   Quality of Service factors enter into elements of cell loss ratio, cell transfer delay, and cell delay variations.

■   A conformance definition is used to decide what traffic the network will accept, that is, the rate and with what burstiness an end point is allowed to send traffic. This definition is enforced by the police function.

■   A definition of a compliant connection, which is a threshold set by the service provider, must be made. Even though a cell may be found to be nonconforming, that does not mean that the connection is noncompliant. The precise definition of a compliant connection is left to the network provider. However, the network provider must define a connection where all the cells are conforming—that is, they are in compliance with the contract—as compliant.

▲   Finally, the service category, constant bit rate for instance, must be decided.

Once these parameters have been agreed, cell transmission can begin. Before we move on, we will take a brief look at the policing function and congestion control.

Policing and Congestion Control

Once the connection has been established, the network monitors the traffic by means of a policing function through the use of an algorithm. This function ensures that the end points hold up their end of the traffic contract. One such policing algorithm that has been defined by the ATM Forum, and is now standardized, is the *Generic Cell Rate Algorithm (GCRA)*, often referred to as the leaky bucket.

Basically, if a cell is found to be nonconforming—that is, if it violates the terms of the contract—it is dropped. The system, however, may be more forgiving than the basic description just given. The following analogy using the concept of the leaky bucket explains the point further. Each cell that passes a leaky bucket drops a token in the bucket that is drained at a specified rate. Cells that arrive at a rate in excess of the drain rate are accepted as long as the bucket does not overflow. The depth of the bucket defines tolerance to jitter. Greater depth implies that long bursts may be sent at a higher rate.

Congestion control has a direct impact on Quality of Service. During periods of congestion, the switch employs a cell discarding policy to control the congestion. When congestion is imminent, it begins discarding the least important cells. If the congestion continues above a certain threshold, only cells that have been defined with a high priority are allowed through the network. The CLP bit in the header is used to make this determination.

ATM networks are less prone to congestion than Frame Relay. The reason for this is that admission control through traffic contracts and policing complements the congestion control mechanism. By being discriminate about what is allowed through and under what circumstances, the network is less prone to congestion than one that allows anything through under any circumstance.

The Journey Through the Network   With the help of Figure 11-27, we will now follow the path of a cell as it enters and traverses the network. Now, assuming that the terms for transport across the network have been negotiated and the end points and the network have reached an agreement on the traffic contract, the cell enters the network. With the traffic contract mapped onto a PHY layer, the cell is ready to begin its journey.

As the cell goes through the network, each network node—ATM switch—only has a need to look at the PHY and ATM layers to perform the switching function, with one exception, SMDS-bearing traffic that uses a connectionless service address in the AAL layer.

As the cell leaves the source end point and proceeds toward its destination, it is given a VPI/VCI pair of 14/7 that identifies the specific destination. The actual values of the VPI and VCI may have been agreed upon beforehand with the service provider, or they may have been allocated though the use of a signaling protocol. When the cell enters the first switch, it performs the following functions:

▼   **Translation**   The switch looks up the received VPI/VCI values and replaces them with a new VPI/VCI 3/11 pair corresponding to the outgoing link or transmission path (TP).

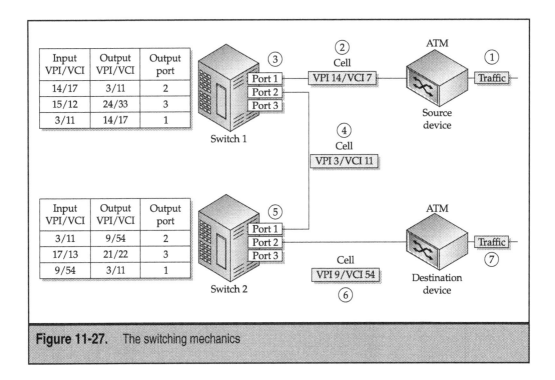

Figure 11-27. The switching mechanics

- ■ **Switching** The correct output port is determined through the use of the translation table.

- ▲ **Buffering** The cell may be buffered for a brief moment if necessary as it waits for the port to become free.

The cell is then sent on its way with the new VPI/VCI value pair to the receiving switch on the other end of the link. The process is again repeated at the second switch, with the cell being sent out with a new VPI/VCI value pair of 9/54.

As is evident from the figure, the VPI/VCI is only relevant on each link. As the cell traverses each switch on the path to its final destination, the switches attempt to live up to the agreements for the connection. The CLP bit becomes a critical component to the survival of the cell within the network during times of congestion.

ATM Signaling

An end device wishing to connect to an ATM network sends a signaling request to its directly connected switch. The request contains the ATM address of the target device and any QoS parameters that are required for the connection. Different signaling protocols are used between a UNI and an NNI. The ATM Forum UNI 3.1 specification, which is based on the Q.2931 public network signaling protocol developed by the ITU, is the current standard for UNI signaling. UNI signaling requests are carried in default connections of VPI = 0 and VCI = 5.

To set up the connection, the request is first sent into the connected switch, which propagates the request through the network. On reaching the final destination, the request is either accepted or denied. The connection request, however, must be managed, and ATM provides a suite of connection management message types to do this. Connection management message types, including setup, call proceeding, and connect and release, are used to set up and tear down an ATM connection. Figure 11-28 shows the logical flow and message types that are used to establish a connection.

The source sends the *setup message,* which includes the address of the target system and QoS service parameters, whenever it wants to establish a connection. The switch to which the source device is connected (the ingress switch) forwards the message through the network, at the same time sending a *connect proceeding* message back to the source in response to the setup message. When the setup message arrives at the ATM switch to which the target device is connected (the egress switch), the switch forwards the setup message across its UNI to the target device. The target device in turn responds by sending back to the egress switch either a *connect message* if the connection is accepted, or a *release message* if the connection is rejected. The egress switch sends the response across the network along the same path that the setup message traversed, to the ingress switch, which passes it on to the source device in acknowledgment of the original request. The ingress switch also sends a *connection acknowledge* message back to the destination to acknowledge the connection. Once the connection has been established, data can flow.

PNNI and ILMI

In a router network, each router builds a very extensive routing table that reflects the topology and the different paths to each device. For an ATM network to function effectively, each switch must also have some concept of the available paths through the network. If a connect setup message is sent into the network, it cannot be effectively handled if the ingress switch has no idea of the devices that are connected to the network and

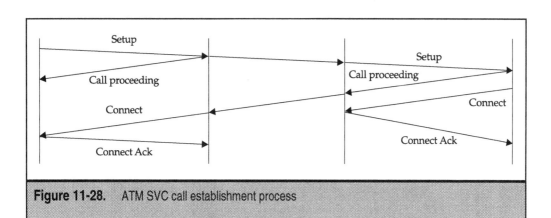

Figure 11-28.   ATM SVC call establishment process

a path to get there. Either this information can be manually placed in each switch or some type of signaling protocol can be used.

PNNI is the ATM routing protocol that enables switches to automatically discover the topology and the characteristics of the interconnection links. It provides two key functions in an ATM network: ATM topology discovery and call establishment. A link state protocol much like OSPF, PNNI tracks the availability of a link and other things such as the bandwidth of the link. As the network changes, this information propagates throughout the network to all the connected switches.

When a setup message is sent to an ingress switch, the PNNI protocol is invoked and a PNNI routing table is referenced to determine a path to the destination that meets the required QoS specified by the source. The ingress switch then builds a *designated transit list (DTL)*, a list that defines each switch hop to support the circuit to the destination. VCI 18 is reserved for PNNI.

The ILMI, or *integrated local management interface*, provides the means by which end devices can determine the status of components at the other end of the physical link and negotiate operational parameters. ILMI is an optional feature that may or may not be enabled by a service provider. Once enabled, it operates over VCI 16, and it allows the connected devices to dynamically learn the statuses of components; negotiate operational parameters; and share information such as addresses, peer interface names, and IP addresses.

In ATM, VCI values 0 through 31 are reserved for management and control functions. Common reserved VCI values and their uses are shown in Table 11-6.

LAN Emulation

LAN Emulation, or LANE as it is frequently called, is a standard that is defined by the ATM Forum to allow the ATM network to operate as a local area network, emulating a particular LAN and providing a similar set of features. When it is operating in this mode, attached devices benefit from the features that are common to 802.3 Ethernet or 802.5 Token Ring LAN.

The LANE protocol defines a service interface, called an *emulated LAN* or ELAN, for higher-layer network protocols that is identical to that of existing LANs. However, it

VCI	Function
5	Signaling from a device to an ingress switch for call management messages
16	ILMI for link parameter exchange
18	PNNI for ATM dynamic routing and topology discovery

Table 11-6. Common VCI Values

does not attempt to emulate the MAC protocol of the specific LAN, meaning it does not implement CSMA/CD of Ethernet or the token passing of token ring networks. From a functional perspective, the basic function of the LANE protocol is to resolve MAC addresses to ATM addresses, with the goal being to enable end devices to set up direct connections between themselves and then forward data. From the perspective of the end devices, they are operating across a normal LAN.

Multiple ELANs may coexist on a single ATM network, since ATM connections are separate entities that do not interfere with each other. A single ELAN emulates either an Ethernet or a Token Ring. FDDI LANs must be mapped into one of these through some type of translational bridging, as LANE does not provide a service interface for FDDI. The benefit of LANE is the fact that devices in remote regions of a building, a city, a state, a country, or the world can all communicate with each other as if they were on the same LAN, as shown in Figure 11-29.

A single ELAN consists of the following entities:

▼ **LAN emulation client (LEC)**   An end system that emulates a LAN interface to higher-layer protocols and performs data forwarding, address resolution, and other control functions. Each LEC is identified with a unique ATM address and is a member of a single ELAN. If the LEC is an end device such as a desktop computer with an appropriate ATM network interface card, the ATM address of the device is associated with a single MAC address. If, however, the LEC is a LAN switch or a router, the ATM address would be associated with the MAC addresses reachable through the ports of that LAN switch or router.

■ **LAN emulation server (LES)**   Implements the control functions for a particular ELAN. Only one LES can exist per ELAN, and for membership in a particular ELAN, a device must have a control relationship with the serving LES. The LES provides joining, address resolution, and address registration services to the LECs in its ELAN.

■ **LANE broadcast-and-unknown server (BUS)**   A multicast server that is used to flood unknown destination address traffic and forward multicast and broadcast traffic to clients within a particular ELAN. Each LEC is associated with a single BUS within an ELAN, but multiple BUSes may exist within the same ELAN, each coordinating and communicating with the others. LANE does not define the configuration of multiple communicating BUSes, which thus remains a vendor-specific implementation. One combined LES and BUS is required for each ELAN.

▲ **LANE configuration server (LECS)**   Bearing an unfortunate name, as it can easily be confused with a LEC, this contains the database that determines which ELAN a device belongs to. Each LEC consults the LECS once when it joins the ELAN to determine which ELAN it should join. The LECS returns the ATM address of the LES for that ELAN. One LECS is required for each ATM LANE cloud.

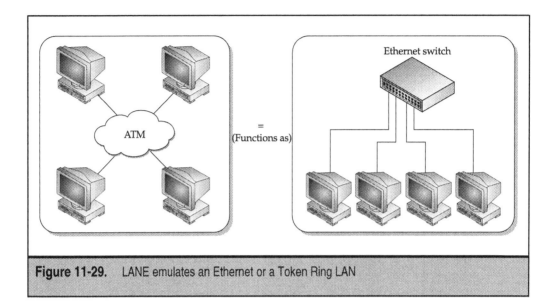

Figure 11-29.   LANE emulates an Ethernet or a Token Ring LAN

The Future of ATM

As a technology that once promised to be the final networking solution that would tran-
scend the local and wide area divide, ATM has lost some of the fervor that it once stirred.
The future of the technology is not as clear now as it was once perceived to be. The advent
of DWDM and the higher speeds that it enables make many of the issues that ATM was
built to address moot. Class of service, for instance, is extremely important when multiple data sources and traffic types are vying for a limited resource—bandwidth. If the limitations are removed and an abundant supply of bandwidth is made available, which is
what DWDM enables, the concept of class of service becomes moot because each traffic
type can be provisioned with the bandwidth that it requires.

The introduction of 1GbE and 10GbE also muddies the water. When ATM first became popular in the early 1990s, it was commonly thought that it would lead to the demise
of Ethernet. ATM was seen as the technology that would replace Ethernet on the desktop
and would be the seamless end-to-end solution. Around that time, 100 Mbps Ethernet
was introduced, and the rest, as they say, is history. Ethernet did not die and is still the
most prevalent local area network technology in use today. 1GbE and 10GbE are already
making waves and are being deployed in the access market and metro area networks. The
combination of Gigabit Ethernet and DWDM technologies raises many questions about
SONET/SDH and ATM.

ATM is now being positioned as a complementary technology rather than one that replaces. On the ATM Forum's web site, it is presented as a technology that interworks with
IP, Frame Relay, Gigabit Ethernet, DSL, wireless, and SONET/SDH. The next-generation
network is described as depending on ATM and complementary technologies to handle

the simultaneous traffic of voice, video, data, and images. We believe this to be true. ATM will most likely not deliver on its promise to be the single networking technology for the local, access, and core networks that it was once hoped to be in conjunction with B-ISDN. However, elements and features of the technology will still be found in many of access technologies that are emerging and those that are already deployed. The stronger features of the technology will also be adopted in new and emerging metro area technologies such as the one the next section describes.

MULTIPROTOCOL LABEL SWITCHING

It goes without question that the Internet has evolved into a ubiquitous network that has inspired a variety of new applications in business and consumer markets. In addition to becoming the network of choice for delivering these new and emerging applications, it also holds the distinction of sealing the IP protocol as the de facto protocol for data networks. Any doubts or questions that may have been present during the period of the late 1980s to early 1990s have now been totally dispelled and IP has become more entrenched than even Ethernet.

This creates a problem, however, because the service demands of these new applications are stretching the capability of the existing infrastructure and the IP protocol. To deliver on the requirements of new services such as voice and multimedia, a few things are required. First, the network must be able to provide different levels of service for different traffic types. Second, the bandwidth of the network must be managed more efficiently, and third, packets need to be delivered and a lot faster.

IP was not designed to deliver a guaranteed level of service for different classes of applications or traffic profiles; the IP protocol treats all traffic the same. Well, to be more accurate, most implementations of IP treat all traffic the same, for while the IP datagram does have a *Type of Service (TOS)* indicator that specifies such things as precedence and is meant to be used to indicate varying levels of importance of different datagrams, most routers ignore it. Implementations of IP assume that the network and its resources are a level playing field that is shared equally by all participants (traffic types). Interestingly, the implementations of IP that do attempt to discriminate between different traffic types have been predominantly vendor specific, and almost all do this through some means other than the TOS field. These features, such as priority queuing in a Cisco router, use a logic that is implementation specific. These approaches provide relief and have been successful in implementation, but even so they are somewhat lacking and have fallen short of providing the true Quality of Service that today's applications require.

The initial deployment of the Internet addressed the requirements of the applications that were prevalent at that time. The network catered to simple file transfer applications, Telnet, and e-mail delivery, among the other applications that were then common. A simple software-based router platform was adequate for this purpose, but as the requirements change, a more efficient platform is required to satisfy the new demand for bandwidth, low latency, and guaranteed service levels that newer application types require.

In the case of voice, further enhancements such as fragmentation were introduced to gain more control over packet size in an attempt to control latency for voice traffic. By fragmenting large data packets that are larger than a certain size while employing some type of priority queuing, more control over bandwidth management is obtained and issues of latency can be better controlled.

However, we are at a point where IP could benefit from some of the features that ATM has to offer, services such as QoS and faster switching decisions. This is where MPLS, *Multiprotocol Label Switching,* fits in, providing a level of bandwidth management for IP-based networks that was previously missing from the IP protocol. MPLS addresses issues of scalability, traffic engineering, and routing on the basis of QoS and service quality metrics.

MPLS Overview

MPLS and IP form a middle ground that combines the best features of IP and the best features of circuit switching technologies such as ATM. Switching as a technology is faster than routing because the decision is made at layer 2 of the OSI model instead of layer 3, as is the case for routing. Limiting the decision to layer 2 means that only the first few bytes of each packet need to be read in order to decide the path. In routing, the router has to read a lot further into the packet and must do a lot more computation to determine the best path.

For this reason, ATM switches are a lot faster than routers. The efficiency and speed of ATM is further enhanced through the use of fixed-length labels instead of the longest-match algorithm that is used by routers. MPLS bridges the gap between these two techniques by adopting principles of routing to do the initial path discovery and principles of label switching for subsequent user traffic across a path that was previously mapped. Through the use of MPLS, a device is able to the same job as a router with the performance of an ATM switch. Furthermore, MPLS simplifies the mapping of IP to ATM by replacing the ATM signaling protocols with IP control protocols.

The history of MPLS is rather short, and rather than the text recounting the details, Figure 11-30 provides a visual timeline that accounts for the significant events leading up to MPLS today, and Table 11-7 lists different RFCs that address the technology.

MPLS is an IETF-specified framework that performs the following functions:

▼ Manages traffic flows at varying granularity; device-to-device and application-to-application, for instance

■ Supports IP, ATM, and Frame Relay layer 2 protocols

■ Functions independently of the layer 2 and layer 3 protocols

■ Maps and maintains a correlation between upper-layer addresses (IP addresses, ports, etc.) and simple fixed-length labels used by different packet-forwarding and packet-switching technologies

▲ Provides an interface to existing routing protocols such as OSPF

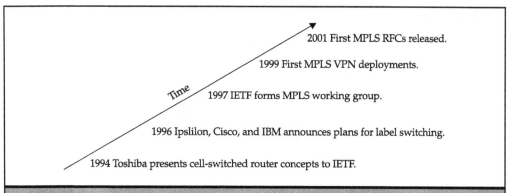

Figure 11-30.   MPLS timeline of events

RFC	Description
RFC 2702	Requirements for Traffic Engineering over MPLS
RFC 3031	MPLS Architecture
RFC 3032	MPLS Label Stack Encoding
RFC 3035	MPLS Using LDP and ATM VC Switching
RFC 3036	MPLS LDP Specifications
RFC 3037	LDP Applicability
RFC 3034	Use of Label Switching on Frame Relay Networks Specification
RFC 3035	MPLS Using LDP and ATM VC Switching
RFC 3036	MPLS LDP Specification
RFC 3037	MPLS LDP Applicability
RFC 3038	VCID Notification over ATM Link for LDP
RFC 3033	The Assignment of the Information Field and Protocol Identifier in the Q.2941 Generic Identifier and Q.2957 User-to-User Signaling for the Internet Protocol
RFC 3063	MPLS Loop Prevention Mechanism
RFC 3107	Carrying Label Information in BGP-4
RFC 3209	RSVP-TE: Extensions to RSVP for LSP Tunnels
RFC 3210	Applicability Statement for Extensions to RSVP for LSP-Tunnels
RFC 3212	Constraint-Based LSP Setup Using LDP
RFC 3213	Applicability Statement for CR-LDP
RFC 3214	LSP Modification Using CR-LDP
RFC 3215	LDP State Machine

Table 11-7.   MPLS RFCs

MPLS Operation and Components

Unlike typical routing, MPLS is based on the concept of flows or *forwarding equivalence classes (FECs)*, a grouping of packets that are all treated the same way by a router. Flows or FECs consist of packets between common end points identified by features such as network addresses and port numbers. Packets belonging to an FEC are forwarded in the same manner and over the same path. The forwarding process is therefore a simple matter of assigning a packet to an FEC and determining the next hop of the FEC. The assignment of a packet to a particular FEC is done by the *label edge router (LER)*, the first router to receive the packet as it enters the network. Once assigned to an FEC, subsequent *label switch routers (LSRs)* need only perform a forwarding function; no further packet classification is required, as is illustrated in Figure 11-31.

In a typical network that is based on routing, a router reads the destination address of an incoming packet and references a forwarding (routing) table to determine the appropriate route for that packet. Each router in the network shares a common understanding of what the network looks like and the different routes between end points by running and sharing in a common routing protocol such as OSPF or RIP. Once the routers within a routing domain converge on a common knowledge of the network, routing of packets is done on a per-packet basis in order to use the best path at the time the packet is received.

In contrast, MPLS calculates the route once on each flow, or FEC, through the provider's network. As a packet enters the MPLS network, a short fixed-length label is affixed to the packet. As the packet travels through the network, each router looks no further than the label to determine the route, thereby reducing processing time and improving scalability. By decoupling the forwarding of IP packets from the information carried in the header of the IP packet, MPLS allows new routing functionality and new forwarding paradigms not available in conventional routing.

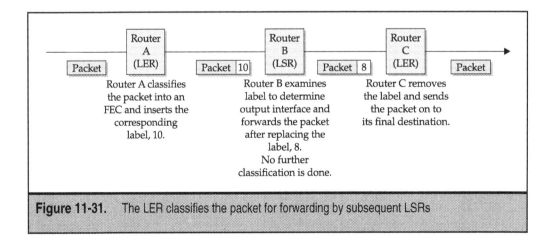

Figure 11-31. The LER classifies the packet for forwarding by subsequent LSRs

FECs and Labels

The FEC, forwarding equivalence class, is a representation of a group of packets that share a common requirement for transport. All packets within an FEC are handled the same way as they traverse the MPLS network. FECs are based on service requirements for a group of packets or are simply based on a common address prefix. Each router in the MPLS network builds a table, called a *label information base (LIB),* consisting of FEC-to-label bindings that define how each FEC is forwarded.

An MPLS label is simply a short fixed-length, 32-bit, locally significant identifier that is used to identify an FEC. The label, which is placed in a packet, represents the forwarding equivalence class and identifies the path along which the packet should travel. The label is placed between the layer 2 and layer 3 headers of the packet, as shown in Figure 11-32. By placing the MPLS label after the link header, MPLS operates independently of the layer 2 transmission protocol and so can be implemented on different layer 2 platforms such as ATM, Ethernet, or Frame Relay.

The label, which is sometimes called the MPLS *shim,* consists of the following fields:

▼   The *Label* field (20 bits) carries the actual value of the MPLS label.

■   The *Class of Service (COS)* field (3 bits), sometimes called the *Experimental* field, is used to determine the type of service that the packet should receive. The value of this field affects such things as queuing and discard algorithms that are applied to the packet as it travels through the network.

■   The *Stack* field (1 bit) supports a hierarchical label stack. It basically allows MPLS to be used simultaneously for routing at a lower level between individual routers within an ISP for instance and at a higher domain-domain level. Each level in a label stack pertains to some hierarchical level, a feature that facilitates tunneling within an MPLS network.

▲   The *Time to Live* or TTL field (8 bits) provides conventional IP TTL functions. This field is used to ensure that a packet does not loop endlessly within the cloud. The value is decremented as it passes through a router and subsequently discarded if the value reaches zero.

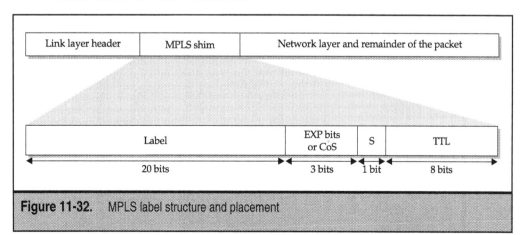

Figure 11-32.   MPLS label structure and placement

A *label switch path (LSP)* is a specific path through an MPLS network. Per RFC 2702, a *traffic trunk* is an aggregation of traffic flows of the same class that are placed in an LSP. This definition has resulted in both terms being used synonymously; it should be noted, however, that the path through which a traffic trunk travels can be changed, and in this respect, traffic trunks are similar to virtual circuits in Frame Relay and ATM networks.

Now that we have defined the concepts of FEC and LSP, the next logical question is, how do LERs and LSRs discover these paths and how are they distributed? Remember, an LSR is a label switch router that resides in the core of the MPLS network. These routers participate in the establishment of LSPs. LERs are label switch routers that reside at the edges of the MPLS network and may be considered LSRs that provide ingress and egress functions. Because of where they are located, they are the first and the last routers to see a packet as it enters and leaves the network; they therefore have additional functions to perform over those of an LSR.

LERs support multiple ports that may be connected to dissimilar networks such as Frame Relay, ATM, or Ethernet and provide functions that classify incoming packets into appropriate FECs for forwarding along an LSP. They are also responsible for removing these labels as they exit the network. LERs also participates in the establishment of LSPs. Because LERs can also function as LSRs, it should be assumed that any references made to functions performed by an LSR are also functions that an LER is capable of performing. However, not all functions that are performed by an LER are relevant to an LSR.

Label Distribution Protocol

The *Label Distribution Protocol (LDP)* is a specification that allows the distribution of labels between label switch routers. LSRs within an MPLS network form a peer relationship. When an LSR assigns a label to a forwarding equivalence class, it needs to let its peers know of this label and its meaning; LDP is used for this purpose. Since a set of labels from ingress and egress LERs in an MPLS domain defines a label switch path, and since these labels are mappings of network layer routing to data link layer–switched paths, LDP helps in establishing a label switch path by using a set of procedures to distribute the labels among the LSR peers.

As in conventional routing, routers participating in a common routing domain are required to share a common knowledge of the network, the links, and the paths between end points. So too in an MPLS network, LSRs must share a common knowledge of the different LSPs, labels, and FECs and LDP is the means by which this is achieved. LDP has the following characteristics:

▼ It provides a mechanism for LSR peers to discover each other and establish communication.

■ It defines four classes of messages: *discovery, adjacency, label advertisement,* and *notification* messages.

▲ It runs over TCP to provide reliable delivery of messages with the exception of the *discovery* message.

Building the LSP   So how are label switch paths built? An LSP is a set of label switch routers and their links, which packets belonging to a specific FEC travel in order to reach their destination. In order for an LSR to build a link switch path, it must make use of the routes learned from the layer 3 IP routing protocol. Other protocols such as RSVP may be used, but they are not required to. A label switch path can be set up through processes called *independent control* or *ordered control*.

Independent control is the process through which each LSR makes an independent decision to assign a label to an FEC and advertise that assignment to its peers. Using this approach allows LSPs to be established at the time of routing convergence. In ordered control, the LERs initiate the assignment, and the label assignment proceeds in an orderly fashion from one end of the LSP to the other across the network. Both independent and ordered control can be supported concurrently without any potential conflicts.

Each of the two models has its strong and weak points. Independent control provides for faster convergence. Any router that hears of a routing change can propagate that information to others. The disadvantage is that there is no single point of control, which makes engineering more difficult.

Ordered control provides more control over the network and provides better traffic engineering, but it is slower at convergence and the label controller is a single point of failure. In ordered control, the distribution of labels is distributed through two methods: *downstream unsolicited (DOU)* and *downstream on demand (DOD)*. Using the DOU method, the label manager (the LER that builds the label) sends the label unsolicited to the downstream LSR. The trigger for this may be something as simple as a timer expiring, say in the case where label refresh messages are sent every 30 seconds, for instance, or it could be some other event such as a change in a routing table. The DOU method is also sometimes called the *push* method.

In DOD, or the *pull* method, as it is sometimes called, an LSR requests a label or a label update.

Interior gateway protocols (IGPs)—routing protocols that are shared among routers in a common routing domain—such as OSPF are used to define reachability of network resources and to bind/map between FEC and next-hop addresses. Figure 11-33 shows the logical steps in discovering and propagating labels throughout an MPLS network. It also shows the action of LERs as a packet enters and leaves the network. As is evident from the figure, the IGP is used to determine and establish reachability to existing networks. Next, LSP maps and propagates label information. As a packet enters the network, the ingress LER inserts a label into the packet and forwards the packet to the next LSR in line, which switches the packet according to the embedded label information. Finally, the egress LER strips the label and sends it to the destination network.

NOTE: By definition, MPLS supports multiple protocols beside IP. At the network layer, it supports multiple protocols such as IPv4, IPv6, IPX, and AppleTalk. At the link layer, it supports Ethernet, Token Ring, FDDI, ATM, Frame Relay, and Point-to-Point links. It can essentially work with not just IP but any control protocol and layer on top of any link layer protocol.

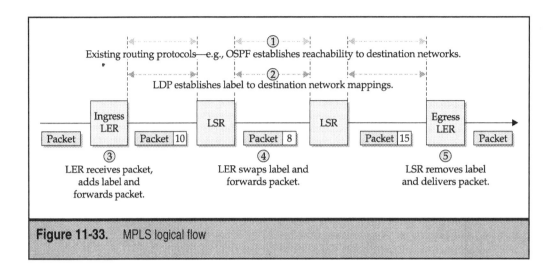

Figure 11-33. MPLS logical flow

Different Traffic Engineering Approaches

Signaling protocols are functionally equivalent to a dynamic routing protocol in a conventional router-based network, and just as a dynamic routing protocol is not a requirement for a router-based network, dynamic signaling protocols are not a requirement in an MPLS network. Also, just as it is possible to create IP forwarding networks without a routing protocol, it is possible to use static mappings of LSPs through an MPLS network by manually supplying every node with the necessary information for packet forwarding, including in and out labels, actions, and next hops. This approach is roughly equivalent to using static routes in an IP network. However, this is just as impractical and difficult to manage in an MPLS network as it is in a large router-based network.

Dynamic signaling protocols have been designed to allow single routers to request the establishment and label binding to FEC for an end-to-end path. The implementation of a dynamic signaling protocol can be as simple as ensuring that a single router is provided with the appropriate label mapping and allowing it to propagate through the network through ordered control. In this simple application, configuring a new LSP over an MPLS core that is signaling enabled does not require anything beyond the configuration in the instantiating router.

In practice, the operation of MPLS and the performance of the network are most certainly enhanced by the ability to create a path using a dynamic signaling mechanism. Three signaling protocols are available for use in MPLS networks today. The first, LDP, has already been discussed and is presented here with a bit more information that is relevant to the context of this discussion. The other two protocols are the subject of the remainder of this section and will be discussed in the context of adding MPLS traffic engineering capabilities to the LDP protocol.

Traffic engineering is the process where data is routed through the network according to a management view of the availability of resources and of the current and expected traffic flows. The class and quality of service required for the data can also be factored into this process. Traffic engineering may be controlled by manual means where the state of the network, the routes, and the traffic are monitored and manually provisioned to compensate for problems as they arise. Alternatively, an automated process that reacts to information fed back through routing protocols or other means may drive traffic engineering.

LDP (RFC3036) defines the protocol specifically designed for the distribution of information required to properly interpret label binding to FEC. It does not represent an end-to-end path; rather, it is a hop-by-hop approach.

As with most new technology in the early stages of deployment, there two camps in support of two different signaling mechanisms for MPLS networks. As a result, separate signaling mechanisms for MPLS traffic engineering have been proposed and are under consideration by the IETF: *Constraint-Based Routing LDP (CR-LDP)* and *Resource Reservation Protocol with Traffic Engineering (RSVP-TE)*. In practice, this means that there are two methods being deployed for the same objective in different MPLS networks today. Table 11-8 shows the major proponents of each.

CR-LDP and RSVP are both signaling mechanisms used to support traffic engineering across an MPLS backbone. *CR-LDP* proposes to extend LDP, which is designed for hop-by-hop label distribution, to support QoS signaling and explicit routing. Like LDP, it uses TCP sessions between LSR peers to send label distribution messages. *Explicit routing*

RSVP-TE	CR-LDP
Cisco	Ericson
Avici	GDC
Argon	Nortel
Ironbridge	
Juniper	
Torrent	

Table 11-8. Proponents of the Different Traffic Engineering Techniques

is the concept that a route is explicitly specified as a sequence of hops rather than being determined solely by conventional routing algorithms on a hop-by-hop basis.

RSVP is a QoS signaling protocol that is an IETF standard and has been in use for some time. RSVP-TE is an extension to the RSVP protocol and supports capabilities that are somewhat similar to CR-LDP but represent a narrower set of service classes defined through a narrower set of explicit route objects. RSVP is a separate protocol at the IP level. It uses IP datagrams for communication between peers and does not require the maintenance of TCP sessions, but as a consequence of this, it must handle the loss of control messages.

The supporters of RSVP-TE hold that there is less pain in introducing a protocol that already has a track record. Adding extensions to the already proved and much under-stood RSVP protocol creates less of a learning curve for implementers and the user com-munity. Adding a new set of functions to the older protocol also provides the potential for backward compatibility. The counterpoint presented by the proponents of CR-LDP is that it is a lot easier to build a new protocol from scratch than to retrofit an existing one.

Both protocols use ordered label switch path setup procedures, and both include some QoS information in the signaling messages to enable resource allocation and the au-tomatic establishment of LSP.

Table 11-9 provides a quick comparison of the two protocols.

Feature	CR-LDP	RSVP-TE
Transport	TCP	Raw IP
Security	Yes	Yes
Multipoint to point	Yes	Yes
Multicast support	No	No
High availability	No	Yes
Rerouting	Yes	Yes
Explicit routes	Strict and loose	Strict and loose
Route pinning	Yes	Yes, by recording path
Traffic control	Forward path	Reverse path

Table 11-9. Comparison of CR-LDP and RSVP-TE Traffic Engineering Protocols

SUMMARY

In the metro area and the network core, copper-based systems are rapidly being replaced by optical fiber cable. Prior to the development of the SONET/SDH standard, many incompatible optical solutions existed that were vendor specific. With the publishing of the SONET/SDH standards, telecommunications companies readily embraced the standard because it ensured interoperability between different vendor systems, increased capacity, and provided an upgrade path for new services.

Around the same time that the SONET/SDH standard was being published, the concept of an intelligent fiber-based network that could efficiently transport different services was also being explored; it became known as a Broadband Integrated Services Digital Network (B-ISDN). The ITU decided that the promise of B-ISDN could be realized through the use of SONET/SDH and of ATM, which was designed to be the final network technology that we would ever need. It was built to function equally well in a LAN, MAN, or WAN environment regardless of the application or traffic type.

ATM never did emerge as the ultimate networking technology, and the development of 100 Mbps Ethernet won the battle for the desktop. Today, ATM and SONET/SDH are facing stiff competition from Gigabit Ethernet and fiber-based optical technologies such as DWDM. In the middle years of the 1990s, label switching emerged. It combines the best features of IP and ATM. Today, it is positioned as a major technology contender for the metro area network along with Gigabit Ethernet. In the core, optically based DWDM solutions appear to be the choice for the long-haul portion of the network.

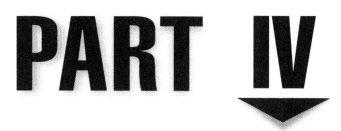

PART IV

Beyond Broadband

n Part 1, we explored the market drivers for broadband and different enterprise and residential network solutions. Part 2 examined current and emerging access technologies. These technologies are used to link enterprise and residential networks to the metro area and the network core. To business and residential customers, these are the technologies that embody the concept of broadband, a convergence of multiple services across a common network. To them, the access network is "The Network" and the only one they will ever see or care about.

Part 3 looked beyond the access network and into the metro area and the core of the network. In this realm, things are a lot different, as this portion of the network hierarchy is more concerned with moving data between different regions from cities to states and countries. Even at this level, there is a logical distinction between the metro area and the core or long-haul network. Within the metro area, things are constantly changing, as it is an evolving entity that is being designed and managed to handle the different types of traffic. At this level, issues of Quality of Service are extremely important, and the different types of traffic are factors influencing its design. In the core of network—the long-haul portion—the emphasis is on moving large amounts of data without regard to the different traffic types.

In this part, we look beyond broadband to the emergence of content services and issues of convergence. We also attempt to raise some of the more frequently asked questions that are typically presented in any discussions of next-generation networks. We present our best interpretation of the issues and our best ideas on how things may play out.

CHAPTER 12

Convergence and Content Services

In this chapter, we look beyond broadband networks to issues of convergence and the developments in content services. To the average user, the concept of the network is limited to that user's access connection to the service provider's network. In fact, the Internet is commonly viewed as the network of the ISP, and so the name of the service—AOL or Roadrunner for instance—is commonly used instead of the Internet. If there is a problem accessing a particular site, it is viewed as local to the service provider.

This chapter explains the challenges associated within delivering content to a global community and some of the ways these challenges are being addressed. But before we address the delivery content, let us review some of the larger issues and questions of convergence.

CONVERGENCE

The convergence of data, voice, and video has raised many issues that have spawned new markets and many questions that have been the subject of debates. In this section, we present some of these questions and explore potential answers. Not having a crystal ball leaves us to look at the issues and make assessments based on existing trends. However, changing market conditions, user requirements, and technological breakthroughs can easily make any such attempt futile. The following should, therefore, be read in the context of the information presented in the preceding chapters and changing trends.

ATM and SONET Versus IP and Ethernet

For the first time, data traffic has eclipsed traditional voice traffic in volume, and as time progresses, voice will represent an even lesser percentage of the overall traffic exchanged on a global basis. This fact presents several questions. First, will future networks be designed for a mixture of data and isochronous traffic such as voice and video in their native format, or will such traffic be packetized and transported as just so many more data packets, as in the case of voice over IP, for example? Second, even if voice and video are packetized, what is the most efficient way to transport them—IP over ATM, pure IP, or pure ATM?

There are two very different views of the best technology for the next-generation networks. These views are centered on two different visions that have evolved from two different evolutionary paths for two different applications: voice and data. The established telephone industry players support ATM and SONET because of their proven capability to transport a mixture of different traffic types with a guaranteed QoS, and because of their management capabilities. However, these systems make inefficient use of bandwidth and can be expensive.

The advocates for IP and Ethernet are typically more data-centric, with a background steeped in LAN-based and packet networks. Here, IP and Ethernet are recognized for their simplicity, ubiquity, economy, and scalability. In this world, QoS is less of a factor, though it is a recognized requirement if other traffic streams are to be supported.

What makes this an interesting debate is that for decades circuit-based traffic has been the predominant traffic type with the most extensive global infrastructure. This meant that the flow of money had, until the early 1990s, chased traditional telephony-based solutions because they represented the largest opportunities and return on investment. But looking ahead, voice traffic looks as if it will represent a fraction of future traffic volume, and the tides have now changed, with investment currency favoring IP-based systems and the build-out of data networks. The path of opportunity has always attracted the biggest investments because that path represents the greatest return on investment.

Today, the path with best opportunity is the path of consolidation. Running multiple infrastructures is expensive, and converging voice, video, and data to a common infrastructure translates to greater cost savings and a better return on investment. Convergence also means the potential for more revenue as new incremental services are launched from a common infrastructure.

Now, this concept is not new; Broadband ISDN was built on the principle of an integrated service. The major difference now is that at the time when B-ISDN emerged, ATM and SONET were seen as the enabling technologies, but today the vision is more data-centric. In the 1980s, the concept of an integrated network had promise, and it still does; the money flow at the time was in telephony-based systems, and so a circuit-based solution was seen as the ultimate solution. But the emergence of the Internet—its growth and its prominence in providing the foundation for a new global economy—happened at a time when the idea of B-ISDN was just getting off the ground. The potential of the Internet was reason enough to rethink the original concept of the integrated network and how it would evolve, all of a sudden the path to the new integrated network—B-ISDN— was not as clear as it once was.

This new network—the Internet—was growing rapidly and was capturing the imagination of different industries. New applications began to emerge at a rapid rate, and the tide of investment dollars began to change course. Investments were being redirected away from voice-based networks and toward the new infrastructure and any new and innovative technology that looked promising.

Investments have since slowed, but their direction has not changed. For this single reason—the flow of investments—it appears that next-generation networks will continue to favor solutions that are data-centric and packet-switched instead of circuit-switched. This trend favors IP, Ethernet, and packet switching technologies, but it also means more development is necessary, especially in areas of the enforcement of Quality of Service. In order to support multiple traffic types, the requirements of each type of traffic have to be satisfied, and while this need is being addressed in emerging packet switching technologies, specifically MPLS, we still have a way to go.

In fact, the days of circuit-switched telephony appear numbered. Service providers and public and private enterprises are spending on infrastructures that will support a converged service. Why waste capital funding an infrastructure that is optimized for telephony when a single infrastructure can be used for both services? Advances in packetized voice technology are quickly removing many of the differentiating qualitative features between circuit-switched voice and packetized voice. Many enterprises already have entire corporate telephone services running across an optimized data infrastructure.

Why MPLS and Not IP over ATM

Okay, so we have made the case for the next-generation networks being data-centric by following the flow of investment capital. We further concluded that the protocol of choice would be IP, as it has become the de facto standard for data networks. We then inferred that the resulting transport technology would be some form of packet/cell switch technology such as MPLS. Here, the logic is not as clear, as it skips an obvious choice—IP over ATM.

So why MPLS, and not IP over ATM? The answer can be summed up in the simple, yet popular quip "been there, done that." In the mid-1990s, several ISPs evolved their networks from router-based cores to an overlay model of running IP over ATM. This was done out of necessity; greater bandwidth, deterministic forwarding performance, and traffic engineering were needed. IP over ATM was seen as a way to satisfy these needs.

The overlay model depended on ATM functionality for the software controls (signaling and routing) and hardware forwarding (label swapping) on every system in the core of the network. The requirements of the applications were met by using layer 3 functionality at the edges of the network and relying on ATM to provide a maximized network throughput. In this model, IP was relegated to the edges of the network because software-based routers were seen as a source of poor network performance.

When viewed in the context of what was available at the time, this approach makes a lot of sense. At that time, networking equipment was not specifically engineered for Internet backbone applications and so was neither as efficient nor as powerful as it is today. Only the ATM switching infrastructure equipment provided the bandwidth and forwarding capacity required at that time. However, the Internet kept growing at an exponential rate, and issues of scalability with ATM became more apparent as growth increased.

The scalability problems included bandwidth limitations of the ATM SAR interfaces, the overhead introduced by the 20 percent cell tax, the complexity of meshing (n*n – 1 issues), the stress imposed on the Interior Gateway Protocol (IGP) because of multiple peering points, and the inability to operate over non-ATM infrastructure. The situation was further compounded by the complexity of operating and managing two disparate technologies that were independently designed for different applications. The architectures of the two technologies were different; each had its own addressing schemes, signaling protocols, and resource allocation schemes that had to be satisfied for an efficient operation.

As these issues became more pronounced, equipment built expressly for backbone routing became available for Internet applications, and it made less sense to continue to perpetuate the problems that the rapid growth was causing in the overlay model. The development of new technologies for the core of the Internet was fueled by the availability of investment capital, which was again chasing the path of opportunity. The Internet was growing at an exponential rate, current technologies were inadequate, the market potential of the Internet was recognized, and so investors were willing to pour capital into any new technology that would facilitate the growth and lessen the growing pains.

The notion of "IP convergence" was also catching the imagination of venture capitalists, and a period of innovation had begun, as new and young start-up companies began pitching technologies that would compete with incumbent vendors. To be successful, these start-ups had to produce a solution that was a combination of the best features of the technologies and approaches that were available then.

By late 1996, a number of vendors began promoting proprietary multilayer switching solutions that integrated ATM switching and IP routing. These approaches went by different descriptions from the different vendors; Cisco coined the term Tag Switching; Toshiba, Cell Switching, to name a few.

Today, the common name for all these approaches is Multiprotocol Label Switching, or MPLS, the latest step in the evolution of technology for the core of the Internet. MPLS delivers a solution that seamlessly integrates the control of IP routing and the speed of layer 2 switching. Tomorrow's network will more than likely be a combination of switching at many different layers. In fact, new layer 4/7 switches are now common; they are addressed separately in a later section.

The Future of Data

One of the more interesting questions raised is whether or not data will follow the path of voice. Since 1996, voice has experienced a rapid decline in its revenue-generating power. This has been attributed to increased competition and aggressive pricing by voice carriers, to lure subscribers from one carrier to the next. Another contributing factor is the emergence of packetized voice over the Internet, specifically VoIP.

As the Internet benefited from an influx of investment, a period of innovation ensued that resulted in the development of many new and innovative applications and services, VoIP being just one. Web sites began offering free long-distance voice calls to attract users to their site, with the hope of using the increased traffic to sell ads to advertisers.

With the potential for free long distance across the Internet, the entrenched voice providers began to see erosion in the number of minutes used. Granted, the performance and quality of voice across the Internet was substantially less than for a circuit-switched telephone call, but a degradation in quality is easily tolerated when the service is free. Today, the quality of voice over the Internet has improved, and now there are service providers that specialize in providing long-distance services across the Internet at a deep discount to circuit-switched solutions.

There is a population that believes the introduction of DWDM and future optical solutions will result in enough bandwidth becoming available that data will eventually see a price decline similar to that of voice. It is further argued that with the abundance of bandwidth, issues of Quality of Service will become moot. After all, QoS is an issue only in those instances in which different traffic competes for a limited resource. By being abundantly available, the resource is no longer limited, and so there is no need for QoS.

This argument has some compelling points. It is true that QoS is an issue only when there is competition for resources. It is also true that DWDM holds a potential that is yet untapped. After all, there have already been demonstrations of optical fiber operating at seven trillion bits per second, and the technology is still in the early stages of development, especially when compared with the long history of wire-based transmission.

When one considers that the fiber in the network core is currently less than five percent utilized, one appreciates the amount of bandwidth that is yet to be tapped. If this is combined with the amount of bandwidth that could be made available through technology improvements, the resulting capacity could indeed create erosion in the revenue-generating potential of data, much as happened to voice when it was used as a loss leader for selling another service.

So how does this affect MPLS? If the main benefit of MPLS is speed and the enforcement of QoS, what is the point of MPLS if speed and QoS are not an issue? If there is indeed an abundance of bandwidth, it could be argued that layer 3 IP routing is all that will be needed, with maybe additional intelligence being applied by reading further into the packet to the headers that contain upper-layer information.

MPLS may be an interim technology, but it will continue to have a place because of how the availability of bandwidth is distributed within the network. The enterprise network and the network core both have an untapped source of bandwidth; the enterprise will continue to benefit from the advancement of Ethernet, and the network core is currently underutilized. The access and metro area networks, however, are still bottlenecks, and as long as there is a bottleneck, there is a place for MPLS.

MPLS will also continue to thrive in areas outside the U.S. Even if fiber delivers trillions of bits per second, not every country will have the resources to upgrade its infrastructure. In countries that cannot, the limited resources will need to be adequately shared between different applications with different requirements, and MPLS will continue to be an efficient technology for those applications.

Security Issues

As consumers and business users take advantage of broadband access, they will be faced with the security challenges of an "always on" connection. Being constantly on is akin to an open-door policy in the roughest of neighborhoods. The odds of an attack are greatly increased, and additional measures must be employed to ensure protection.

Pranksters and malicious users are constantly looking for unsuspecting users on whom to wreak havoc. While there have been instances of theft of personal information such as credit card numbers, the majority of attacks are centered around disrupting the target computer system through the use of a virus, or using the target computer to wreak havoc on some other unsuspecting computer system. In the latter example, the simplest and slowest of computers can be used as a weapon to assist in creating a disruption of service to high-profile computer systems such as government and financial systems.

Security breaches are now so numerous and varied that an entire industry has been created to help address the problem. Hackers are constantly discovering new vulnerabilities, and the security vendors must keep up with these new discoveries. The products of security vendors usually come in the form of a firewall solution that attempts to prevent intrusion. However, in addition to the firewall market, there other solutions that are devoted to repairing the damage done by an intruder and the removal of programs that infect and create problems on a target machine.

Firewalls

Firewalls come in two flavors, dedicated and integrated solutions. Dedicated security systems are a combination of hardware and software that is placed at the "entrance" to the network and used to enforce rules that are set by the user. These solutions provide the benefit of flexibility through strategic placement. Depending on where in the network they are placed, they provide the potential to protect the entire enterprise or residential network or a portion of the network. The usual reason for wanting to protect portions of the network instead of the entire network is that a portion of the network has to be left open for public access.

Integrated solutions are software-only approaches that are typically enabled on the devices they are meant to protect. In some instances, software solutions may be used to protect the entire network when the device on which it is installed forms the only entry to the network. In this configuration, they serve both functions of an integrated system and a dedicated one.

Firewalls are effective only against specific types of threats; others can present themselves through applications that are allowed to pass through, e-mail, for example. Mail messages can be used as transports for different types of viruses. In these cases, care has to be taken in opening any mail from an unknown sender. This is also true when introducing new diskettes to a computer. A good virus scan program is highly recommended.

The scope of this topic is not to provide details of different solutions, as there are many available references on the Web and in bookstores to do that. Our intent is to create an awareness of the challenges and to provide a basic understanding of how these security threats can present themselves.

Convergence and Its Wider Security Implications

Mention of hackers, viruses, and computer security breaches usually evokes the image of a computer system—a personal computer, server, or mainframe—being attacked. Missing from this image are those devices that are becoming more prevalent as a result of convergence. A digital set-top box, for instance, or any IP-enabled appliance in the home for that matter, are all subject to the same security exposures as a computer.

The convergence of different services and traffic types—the basic intent of broadband—is eroding any clear distinction in function between devices that were used exclusively for a particular application. These devices are rapidly being equipped with intelligent chips to enable additional functions and to provide remote management through the use of the IP protocol. As a result, these devices are subject to the same security concerns that are posed to any other IP-enabled computer.

In the enterprise or residential network, IP-enabled devices will fall in two categories: those that become part of the LAN and those that are part of the service provider's domain or VPN. The IP address in a digital set-top box is an example of an IP-enabled device that is part of the service provider network. A personal computer or an IP-enabled appliance that controls the lighting of a building is an example of a device that is part of the local LAN. Those devices that are controlled by the service provider must be protected by the provider and usually will benefit from whatever security policies the service provider employs for its VPN. However, these devices are not protected from malicious intent arising from within the provider's network.

Devices that are part of the local LAN must be protected by whatever security policies are employed by the LAN administrator. The security policies of the local environment must therefore be mindful of any device—large or small—that may be subject to a security breach, and this information must influence the choice of security system. This issue will become more complex and relevant as more IP-enabled appliances begin popping up in the home. The vision of the intelligent refrigerator carries with it the added responsibility of security in addition to the benefits it was meant to provide.

Types of Attacks

In an attempt to create awareness of the different types of security attacks, we will examine some of the more common forms. The speed at which new threats are being developed and the creativity and ingenuity behind the development of these threats make it virtually impossible to have a single published source of information on the most current threats. Entire books have been devoted to the topic, and even those are quickly outdated. The information here is for general awareness, and it is highly recommended that other sources be used for more detailed information and for information on more current threats.

Trojan Horse Programs   Trojan horse programs are back-door programs that are installed on your computer without your knowledge. These programs allow intruders easy access to your computer without your knowledge. These apparently useful programs contain hidden functions that can exploit the privileges enjoyed by the owner of the computer to change system configurations or infect other computers. A Trojan horse relies on a user to install it, or else an intruder who has gained access through some other means can install it.

Denial of Service (DoS)   The DoS attack is a type of attack that creates churn on a target computer by overwhelming it with things to process. This causes the computer to use most of its resources to process unnecessary and redundant information, thereby rendering it too busy to perform the functions it is meant to perform.

The average user may be the subject of a denial of service attack, but it is more common to target a high-profile computer, as was the case in the highly publicized Yahoo attack. The web site became so overwhelmed with traffic that it was rendered useless in processing requests from valid users. When a public, high-profile computer is the target, the threat to the average user is that his or her machine may be used as an intermediary source to generate the redundant traffic.

The whole point of DoS is to get as many machines accessing the target as possible. To do this, the hacker must gain access to multiple machines. Once these machines have been breached, they are used as the source for bombarding the target server with unnecessary traffic, which may consist of something so basic as ICMP pings.

In the always-on world, your computer becomes just another "always available" resource to launch these attacks. A Trojan horse program could be used as a dormant agent that awaits a remote signal to awaken a process that was designed to generate excessive traffic to a target machine. Unprotected Windows networking *shares* can be a source of DoS vulnerabilities. When exploited, Windows shares allow a single machine to propagate a malicious program to connected machines.

HTML Tags   Web browsers have the capability to interpret scripts embedded in web pages. These scripts are downloaded to a user's browser when he or she accesses the web page, and most browsers by default are enabled to run scripts. The exact problem or mischief caused by the script varies by design. The script may, for instance, have additional interaction with the web site without the user's knowledge, or a malicious script may read fields in a form provided by the web server, then send the user input to the attacker.

A user can unintentionally expose his or her browser to a malicious script by doing any of the following:

▼ Following links in web pages, e-mail messages, or newsgroup postings without knowing the ultimate destination

■ Using interactive forms on an untrustworthy site

▲ Viewing online discussion groups, forums, or other dynamically generated pages where users can post text containing HTML tags

These threats are a lot more difficult to address. The onus is really on the web site developer to ensure that the site is free of any such threats. From a user perspective, a user has the option of turning off the execution of scripts in their browser, but this disables a useful feature. Another option is to be more selective about the sites that are visited. Yet another way is to enable a warning feature in the browser that warns before any scripts are executed, but this may not be an option with some browsers.

E-mail Attachments and Spoofing   *Spoofing* is usually an attempt to trick the user into believing that a piece of mail has been sent from a recognized source. The source e-mail address, for example, may have been obtained from the user's address book. The intent of this attack is usually either to get the user to open an attachment that infects the user's machine, or to trick the user into sending sensitive information such as a password.

Viruses and other types of malicious code are often spread via e-mail. The best advice is to not open anything for which you do not know the source.

Instant Messenger and Chat Applications   These applications usually provide options for transfer of files and other features that are common to e-mail such as active web links. These systems present the same risks as e-mail-borne attacks, just in a more real-time fashion. The best advise is to not exchange files with strangers and to limit your sessions to those you know.

Sniffer Applications   Now here is a really sophisticated threat. *Sniffers* are devices that are built to capture data as it traverses the physical medium. These devices are designed for troubleshooting; they enable a technician to see what is happening at the bit level of a connection. However, they are just as effective for creating mischief, spying, and stealing data.

Access to the physical infrastructure is a requirement for this type of device. Cable and DSL access networks have this potential risk. Wireless systems in enterprise and residential networks are especially vulnerable. When data traveling over such a system is not adequately encrypted, it is very possible (and without much hassle) to capture it as it traverses the air. Wireless sniffers are now common, and sales teams have been known to sit outside an office building, capture the transfer of unencrypted data, and present the data to the client to highlight security exposures and maybe sell their security services. It is highly advisable that wireless transmissions be encrypted.

Flash Crowds and Bottlenecks

A broadband connection leads users to expect a much-improved level of performance relative to a simple dial connection. Although the broadband connection carries data at a much higher rate, that is only one of many factors that impact the speed of a service, but this fact is lost on the average user.

Having a high-speed link is no guarantee that the user will realize high-speed delivery of a service. Other factors that play a part in the Quality of Service enjoyed by the user include:

▼ The first-mile link (the link that connects the source of the service to the network)

■ Server loads and capacity

▲ The state of the network and its resources between the user and the target service

If any one component is bogged down, the user experience is adversely affected.

The design and performance of the serving web site and its links to the network cannot therefore be divorced from discussions about service quality. This is also true of the network path between the user and the originating server.

Bottlenecks Within the Network

The Internet is a network of many networks that are all linked. Common points of congestion are points of network peering—the interconnection points between the independent ISPs that provide services to end users. Within this mesh of networks, there is a hierarchy of providers; smaller providers peer to larger ones, and larger providers peer to each other.

ISPs first optimize packet flows inside their own networks and then interconnect at one of many possible peering points. Border Gateway Protocol version 4 (BGP4) is the routing protocol that ISPs use to implement their peering policies and to control the traffic exchange at the peering points. These policies and routing practices have important implications for both the cost and the performance of first- and last-mile networks.

Smaller ISPs are much more numerous than larger ones, and thus much of the peering occurs when a smaller ISP buys connectivity from a larger provider. The size of a link between the networks becomes the weakest link between the user and the serving web site. While a smaller ISP has to buy a connection from a larger provider, the same is not necessarily true for links between two large ISPs. Because both mutually benefit from the peering, and because peering carries considerable setup costs, larger ISPs tend to split the cost of the peering but limit the number of peering points. As a result, the limited number of peering points translates into potential bottlenecks in the network. The same is true of the peering points of smaller networks, except in this case the issue is not only the number of peering points but also their size.

Flash Crowds and Issues of the First Mile

The bandwidth capacity of the serving web site plays an important role in the user experience. The first mile, as it is called, must be designed to accommodate the traffic requirements of the busiest period of the day. Having knowledge of trends and traffic profiles, such as the type and size of the data commonly requested, helps in planning and main-taining acceptable levels of performance.

The term *flash crowds* has been used to describe the problem in which a crowd of users converges on a single web site because of an event, such as breaking news or a product sale. The convergence of traffic on a web site covering a breaking news event has the po-tential to cripple the site by causing server crashes and bottlenecks across its access links. Capacity planning in this case is a lot more difficult because transient spikes are more dif-ficult to project. The extent of the spike would have to be predicted in order to do ade-quate planning, and at best this would be a guess. Planning based on guessing has one of two results: failure or an expensive, overly engineered solution. The remainder of this chapter is devoted to the techniques and technologies that have emerged to address prob-lems of flash crowds: efficient and fast content delivery and technologies that improve the user experience.

CONTENT DELIVERY NETWORKS (CDN)

In support of the growing population of Internet users, content providers must find ways to distribute the same information to thousands of users in a reliable and efficient man-ner. The distribution of information—page views, graphics, text, and rich media such as streaming audio and video—is not even, nor will it ever be. Popular web pages create hot spots of network load, with the same data being transmitted over the same network links thousands of times per day.

These hot spots or network bottlenecks also move around, from web page to web page, service provider to service provider, and region to region. On a particularly heavy news day, a news service web site may be the flavor of the day, and if the news is regional in context, most viewers may be from a particular region. On another day, users from all over the world may be drawn to a financial service web site because the Nasdaq is in free fall; in this in-stance, the hot spot becomes the point of convergence, the server serving the content.

Content delivery networks (CDNs) were created to address this problem. A CDN is a comprehensive, end-to-end solution for optimizing global networks for web content de-livery. Users requesting information from a popular web site may well have those re-quests served from a location much to closer to them than the original server on which it is generated. By serving content from points a lot closer to the user, a CDN reduces the likelihood of hot spots by dispersing the different points of convergence and by distribut-ing the workload between multiple servers.

Delivering content from the edge of the network instead of the original server also has the added benefit of added reliability. The probability of lost packets is decreased, pages load faster, and the performance of streaming audio and video clips is improved. The user experience overall is improved.

CDN Service Description

CDNs employ a network of web and caching servers that are less expensive than the host-ing server but are efficient at delivering web content. The typical arrangement is for an owner of a web site to outsource the content of their site to an external content delivery network. The CDN provider takes the content and distributes it across the network to all edge servers, a process that can take from minutes to two to three hours, depending on the size of the CDN. Once loaded, the CDN intercepts all IP requests and serves the content from the available cache that is physically closest to the user.

Akamai is an example of a CDN provider with a network that spans 63 countries and approximately 13,000 servers. The deployment and management of a CDN involves complex challenges that extend beyond the deployment of edge servers. An entire suite of services and capabilities must be built into the solution in order to provide a robust infrastructure and a reliable service. The objective is to improve the user experience with the hope of not distracting users with performance-related problem; by doing this, the content owner has a much greater chance of a return visit. Factors that need to be considered in the design and deployment of a CDN include the following:

▼ The content must be delivered to the edge servers if the service is to be of any use to the users. This is a fundamental requirement of a CDN.

■ A basic service requirement is the ability to identify the location of the requesting user and respond from the edge server that is closest to the user.

■ The performance of the edge server must be monitored for performance degradation, and the CDN must be able to load-balance in the event a server becomes overloaded.

■ The Internet is a network of networks, and so the CDN provider must deploy enough servers to address users from different service provider networks. Issues of content management across different service provider networks must also be addressed to ensure the reliable delivery of content from the original server to the edge servers in different service provider networks.

■ Consideration must also be given to geographic placement. Pockets of population by city, region, and country must be considered in the deployment of edge servers. Because the objective is to improve the user experience by placing the content closer to the user, edge servers must be deployed on a global basis.

■ The content must also be kept fresh and synchronized with the serving server. This can be a very challenging task, and the more frequently the source data changes, the more complex the task becomes.

■ Fault tolerance is a fundamental requirement for the delivery of a reliable service. The solution must factor in failure and automatically adjust to prevent service interruption.

■ A well-designed CDN must track real-time conditions of the network and avoid hot spots. If congestion is detected, it should have the capability to quickly reroute content in response to the congestion and outages.

▲ Reporting and billing are both requirements. Real-time logging and billing and detailed reporting on traffic patterns, user location, Internet conditions, and other reporting elements must be tracked.

From this list, it is evident that the efficient operation of a CDN and its ultimate usefulness to a distributed user base are dependent on multiple factors. It is virtually impossible for a single CDN to meet all these requirements without a significant investment, and even then, the likelihood of that CDN having enough edge servers to satisfy a global user base is very low.

If no single CDN is able to meet all the requirements, this raises a few questions. The first is from the perspective of the content owner. Will the content owner need to contract with more than one CDN in order to gain adequate coverage for users within multiple geographic regions? This issue becomes particularly relevant when a single CDN may not have the coverage within the target regions of the content owner. Furthermore, it is likely that the content owner may not be able to predict where the users of the service will be connecting from. Multiple contracts with multiple vendors' CDN services place an additional cost burden on the content owner.

From the perspective of the user and that user's ISP, the question arises whether or not the user will be able to gain the benefit of fast access to an edge server if the only CDN provider in the region is affiliated with a competing ISP. Obviously, these questions raise issues that cannot be addressed entirely by technology. At the core of these questions are contractual issues that can be addressed only through a business arrangement between the different ISPs, CDN providers, and content owner. These questions, technical and business, are being addressed through the formation of the Content Alliance.

The Content Alliance

In August 2000, Cisco Systems announced the formation of the Content Alliance to speed the adoption of compatible CDN technology. At the time of the announcement, the charter members of the alliance consisted of Cable & Wireless, Digital Island, Genuity, GlobalCenter, Mirror Image Internet, NaviSite, PSINet, ServInt and Network Appliance.

Because the Internet is really a collection of independent networks, the Alliance was formed to help develop standards and protocols to advance content networking. The Alliance is not a standards organization; it generates proposals for standards and depends on traditional standards bodies such as the IETF to gain broad industry acceptance. The initial focus of the group was on *content peering*, a term that describes the process that enables the CDNs of multiple independent service providers to work in cooperation. In addition to the development of technologies, the Alliance is also focused on defining specifications to address issues of authorization for the use of content between networks, and the sharing of logging or billing information for charge settlement.

Content Peering

Content delivery networks accelerate web sites by intercepting user requests and redirecting them to devices located near the user. These devices may be located on the networks of many service providers. As previously discussed, a CDN approach is only as effective as its coverage. If a user requests a web page from a location that is not served by an edge server, the request must travel the full path to the original server, which may be in the same city but could just as well be in a different country.

However, no individual CDN has the reach and coverage to span all the geographies and networks across the Internet. Likewise, a given CDN cannot take advantage of all the value-added functionality—speed, security, Quality of Service, on-premise delivery, and other features—that an individual service provider may implement. Given the collection of independent networks that form the Internet, content peering allows CDNs to interoperate, ensuring fast performance by delivering web content from devices located close to the viewing audience.

With *content peering,* a web site owner can work with a preferred hosting service provider but gain the reach of the combined peering networks. By leveraging the reach, power, and features of different CDNs, content peering creates the ability to deliver the benefits of content delivery networks to a global user base regardless of where the server is hosted and by whom. Content peering requires the CDNs to share information in three areas:

▼ **Content distribution**   The process of moving files to the remote delivery devices

■ **Content request-routing**   The process whereby a viewer's page request is redirected to the appropriate delivery device

▲ **Accounting**   The process for collecting usage and billing data

Features of CDN Operation

CDNs provide service providers and independent CDN operators with a platform for delivering value-added services to content owners. The ability to offer additional value represents an additional revenue stream to service providers and forms a fundamental source of revenue for companies like Akamai who specializes in such services.

The value proposition to the content owner is an improved user experience that has less to do with the initial attraction of a user to the web site than with the initial impressions of the site as they affect the decision to make subsequent visits. An initial experience of long waits and aborted page loads may bias the user against returning. If, however, the user experience is satisfactory, it gives the user time to focus on the content, which should be the deciding factor on whether or not the user pays a return visit.

Today's CDNs are a lot more functional than their original versions. Once limited to serving static web pages, CDNs now have the capability to serve dynamic and personalized content. This is quite an accomplishment when one considers that the fundamental operation involves the distribution of content from a central point to distributed edge servers, a process that could take hours. Dynamic content implies content that changes at the time of the request. This appears to be impossible without storing all the content-generating applications, such as databases, at each edge server. But before we look at some of the ways this is accomplished, a closer look at the components of CDNs is in order.

Evolution of Content Delivery

In the earlier years of the World Wide Web, the Web was seen as a means by which an organization could publish information for public consumption, or to a closed user group. Users accessing the content would connect directly with the server that served the content.

As the web sites grew and more functions were added, the problems of content management and the addition of dynamic features became more and more challenging.

Application servers emerged and were used for content management, as shown in Figure 12-1. These servers were placed between the user and the back-end business systems—applications, databases, and modern and legacy servers. With this approach, the back-end systems could continue to operate in the way they were designed, and continue to address the functions for which they were intended, without having to understand HTTP requests and HTML functions. The application server was used as a translator; on one side, it understood the web-based structure of the user request, and on the other, the native structure of the serving database or server.

In recent years, this intermediate layer—occupied by the application server in Figure 12-1—has grown in function and components. This layer—the components of which are often referred to as middleware—has seen significant development in an attempt to minimize the complexity of client programs and improve performance. Functions of security were also added to this layer to ensure the security of data and user traffic.

Contemporary data centers are a lot more complex than their original counterparts. The growth of e-business and other e-enabled services has placed demands on content hosting and delivery that have resulted in a more complex infrastructure, built to deliver personalized and dynamic content. Within a typical content hosting data center, it is now common to find components that include routers, switches, firewalls, reverse proxy caches, devices that do load balancing, DNS servers, web servers, application servers, database servers, and storage appliances. And each component may be duplicated to ensure redundancy, provide adequate capacity, and improve reliability and availability. The entire site is usually homed to the Internet through different access points and sometimes through different service providers, where possible.

The architecture of this setup is depicted in Figure 12-2. Requests coming in from the Internet are first directed to the policy server, which validates the request against a set of rules and authorized user lists. Valid requests are then sent to the load balancer for distribution to a web server. These web servers interact with application servers, which in turn interact with the appropriate content source: a database server, a transaction server, or some other server type.

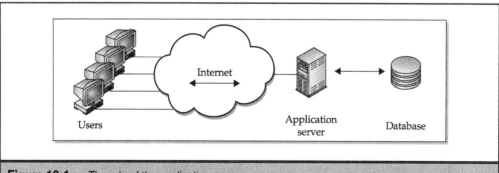

Figure 12-1.   The role of the application server

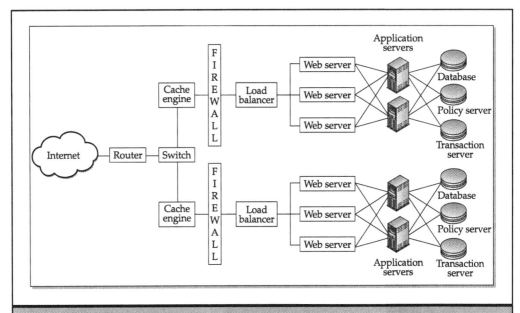

Figure 12-2. Typical architecture of a Web hosting service

This architecture provides a level of redundancy and, when viewed as a whole, is just an extension of a client/server approach, where the server function is now being delivered through the combination of many components acting as a single resource to the client. Taking this view of the system, one can easily see that it is still a centralized approach to the delivery of content, and a centralized approach is subject to many weaknesses, a few of which are listed here:

▼ A centralized approach is not scalable. Issues of performance and capacity are addressed by adding more components, further complicating operation and management.

■ A centralized approach does not address latency introduced within the network. A user request and the response must still traverse the full path across the network and be subject to whatever problems the network may be experiencing.

■ A centralized approach limits global reach. Users in other parts of the world may have adequate infrastructure for accessing local content, but the links back to the country of the hosting service may be limited.

▲ A centralized approach does not solve the problems created by flash crowds or denial of service attacks without having additional capacity that goes unused during normal operation. On a day like September 11, 2001, the added surge of traffic could easily crash servers at the time when the content was needed most.

Web Site Mirroring

Mirroring a web site across different geographically dispersed networks addresses many of the weaknesses of a centralized approach. However, it is expensive to set up a completely separate but identical infrastructure, and though this approach addresses issues of availability, it introduces new challenges such as data synchronization.

Mirroring improves availability and is useful for server load distribution but does little to address network congestion problems between the user and the serving site. Users that are directed to one site may experience an improvement in performance, but the body of users that remain with the original may be not be so lucky.

Mirroring also does not fully address the problems created by flash crowds; the convergence of traffic to a single point within a region creates the same problems as in the centralized model. Tons of spare capacity would be needed to address this problem, which would be a waste during times of normal operation. Finally, mirrored sites further complicate the administration and management of content.

Logical Functions of a CDN

A CDN addresses the shortcomings of both the centralized approach and the model that employs mirroring by using a distributed model that separates the content generation tier from the tier that is responsible for assembly and delivery, as shown in Figure 12-3. This model logically divides the functions of the centralized model into three tiers, a *content generation* tier, an *integration* tier, and an *assembly and delivery* tier.

Content Generation   The *content generation* tier is usually centrally located in a fully redundant data center. The primary function of this tier is the generation of the requested information from legacy systems, databases, and transaction servers. Within this tier, a policy-based server may also be used to administer rules and security policies, for users and user requests. The policy server works in conjunction with the firewall to administer these security policies. The following paragraphs look at some of the key features, components, and functions of this tier in more detail.

The *application server* understands the different data structures of the different servers that are used to serve content. A user request received by an application server is translated into a set of requests that must be sent to the appropriate content servers, in a structure that these servers understand. The response must then be converted from the native structure of the serving platform to one that can be processed by the web-site functions.

In the centralized model, the application server may also be involved in the page assembly and user handling processes. In the distributed model of a CDN, these functions are offloaded to the content delivery and page assembly tiers, greatly enhancing the performance of the network.

Previously, we mentioned that earlier versions of CDNs were capable of handling only static pages. The current versions of CDNs are now capable of serving personalized and dynamic pages through the use of a specification called *edge side includes (ESI)*. Akamai Technologies, ATG, BEA Systems, Circadence, Digital Island, IBM, Interwoven, Oracle, and Vignette jointly developed the specification. ESI is an open language for creating a uniform programming model that facilitates interoperation of ESI-complaint systems from different vendors.

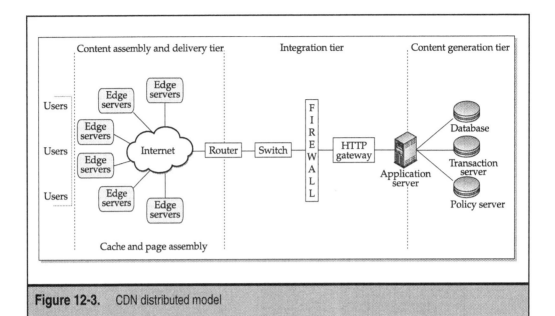

Figure 12-3. CDN distributed model

ESI enables a web site developer to break down web pages into fragments of different cacheable profiles. Maintaining these fragments as separate elements on the content delivery network enables dynamically generated content to be cached, then reassembled and delivered from the edge of the network. The freshness of the information is managed through a process of sending invalidation messages from the application server to the edge servers informing them to overwrite outdated objects residing on them. In this way, changing content can be controlled much as it was when only static pages were served.

To provide a basis for understanding how dynamic pages are created, we will briefly compare how a user request is handled in a centralized solution with the same process on a CDN. In the centralized model, the request for information from a user is sent to the application server. The application server translates the request and sends a database query to a database server, if the request was for a database record. The target server does not have to be a database server; it could be any back-end system, including a transaction server or a legacy mainframe.

The application server combines the returned values with other components such as advertising and menu navigation into a fully assembled page that is sent back to the user. Subsequent requests for the same information are handled essentially in the same manner, with the difference lying in where the components are gathered. The second request, for instance, may be satisfied from information that is held in cache as a result of the first request.

In a CDN, the initial request is always sent to the edge server, which checks its internal cache to see if it contains the page as a result of a previous request. What is cached, how long it is stored in cache, and other general rules for caching content are defined by the content provider via a metadata configuration file. A component of this rule is the *time to live (TTL)*. This value determines how long a page fragment stays in cache and is based on how frequently the information on the fragment changes. A particular piece of information may be updated every 6 hours; another may be updated every 24 hours; each would have a different TTL value.

The fact that a different user requested a page has nothing to do with its relevance to the next user that requests it. Having a local copy on the edge server speeds future requests for the same information by other users. Fragmenting the page allows different parts of the page to have different refresh values. An image, for instance, may have a longer TTL than another fragment.

Not all information is cacheable; in fact, some is deemed uncacheable. Personalization of content is a lot harder to manage using techniques of caching. The look and feel of a page based on a user's preferences and location tend to rely more on the application server, content management systems, and cookies to personalize the user experience. In these instances, uncacheable fragments can still be speeded up through the use of persistent connections between the original site and a number of edge servers. Keeping the connection open means that a path for communication is always open back to the originating server and performance is enhanced by not having to establish a new connection for every request. Compression techniques further aid in the delivery of content across the persistent connection.

The ESI markup language includes the following features:

▼ **Inclusion**   ESI provides the ability to fetch and include files that are used to make up a web page. Each file is subject to its own properties, such as TTL, configuration and control, and revalidation instructions, among others. Each included file may include other files up to three levels of recursion.

■ **Conditional inclusion**   ESI supports conditional processing based on Boolean comparisons or environmental variables.

■ **Environmental variables**   ESI supports the use of a subset of standard CGI environment variables such as cookie information. These variables can be used inside ESI statements or outside of ESI blocks.

▲ **Exception and error handling**   ESI enables developers to specify alternative pages and default behavior, such as serving a default HTML page in the event that an original site or document is not available. Further, it provides an explicit exception-handling statement set. If a severe error is encountered while processing a document with ESI markup, the content returned to the end user can be specified in a "failure action" configuration option associated with the ESI document.

Detailed information on ESI and its capabilities can be found at www.esi.org.

▼ **Directory and policy servers**   Security solutions vary from basic user ID and password pairing to sophisticated solutions that employ elements of authorization, authentication, and accounting. Policy servers work in conjunction with directory servers and firewalls to validate user requests against predefined rules.

- **Database and transaction servers and legacy systems** The repository of data and the handlers of requests. These systems are usually not web-aware and function in ways that are native to their own internal architecture. These back-end systems work in conjunction with other back-end systems to provide a consistent data repository and transaction management.

- ▲ **Storage management** Extremely critical to any operation. Once a centrally managed function, storage systems are taking on a new structure and architecture of their own. The introduction of terabit systems and storage appliances has enabled storage to be deployed at any point in the network. We have depicted it here in the content generation tier because this is where the most critical storage function resides. However, storage is also a function that resides within the content assembly and delivery tier.

Integration Tier The *integration* tier lies between the components that provide content generation and those that do page assembly and delivery. The two main components of this tier are the HTTP gateway and the firewall. The HTTP gateway is just a fancier version of the web server of the centralized model depicted earlier in Figure 12-2. The gateway's basic functions are to forward requests from the edge servers—for fragments and files—to the application and then send back the results. The gateway supports a multitude of different web-based languages, such as the Common Gateway Interface (CGI), Java Server Pages (JSP), and Microsoft's Active Server Pages (ASP).

Content Assembly and Delivery The *content assembly and delivery* tier embodies the major benefit of a CDN: moving the content closer to the user, resulting in a much improved user experience. This tier forms a distributed and fully managed edge network; the bigger the network, the better the user community is served. The edge network may be further divided into a logical hierarchy consisting of edge servers and regional edge servers forming a core. The edge servers generate and serve content, and if they do not have a particular file in cache, they may request copies from the core servers, which in turn may request the information from other regional servers. Using this method, the traffic to the original site is further reduced and issues of flash crowds become less a factor.

Some of the functions performed with the edge network are discussed in the following paragraphs.

- ▼ **Traffic analysis and monitoring** The performance of the edge network is dependent on knowing the network health and performance of the surrounding Internet. Using real-time information and statistics, optimal routing for content can be performed for each user request.

- **Load balancing** The performance of each edge server is monitored to ensure adequate load distribution. Through the use of mapping software, the user is routed to the server in the best position to service a user. By having detailed knowledge of traffic statistics and server loads, the mapping software is able to combine information about the user's local name server to direct the user request to the optimum edge server. Once determined, the chosen edge server proceeds to fulfill the request by delivering content from its cache. If the content is not in its cache, the edge server requests it from, in order of priority, another edge server within the region, the core hierarchy, and then the original site.

- **Caching** Frequently requested content, from text files to video clips, is stored locally by the edge server. This eliminates the latency involved in transferring these objects from a central location and optimizes the consumption of bandwidth and server resources.

- ▲ **Content assembly** Edge servers utilize the cached fragments to build web pages to satisfy user requests. When a request first comes into an edge server, it looks within its cache to see if the request can be satisfied with static contents already stored in cache. If so, it just returns the static page to the user. If not, the fragments necessary for the creation of the page are gathered from the local cache, other regional edge servers, the core hierarchy, and/or the original site. Once the data have been received, the page assembly engine of the serving edge server uses the information contained in a metadata file to assemble the page. The output of the process is cached and then sent to the user.

CDN Summary

CDNs are highly scalable and provide improved performance and reliability that directly benefit the user. The distributed model of their architecture protects against issues of flash crowds and network hot spots. The formation of the Content Alliance has helped lay the groundwork for a cooperative model between independent CDN operators, which has made the content owner's life a lot simpler. Through a single contract with a single CDN provider, the content owner gains the benefit of the reach and coverage provided by their CDN provider and the reach and coverage of any other CDN with which there is an agreement.

LAYER 4 SWITCHING

A new breed of switching technology has been developed as a result of the requirement for efficient content delivery. These products, known as web switches, e-commerce switches, or content-aware switches, are typically targeted at a specific function, such as load balancing.

These switches differ from layer 2 or 3 switches in that they can look further inside packets and make sophisticated forwarding decisions according to information contained in the HTTP header. A new technology, first introduced by Alteon WebSystems (now a part of Nortel Networks), layer 4 switching has quickly become an integral technology in enterprise and content provider networks.

Since the introduction of the first layer 4 switch, many vendors have introduced products that bear that designation but in fact operate in a way that conflicts with the true definition of layer 4 switching. Layer 4 switching implies that a decision regarding a path is being made according to the contents of the transport layer of a packet. This means that the switching decision is being made at a TCP or UDP level in the case of IP or layer 4 of the OSI model, but not all switches transfer packets according to this information. The generic term *Web switches* is therefore more appropriate in describing this new breed of switches. These switches have also been termed layer 4-7 switches in an effort to describe their functions more appropriately. We have chosen to use the term layer 4 switch, even though it is not technically correct in all its applications. However, the description has become a standard way of describing these products.

Why Layer 4 Switching

The original term, layer 4 switch, was coined by Alteon WebSystems and was used to describe their product, the ACESwitch, that performed layer 2 switching functions based on the MAC address of the target device. The ACESwitch was also capable of making routing decisions based on the values of TCP and UDP port numbers in the packet, a capability that facilitated service-based load balancing and was the basis for the description.

Placing the switch in front of multiple web servers meant that the traffic could be load-balanced across them without the user having any knowledge that this was taking place. In this configuration, the web servers all have identical content and must employ a replication scheme to ensure that the content remains in sync. If a web server fails, the web switch knows enough to stop forwarding packets to the failed server, thereby adding a layer of redundancy.

There are now many competing products that provide the same load-balancing functions along with many more functions. Many of these newer devices add a level of intelligence that monitors server load and response time, and that switches incoming requests to the server with the least load.

Other enhancements include the capability to add levels of priority to the packet to enforce Quality of Service. On receipt of a packet, the switch looks inside the packet to determine its TCP or UDP port number and compares that number with a list of UDP and TCP port numbers for which it has rules. If the packet matches a number, it is passed on to another process for further processing; if not, it is sent to be switched to the appropriate output interface. Once a packet has been identified as needing priority, a layer 2 priority tag such as an 802.1p priority tag may be inserted in the layer 2 header of the packet, which is then sent on its way. Or, an IP ToS may be assigned.

Functions of Layer 4 Switches

Many of the features provided by layer 4 switches can also be performed in a router through the use of extensive filtering combined with policy-based routing. However, these functions are provided in software, and functions performed in software are a lot slower than those performed in hardware. Layer 4 switching performs the switching functions in hardware, making it possible to make decisions at close to full wire speeds. These switches can be classified into several types of products.

Server/Web Load Balancing   Defining virtual server groups enables incoming requests to be switched to the least loaded server. Load balancing schemes are also used when hardware and path defects are detected. The load balancing decisions are usually based on the TCP or UDP port number, which identifies a specific application. With a single server running multiple applications, each application has an identifying port number. If multiple identically configured servers are running the same number of applications, it becomes possible to control the load on each server by carefully routing each request to the most available server. If a server goes down, the layer 4 switch reroutes the traffic away from the failed server to one that is functioning.

Access Control   An *access control list (ACL)* is a table that stores rules governing traffic that flows through a particular device. This is where the layer 4 description becomes inaccurate. ACLs are able to manipulate any data within the packet, not just those that are in the layer 4 headers. Information such as source IP address may be used to block a specific device from using the network, and in theory, an ACL can just as effectively use any information above layer 4 to make decisions, if it was designed to do so. ACLs are not unique to layer 4 switches, they are also found in routers. But as opposed to software-based routers, ACLs are implemented in the switches' hardware, which means minimum degradation of performance while switching.

Quality of Service   Layer 4 information can be used to establish application-based priority. Without layer 4 switching, the rules regarding QoS and CoS are restricted to information contained in the layer 2 and 3 headers. MPLS, for instance, uses labels that sit at the data link layer to enforce switching decisions. Layer 4 switching enables more control, and additional layers of QoS and CoS may be applied to specific applications. This can be especially useful in Voice over IP applications, where packets sent by these applications must be granted the highest priority in the network.

Performance Measurements

Unlike layer 2 or 3 switches that measure performance in packets per second, the performance of web switches is measured in transactions per second, with a single transaction consisting of a request and a response. Other metrics include:

- ▼ The number of HTTP connections per second the switch is able to set up
- ■ The number of concurrent connections the switch can handle
- ▲ The error rate, or the rate of transaction requests that aren't completed

SUMMARY

The years from 1990 till now have seen many changes. The rate of technological achievements, innovation, and expansion is unprecedented. During this time, the way we traditionally viewed the world and its telecommunication, data, and broadcasting infrastructure has significantly changed. For the first time, there is a distinct blurring of lines between the telephone, data, and audio/video broadcast network industries. The providers of services, who once operated within the confines of their respective industries, now compete across industries to deliver all three services.

As a result of this change, there is evidence of convergence. It has brought us a lot closer to the promise of an integrated network than we have ever been, and to all the benefits to be gained from this arrangement. For the consumer, it means a single provider will be able to provide an integrated service. It also means the potential for lower prices for a combined service, as a result of competition. To the provider, it means additional revenues and a single infrastructure from which multiple services can be delivered.

As we continue along this journey of convergence, new technologies and services will continue to evolve, which makes the future unpredictable but instills a sense of anticipation. The access and metro area networks are the two networks that will see the most change in the coming years because these are the networks that are furthest from their full potential. The metro area networks will eventually be upgraded, and issues of bandwidth will be addressed as more fiber is deployed. Today's dependence on ATM and SONET/SDH will eventually yield to newer and improved schemes. Today it is MPLS; tomorrow it may be new and improved versions of packet/cell switching that are a lot more application aware. Copper will eventually be replaced by fiber, and issues created by limited bandwidth will become less of a concern.

Today there are many competing access technologies, and the reality is that we may never get to a single solution. The only thing that appears certain is the inevitability of wireless playing a more significant role and the assurance that fiber will be deployed a lot closer to the home.

But even so, we can only speculate, and your guess is as good as ours. So our best advice—if you would care to attempt to predict the future of broadband—is to follow the money, because the flow of investment capital is our best indicator of our technological future.

Index

▼ **E**

F

▼ I

N

O

INTERNATIONAL CONTACT INFORMATION

AUSTRALIA
McGraw-Hill Book Company Australia Pty. Ltd.
TEL +61-2-9417-9899
FAX +61-2-9417-5687
http://www.mcgraw-hill.com.au
books-it_sydney@mcgraw-hill.com

CANADA
McGraw-Hill Ryerson Ltd.
TEL +905-430-5000
FAX +905-430-5020
http://www.mcgrawhill.ca

**GREECE, MIDDLE EAST,
NORTHERN AFRICA**
McGraw-Hill Hellas
TEL +30-1-656-0990-3-4
FAX +30-1-654-5525

MEXICO (Also serving Latin America)
McGraw-Hill Interamericana Editores S.A. de C.V.
TEL +525-117-1583
FAX +525-117-1589
http://www.mcgraw-hill.com.mx
fernando_castellanos@mcgraw-hill.com

SINGAPORE (Serving Asia)
McGraw-Hill Book Company
TEL +65-863-1580
FAX +65-862-3354
http://www.mcgraw-hill.com.sg
mghasia@mcgraw-hill.com

SOUTH AFRICA
McGraw-Hill South Africa
TEL +27-11-622-7512
FAX +27-11-622-9045
robyn_swanepoel@mcgraw-hill.com

**UNITED KINGDOM & EUROPE
(Excluding Southern Europe)**
McGraw-Hill Education Europe
TEL +44-1-628-502500
FAX +44-1-628-770224
http://www.mcgraw-hill.co.uk
computing_neurope@mcgraw-hill.com

ALL OTHER INQUIRIES Contact:
Osborne/McGraw-Hill
TEL +1-510-549-6600
FAX +1-510-883-7600
http://www.osborne.com
omg_international@mcgraw-hill.com

www.ingramcontent.com/pod-product-compliance
Lightning Source LLC
Chambersburg PA
CBHW080134060326
40689CB00018B/3787